D1061460

GROUP THEORY

RUDOLF KOCHENDÖRFFER

PROFESSOR OF MATHEMATICS, UNIVERSITY OF TASMANIA

McGRAW-HILL · LONDON

NEW YORK · SYDNEY · TORONTO · MEXICO · JOHANNESBURG
PANAMA · SINGAPORE

Published by
McGRAW-HILL Publishing Company Limited
MAIDENHEAD · BERKSHIRE · ENGLAND

07 094089 4

Authorized translation from the first German-language edition of Lehrbuch der Gruppentheorie unter besonderer Berücksichtigung der endlichen Gruppen, first published by Akademische Verlagsgesellschaft Geest & Portig K.-G.

PRINTED AND BOUND IN THE GERMAN DEMOCRATIC REPUBLIC

PREFACE

The last two or three decades have brought about rapid progress in the theory of groups. In particular, after a period of stagnation, the theory of finite groups has been enriched by important and even sensational results. Thus, an increasing number of students may wish to become familiar with the basic facts of group theory as early as possible. The present book is intended to serve this purpose.

The theory of groups is a wide and highly developed branch of mathematics. For this reason, an attempt to give an introduction to all parts of the theory in a book of reasonable size would very likely result in leaving the reader almost everywhere without the knowledge that is required to study research papers. Thus, it is necessary to select the topics to be treated, and in doing so one will, of course, be guided by the main lines of current research. The present book is mainly devoted to the theory of finite groups. However, it would not be reasonable to confine oneself exclusively to the finite case, even if finite groups were the only goal. Therefore, the basic concepts are treated without any finiteness conditions so that the book also provides an introduction to the general theory of groups. The fact that the German edition has been well received seems to justify the attempt to present the rudiments of group theory in a textbook that is accessible to younger students and, as regards finite groups, will enable the reader to proceed to more advanced study of current problems.

The prerequisites, apart from some general experience in mathematics, are no more than the material usually taught in the first year course at a university. Indeed, with very few exceptions, the reader is only assumed to be familiar with the basic facts of linear algebra and the rudiments of elementary number theory, such as congruences and the properties of the greatest common divisor. Any gap in the student's knowledge will easily be filled by reference to any elementary textbook on these subjects, for example [49] and [52].[1]

For an introductory course on group theory the following material may be used: chapters 1 and 2, sections 3.1, 3.2, 3.4, 4.1, 4.2, and 4.3, Theorem 4.4.1, section 4.5, chapter 5, sections 6.1, 7.1, 7.2, 7.4, and 7.6.

[1] Numbers in brackets refer to the Bibliography at the end of this volume.

A subsequent course on finite groups might include section 3.3, the remaining sections of chapter 6, section 7.5, chapter 8, and selected topics from chapters 9 to 13.

I have tried to make chapter 13 on group representations as self-contained as possible. Only the basic material of sections 1.1, 1.2, 1.3, 2.1, 2.2, 2.3, 2.4, 2.6, and 2.7 will be used.

The Suggestions for Further Studies at the end of the volume contain references to literature on subjects that are not treated in the present book, and hints for a more detailed study of some topics of current interest.

I wish to acknowledge the kind cooperation of both the Akademische Verlagsgesellschaft Geest & Portig K.-G., Leipzig and the McGraw-Hill Publishing Company. I also thank my wife for her patience and skill in preparing the typescript.

RUDOLF KOCHENDÖRFFER

CONTENTS

CHAPTER 1

GROUPS AND SUBGROUPS

1.1 Definitions and Examples

Let S be a non-empty set. A *binary operation* f on S is a rule that assigns to every ordered pair x, y of elements of S a uniquely determined element $f(x, y)$ of S. It is important to consider ordered pairs, because $f(x, y)$ and $f(y, x)$ may be distinct elements of S. A non-empty set S, together with a binary operation defined on S, is called an *algebraic system with a binary operation*. The various names given to such systems refer to additional axioms that are postulated.

Since we are concerned only with a single operation on S, we may simplify the notation and write xy instead of $f(x, y)$. This notation suggests calling xy the *product* of x and y. Accordingly, x and y are called, respectively, the left and the right factor. Occasionally, however, we prefer the additive notation, i. e. we write $x + y$ instead of xy and speak of the sum of x and y.

An algebraic system H with a binary operation is called a *semigroup* if the *associative law* holds,

$$(xy)z = x(yz) \quad \text{for arbitrary elements } x, y, z \quad \text{of} \quad H.$$

There are two ways of forming products whose factors are three given elements x, y, z, taken in this order, namely $(xy)z$ and $x(yz)$. By the associative law, these two products coincide. Thus, we may omit the brackets and refer without ambiguity to the product xyz of the elements x, y, z, taken in this order. (In general, however, the product depends on the order of the elements.)

Now, let n be a natural number, $n \geqq 3$, and let $x_1, x_2, \ldots, x_n$ be n elements of a semigroup H. There are various ways of forming products whose factors are these n elements taken in the given order. These ways are described by brackets indicating the successive steps in which the multiplication is to be carried out. We shall prove that all these ways lead to the same final result, in other words, *the product of n elements of a semi-group is uniquely determined* *solely by these elements and their order.* This result permits us to denote the product simply by $x_1 x_2 \cdots x_n$ without brackets.

For $n = 3$, our proposition is the associative law. We assume that any product of fewer than n elements is uniquely determined by the elements and their order. We consider a certain method of forming the product of $x_1, x_2, \ldots, x_n$. At the last step of this method, we have to carry out a multiplication

$$a_1 a_2 = (x_1 \cdots x_k)(x_{k+1} \cdots x_n).$$

Owing to the inductive hypothesis, the products

$$a_1 = x_1 \cdots x_k, \quad a_2 = x_{k+1} \cdots x_n$$

are uniquely determined. The final step in some other method of computing the product of $x_1, x_2, \ldots, x_n$ is a multiplication

$$b_1 b_2 = (x_1 \cdots x_l)(x_{l+1} \cdots x_n).$$

We have to show that $a_1 a_2 = b_1 b_2$. Clearly, we may assume that $k < l$. By our inductive assumption, $c = x_{k+1} \cdots x_l$ is uniquely determined, and we have $a_1 c = b_1$, $c b_2 = a_2$. The associative law gives

$$a_1 a_2 = a_1(c b_2) = (a_1 c) b_2 = b_1 b_2,$$

and this completes the proof.

The associative law enables us to define powers x^n for natural numbers n, namely

$$x^n = x x \cdots x \quad (n \text{ factors}).$$

Clearly, we have the familiar laws

$$x^m x^n = x^{m+n}, \quad (x^m)^n = x^{mn}. \tag{1.1}$$

If a semigroup H contains an element e such that

$$xe = ex = x \quad \text{for every} \quad x \in H,$$

then e is called a *unit element* of H. A semigroup H cannot contain more than one unit element. Indeed, for two unit elements e_1, e_2, we have $e_1 = e_1 e_2 = e_2$.

Suppose that H is a semigroup with unit element e. For a given element x of H, there may exist an element x^{-1} of H such that

$$x x^{-1} = x^{-1} x = e.$$

Such an element x^{-1} is called an *inverse* of x.

A *group* is, briefly, a semigroup with unit element in which every element has a unique inverse. It turns out, however, that a group can be defined by slightly weaker postulates.

Definition. *An algebraic system G with a binary operation is called a* group *if the following conditions are satisfied:*

(A) $(xy)z = x(yz)$ *for arbitrary elements* x, y, z *of* G *(associative law).*

(U_r) *G contains at least one right unit element, i. e. an element e with the property*

$$xe = x \quad \text{for every} \quad x \in G.$$

(I_r) *For a fixed right unit element e and arbitrary element x of G, there is at least one element* x^{-1} *in G such that*

$$xx^{-1} = e.$$

We call x^{-1} *a right inverse of x with respect to e.*

These postulates already imply the existence and uniqueness of the (two-sided) unit element and of (two-sided) inverses:

1.1.1 *A group G contains one and only one unit element, i. e. an element e such that*

$$xe = ex = x \quad \text{for every} \quad x \in G.$$

For every element x of G, there is one and only one inverse x^{-1} *in G such that*

$$x^{-1}x = xx^{-1} = e.$$

Proof. Let e be a right unit element such that every $x \in G$ has at least one right inverse x^{-1} with respect to e. We have $xx^{-1} = e$ and $x^{-1} = x^{-1}e = x^{-1}xx^{-1}$. Multiplying both sides of the last equation on the right by a right inverse $(x^{-1})^{-1}$ of x^{-1} (with respect to e), we obtain

$$e = x^{-1}(x^{-1})^{-1} = x^{-1}xx^{-1}(x^{-1})^{-1} = x^{-1}xe = x^{-1}x.$$

This shows that every x^{-1} with $xx^{-1} = e$ also satisfies the equation $x^{-1}x = e$; in other words, that every right inverse is a left inverse. Thus, we may speak of inverses without referring to the side. Moreover, it follows that x is an inverse of x^{-1}. From $xx^{-1} = e$ and $x^{-1}x = e$, we obtain

$$ex = xx^{-1}x = xe = x.$$

Consequently, e is not only a right but also a left unit element so that e has the properties of a unit element as defined above. Since even a semigroup has at most one unit element, we conclude that e is the only unit element of G. Finally, let x^* be an arbitrary element such that $xx^* = e$. Multiplying both sides of this equation on the left by x^{-1}, we obtain $x^{-1}xx^* = x^{-1}$ or $x^* = x^{-1}$. Thus, the inverse is unique. This completes the proof of 1.1.1.

For two arbitrary elements x, y of a group, we have $xyy^{-1}x^{-1} = e$. Since the inverse is unique, we conclude that

$$(xy)^{-1} = y^{-1}x^{-1}.$$

Similarly, we obtain for more than two factors

$$(x_1 x_2 \cdots x_{n-1} x_n)^{-1} = x_n^{-1} x_{n-1}^{-1} \cdots x_2^{-1} x_1^{-1}.$$

For a negative integer n, we define

$$x^n = (x^{-1})^{|n|} \qquad (n < 0),$$

and by x^0 we understand the unit element for every x. It is easily verified that in view of these definitions, eqs. (1.1) hold for arbitrary integers m, n.

The following theorem gives another characterization of a group.

1.1.2 *An algebraic system G with a binary operation is a group if and only if the following conditions are satisfied:*

(A) *The associative law.*

(Q) *For any two elements a, b of G there exists at least one element x such that*

$$ax = b$$

and at least one element y such that

$$ya = b$$

(existence of right and left quotients).

Proof. In a group G, condition (Q) is obviously satisfied by $x = ab^{-1}$ and $y = ba^{-1}$.

Conversely, suppose that (A) and (Q) hold. By (Q), the equation $ax = a$ has at least one solution $x = e_a$. Using (Q) again, we see that, for an arbitrary $b \in G$, we can find an element $y \in G$ such that $ya = b$. By (A), we obtain

$$be_a = (ya)e_a = y(ae_a) = ya = b.$$

Thus, e_a is a right unit element in G. By (Q), for every $c \in G$ the equation $cx = e_a$ has at least one solution, i. e., every element of G has at least one right inverse with respect to e_a. Thus, (U_r) and (I_r) are satisfied; hence, G is a group.

We remark that in a group $x = a^{-1}b$ is the only solution of $ax = b$. This follows immediately by multiplying $ax = b$ on the left by a^{-1}. Similarly, we can verify that $y = ba^{-1}$ is the only solution of $ya = b$. In other words, in a group, right and left quotients are unique.

For a finite algebraic system with an associative operation, the existence of right and left quotients already follows from their uniqueness.

1.1.3 *Let G be an algebraic system with a binary operation that contains only a finite number of elements. Then, G is a group if and only if the following laws hold:*

(A) *The associative law.*

(C) $ax = ay$ *implies that* $x = y$,

$xa = ya$ *implies that* $x = y$

(left and right cancellation law).

Proof. Multiplication of $ax = ay$ and $xa = ya$ by a^{-1} on the left or on the right, respectively, shows that (C) holds in every group even if it contains infinitely many elements.

Conversely, let $a_1, a_2, \ldots, a_n$ be the elements of G. For an arbitrary element a_i of G, the first part of (C) implies that all the products

$$a_i a_1, a_i a_2, \ldots, a_i a_n$$

are distinct. Thus, these products coincide with $a_1, a_2, \ldots, a_n$ but for their order. It follows that every $a_k \in G$ can be written in the form $a_k = a_i a_m$ with a suitable element a_m. This shows that the first part of (Q) in 1.1.2 is satisfied. The second part of (Q) can be established in the same way. Hence, G is a group.

We do not intend to enter into a closer investigation of group axioms, but we refer to [6].

A group G is said to be *finite* if it contains only a finite number of elements. The number of elements in G is called the *order* of G and will be denoted by $|G|$. If G contains infinitely many elements, then $|G|$ means the cardinal number of G.

A semigroup H is called *commutative* or *abelian if*

$$xy = yx \quad \textit{for arbitrary elements} \quad x, y \textit{ of } H \quad \text{(commutative law)}.$$

It is easy to prove that, in an abelian semigroup, any product of a finite number of elements remains unchanged under an arbitrary permutation of the factors. In an abelian semigroup, we have

$$(xy)^n = x^n y^n \tag{1.2}$$

for any natural number n.

An *abelian group* is a group in which the commutative law holds.

In an abelian group, (1.2) obviously holds for any integer n. It is clear that for an abelian group the two parts of condition (Q) in 1.1.2 and of (C) in 1.1.3 coincide.

Two elements x and y of an arbitrary semigroup are called *permutable* if $xy = yx$. We also say that x and y *commute* with one another. In an arbitrary semigroup, for example, every element commutes with its powers. The unit element of a group commutes with every element, and each element commutes with its inverse.

Occasionally, it is expedient to write the operation in a semigroup or a group as addition instead of multiplication. In particular, we prefer the additive notation for abelian groups. In an additively written group, the associative law, of course, reads as follows:

$$(x + y) + z = x + (y + z).$$

Instead of powers x^n, we have the *multiples*

$$nx = x + x + \cdots + x \qquad (n \text{ summands})$$

satisfying the rules[1]

$$mx + nx = (m + n)x, \; n(mx) = (nm)x.$$

We speak of the null element instead of the unit element, and denote it by 0. It is defined by

$$x + 0 = 0 + x = x \quad \text{for every } x.$$

The additive inverse of an element x is denoted by $-x$, and instead of $x + (-y)$ we simply write $x - y$.

A group G is completely determined when a rule is known by which the product of any two elements can be found. In the case of a finite group, this rule can be expressed by the Cayley group table. This is a square table with $|G|$ rows and $|G|$ columns. The rows and columns are labelled by the elements of G. At the intersection of the row labelled x and the column labelled y there stands the product xy:

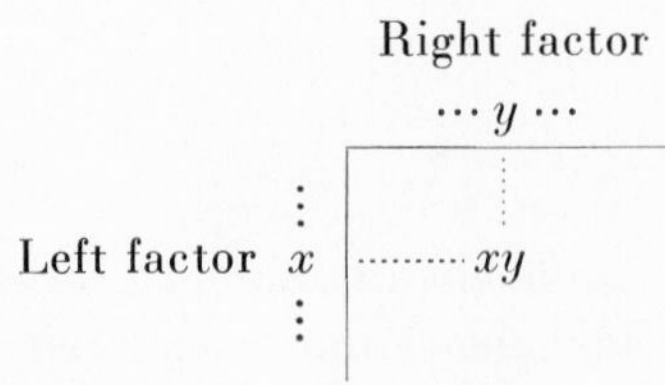

In each row and each column of a group table every element occurs exactly once. For the row labelled x contains all the elements xy, $y \in G$, and for given

[1] Note that on the left-hand side of the first equation the sign + refers to the group operation, whereas on the right-hand side $m + n$ means the sum of the integers m and n.

x and z, there is a unique y such that $xy = z$. A similar argument applies to the columns. A square table such that in every row and in every column each element of a certain set occurs exactly once is called a Latin square. However, not every Latin square may be regarded as a group table. For if we use a Latin square to define a binary operation, then (C) of 1.1.3 is clearly satisfied, yet the associative law need not hold. Since group tables are of minor importance, we refrain from indicating the additional property of a Latin square that implies the associative law (cf. [83]). Provided that the rows and columns are labelled in the same order, a group table is obviously symmetric with respect to its principal diagonal if and only if the group is abelian.

We briefly mention two other kinds of algebraic systems with a binary operation.

A *quasigroup* is an algebraic system with a binary operation such that for any pair a, b the equations $ax = b$ and $ya = b$ have unique solutions x and y, respectively. Thus, the Cayley tables of finite quasigroups coincide with the Latin squares.

A *loop* is a quasigroup with a unit element.

The concepts introduced so far can also serve to define some important algebraic systems with two binary operations.

An algebraic system R with two binary operations

$$x + y \quad \text{(addition)}, \; xy \text{ (multiplication)}$$

is called a *ring*, if the following conditions are satisfied:

(a) The elements of R form an abelian group with respect to addition.

(b) The elements of R form a semigroup with respect to multiplication.

(c) $(x + y)z = xz + yz, \;\; z(x + y) = zx + zy$

for arbitrary $x, y, z \in R$ (distributive laws).

The null element of the additive group is called the null element of the ring. If the multiplicative semigroup contains a unit element, it is called the unit element of R. We say that R is a *commutative ring* if the multiplicative semigroup is abelian. R is said to be a *skew field* if all elements other than the null element form a group with respect to multiplication. If, in addition, this multiplicative group is abelian, then R is called a *field*.

We now consider some examples of groups.

Example 1. Let Ω be an arbitrary non-empty set. A one-to-one mapping of Ω onto itself is called a *permutation* of Ω. The image of an element α of Ω under the permutation x is denoted by αx. If we carry out two permutations x and y of Ω in succession, first x and then y, the result is also a permutation of Ω.

We denote this permutation by xy and call it the product of x and y. This definition of xy can be expressed as follows:

$$\alpha(xy) = (\alpha x)y \quad (\alpha \in \Omega). \tag{1.3}$$

If we perform three permutations x, y, z in this order, then an arbitrary element α of Ω is carried into $((\alpha x)\,y)z$. By (1.3), we have

$$((\alpha x)\,y)z = (\alpha x)(yz) = \alpha(x(yz))$$

and

$$((\alpha x)\,y)z = (\alpha(xy))z = \alpha((xy)z).$$

Since this holds for every $\alpha \in \Omega$, we conclude that the permutations $(xy)\,z$ and $x(yz)$ coincide. Thus, the multiplication of permutations is associative. The identity permutation, which carries every $\alpha \in \Omega$ into itself, plays the role of the unit element. Finally, if x is any permutation of Ω, then the inverse mapping x^{-1} exists and is again a permutation of Ω. This shows that the setof all permutations of Ω forms a group under multiplication. It is obvious that the structure of this group depends only on the cardinal number $|\Omega|$ of the set Ω. We denote this group by $S_{|\Omega|}$ and call it the *symmetric group of degree* $|\Omega|$. The permutation x that carries α into αx is also denoted by

$$x = \begin{pmatrix} \alpha \\ \alpha x \end{pmatrix} \qquad (\alpha \in \Omega).$$

If $|\Omega|$ is finite, $|\Omega| = n$ say, and if $\alpha_1, \alpha_2, \ldots, \alpha_n$ are the elements of Ω, then the symbol

$$x = \begin{pmatrix} \alpha_1 & \alpha_2 & \cdots \alpha_n \\ \alpha_1 x & \alpha_2 x & \cdots \alpha_n x \end{pmatrix}$$

explicitly exhibits the permutation x. The second row is a permutation of the first row in the sense of combinatorics. Since there are $n!$ distinct permutations of n elements, the symmetric group S_n of degree n is of order $n!$. The inverse of the permutation x is

$$x^{-1} = \begin{pmatrix} \alpha x \\ \alpha \end{pmatrix} \quad \text{or} \quad x^{-1} = \begin{pmatrix} \alpha_1 x & \alpha_2 x & \cdots \alpha_n x \\ \alpha_1 & \alpha_2 & \cdots \alpha_n \end{pmatrix}.$$

The six permutations of the set $\Omega = \{1, 2, 3\}$ can be expressed as products of two of them, namely

$$a = \begin{pmatrix} 1 & 2 & 3 \\ 2 & 3 & 1 \end{pmatrix}, \quad b = \begin{pmatrix} 1 & 2 & 3 \\ 1 & 3 & 2 \end{pmatrix}, \quad a^2 = \begin{pmatrix} 1 & 2 & 3 \\ 3 & 1 & 2 \end{pmatrix}$$

$$ab = \begin{pmatrix} 1 & 2 & 3 \\ 3 & 2 & 1 \end{pmatrix}, \quad a^2 b = \begin{pmatrix} 1 & 2 & 3 \\ 2 & 1 & 3 \end{pmatrix}, \quad a^3 = e = \begin{pmatrix} 1 & 2 & 3 \\ 1 & 2 & 3 \end{pmatrix}.$$

It is easily checked that, besides $a^3 = e$, the relations $b^2 = e$ and $ba = a^2b$ hold.

Example 2. For an arbitrary field F, the set of all regular $n \times n$ matrices with elements in F form a group with respect to matrix multiplication. This group is called the general linear group of degree n over F and is denoted by $GL(n, F)$.

Example 3. Let A be an additively written abelian group and Λ an arbitrary ring. Suppose that to an arbitrary $\lambda \in \Lambda$ and any $a \in A$ there is assigned a unique element λa of A. We say that A is a *Λ-module* if the following laws hold for arbitrary $a, b \in A$ and $\lambda, \mu \in \Lambda$:

(a) $\lambda(a + b) = \lambda a + \lambda b$

(b) $(\lambda + \mu) a = \lambda a + \mu a$

(c) $(\lambda\mu) a = \lambda(\mu a)$

If we wish to stress that the elements of Λ are written on the left, we speak of a *left Λ-module.*

If Λ contains a unit element ε and if $\varepsilon a = a$ for every $a \in A$, then A is said to be a *unital* Λ-module.

In case Λ is a skew field, a unital Λ-module is called a *vector space* over Λ.

If a vector space A over a skew field Λ contains a system of n elements $a_1, a_2, \ldots, a_n$ such that every $a \in A$ can uniquely be expressed in the form

$$a = \lambda_1 a_1 + \lambda_2 a_2 + \cdots + \lambda_n a_n \qquad (\lambda_i \in \Lambda),$$

then A is said to be a vector space of *dimension* n over Λ, and the system $a_1, a_2, \ldots, a_n$ is called a *basis* of A.

The vectors of the three-dimensional real euclidean space may serve as a familiar example.

Example 4. Let α be a real number other than 0, 1, -1. Then all powers α^n with integral exponents n are distinct and form an abelian group with respect to multiplication. Every group that consists of the powers of a single element is called a *cyclic* group. In the present case, the group contains infinitely many elements; hence, we speak of an infinite cyclic group.

Example 5. The integers form an abelian group with respect to addition. The null element is 0, and the inverse of n is $-n$. This group has the same property as the group in the preceding example, translated into the additive notation: the group consists of the multiples of the single element 1 (or -1). So, it is an infinite cyclic group in additive notation.

Example 6. Let m be a natural number and $\varepsilon = \cos(2\pi/m) + i \sin(2\pi/m)$. The distinct powers of ε are $\varepsilon^0 = 1, \varepsilon, \varepsilon^2, \ldots, \varepsilon^{m-1}$. They form an abelian group of order m under multiplication. Since $\varepsilon^m = 1$, we have for $k, l = 0, 1, \ldots, m-1$

$$\varepsilon^k \varepsilon^l = \begin{cases} \varepsilon^{k+l} & \text{if } k+l < m, \\ \varepsilon^{k+l-m} & \text{if } k+l \geqq m. \end{cases}$$

This is a cyclic group of order m. For $m = 1$, we obtain a group that consists of the unit element only. This example shows that there are groups of any given finite order.

Example 7. Let m be a natural number. For $k = 0, 1, \ldots, m-1$ let $[k]$ denote the residue class mod m that contains k. The m residue classes $[0]$, $[1], \ldots, [m-1]$ form an abelian group under addition. For $k, l = 0, 1, \ldots, m-1$, we have

$$[k] + [l] = \begin{cases} [k+l] & \text{if } k+l < m \\ [k+l-m] & \text{if } k+l \geqq m \end{cases}$$

Example 8. For a given natural number m and any integer a relatively prime to m, the congruence

$$ax \equiv 1 (\mathrm{mod}\ m)$$

has a solution, which is uniquely determined mod m. Thus, the $\varphi(m)$ prime residue classes mod m, i. e. the residue classes $[k]$ with $(k, m) = 1$ form an abelian group with respect to multiplication, the prime residue class group mod m. Here $\varphi(m)$ denotes the Euler function.

Example 9. For two real numbers x, y, with $0 \leqq x < 1$ and $0 \leqq y < 1$, we define the addition mod 1 as follows:

$$x \dot{+} y = \begin{cases} x+y & \text{if } x+y < 1 \\ x+y-1 & \text{if } x+y \geqq 1. \end{cases}$$

It is obvious that all real numbers x with $0 \leqq x < 1$ form an abelian group under addition mod 1. The subset of the rational numbers of this interval also forms a group. Another important group is obtained by again taking a subset, namely 0 and all those rational numbers of the interval whose denominators are powers of a given prime number p. This group is called a *group of type* p^∞.

Example 10. The Cayley table

	e	a	b	c
e	e	a	b	c
a	a	e	c	b
b	b	c	e	a
c	c	b	a	e

defines a group of order 4 which is called the *four-group*.

1.2 Subgroups

A subset of a group is sometimes called a *complex*. The number of elements in a complex C is denoted by $|C|$; if C contains infinitely many elements, then $|C|$ stands for the cardinal number of C.

A non-empty complex U of a group G is called a *subgroup* if U is a group with respect to the operation defined on G.

Every group G contains two *trivial* subgroups, namely G itself and the subgroup that consists of the unit element e only. The latter is also denoted by the letter e. If G contains subgroups other than G and e, then they are called *nontrivial* subgroups. A *proper* subgroup of G is any subgroup distinct from G.

The notation $C \subseteq G$ means that C is a complex or a subgroup of G. If we wish to stress that C is a proper subset of G we write $C \subset G$.

We call U a *maximal* subgroup of G if U is a proper subgroup and if there is no subgroup T of G such that $U \subset T \subset G$. A group need not contain maximal subgroups. A subgroup of G that is distinct from e and contains no proper subgroup other than e is called a *minimal* subgroup. Not every group contains minimal subgroups.

A non-empty complex U of a group G is a subgroup if and only if the following two conditions hold:

(a) Whenever $x \in U$ and $y \in U$, then $xy \in U$,

(b) Whenever $x \in U$, then $x^{-1} \in U$.

For since U is not empty the two conditions imply that $xx^{-1} = e$ is contained in U.

The conditions (a) and (b) can be replaced by a single condition:

1.2.1 *A non-empty complex U of a group G is a subgroup if and only if the following condition is satisfied:*

For any two elements x, y of U, the product xy^{-1} belongs to U.

Taking $x = y$, this condition implies that $xx^{-1} = e$ belongs to U. From $e \in U$ and $x \in U$, we conclude that $ex^{-1} = x^{-1}$ is in U, i. e. condition (b) is satisfied. Finally, let x and y be any two elements of U. It follows that $y^{-1} \in U$, so that $x(y^{-1})^{-1} = xy$ belongs to U. Thus, condition (a) is satisfied. It is obvious that, conversely, (a) and (b) imply the condition in 1.2.1.

From 1.2.1 we see immediately that *the intersection of an arbitrary set of subgroups of G is again a subgroup of G.*

Let Z be the complex of all those elements of G that commute with every element of G. In other words, $z \in Z$ if and only if $zx = xz$ for every $x \in G$.

Multiplying both sides of the last equation on the left and on the right by z^{-1}, we obtain $xz^{-1} = z^{-1}x$. Thus, $z \in Z$ implies that $z^{-1} \in Z$. For $z_1 \in Z$ and $z_2 \in Z$, we have

$$z_1 z_2 x = z_1 x z_2 = x z_1 z_2,$$

hence, $z_1 z_2 \in Z$. Finally, Z is not empty, because at least e is contained in Z. Thus, Z is a subgroup of G. We call Z the *centre* of G and denote it by $\mathsf{Z}(G)$.

Clearly, G is abelian if and only if $\mathsf{Z}(G) = G$. On the other hand, there are groups whose centre consists of the unit element only, for instance the group S_3 (cf. Example 1). Such groups are usually called *groups without centre*, though this name is not quite accurate and one should rather speak of groups with trivial centre.

Taking a fixed element a of G instead of all elements of G, we can show in the same way that all elements of G that commute with a form a subgroup of G. This subgroup is called the *normalizer* of a and is denoted by $\mathsf{N}(a)$. It is obvious that

$$\mathsf{Z}(G) = \bigcap_{a \in G} \mathsf{N}(a).$$

Let A be any non-empty complex of G. There are subgroups of G that contain all the elements of A, for example the group G itself. The intersection of all subgroups of G that contain A is denoted by $\langle A \rangle$. We call $\langle A \rangle$ the subgroup *generated by* A. It is clear that every subgroup of G that contains A also contains $\langle A \rangle$. In this sense, $\langle A \rangle$ is the smallest subgroup of G that contains A. We consider the complex C of all finite products

$$a_1^{m_1} a_2^{m_2} \cdots a_r^{m_r} \qquad (r = 1, 2, \ldots)$$

where the a_i are arbitrary, not necessarily distinct elements of A, and the m_i are integers. It is easy to see that C is a subgroup that contains A. On the other hand, every subgroup that contains A also contains all these products. Consequently $C = \langle A \rangle$. Note that distinct products may represent one and the same element of G; for example, in the group S_3, we have $ba = a^2 b$ (cf. Example 1).

For an arbitrary set $A, B, C, \ldots$ of complexes of the group G, we denote by $\langle A, B, C, \ldots \rangle$ the subgroup generated by the union of $A, B, C, \ldots$.

If $\langle A \rangle = G$, then the complex A is said to be a system of *generators* of the group G. We call G *finitely generated* if $G = \langle A \rangle$ for some finite complex A.

A *cyclic* group is one that can be generated by a single element. If $G = \langle a \rangle$, then G consists of all distinct powers of the element a. We have to consider two cases according as all the powers of a are distinct or not. If all the powers a^k, $k = 0, \pm 1, \pm 2, \ldots$ are distinct, then $\langle a \rangle$ is an *infinite cyclic* group (cf.

Example 4). In the second case, we have $a^k = a^l$ for some distinct integers k, l. We may assume that $k > l$, and write $a^{k-l} = e$. Thus, there exist natural numbers n for which $a^n = e$. Let m be the smallest natural number with this property. Then, we have $a^m = e$, and the powers $a^0 = e, a, a^2, \ldots, a^{m-1}$ are distinct. For $a^r = a^s$, with $0 \leqq r < s \leqq m - 1$, would imply that $a^{s-r} = e$ where $1 \leqq s - r \leqq m - 1$, which contradicts the choice of m. This shows that $\langle a \rangle$ is a cyclic group of order m (cf. Example 6).

For an element a of G, we call $\langle a \rangle$ the cyclic subgroup generated by a. If $\langle a \rangle$ is an infinite cyclic group, then a is said to be an *element of infinite order*. If, however, $\langle a \rangle$ is a finite cyclic group, $|\langle a \rangle| = m$ say, then a is called an *element of order m*. It is clear that in any group the unit element is the only element of order 1.

Of course, in a finite group, every element is of finite order. In infinite groups, all possibilities concerning the orders of the elements actually occur. There are infinite groups in which every element other than the unit element is of infinite order. Such groups are called *torsion-free*. On the other hand, there are infinite groups all of whose elements have finite order. In this case, we speak of *periodic* groups. Finally, it is easy to find infinite groups that contain elements both of infinite and of finite order other than the unit element.

By the *commutator* of the elements x, y of the group G, we understand the element $[x, y] = x^{-1} y^{-1} xy$.

We have

$$yx[x, y] = xy,$$

so that x and y commute if and only if $[x, y] = e$. Let K be the complex of the commutators of any two elements of G. The subgroup $\langle K \rangle$ is called the *commutator subgroup* or the *derived group* of G. The complex K itself need not be a group. The commutator subgroup plays an important role in group theory. It is clear that G is abelian if and only if $K = e$. On the other hand, there are groups that coincide with their commutator subgroup.

By the product AB of two complexes A and B, we mean the complex whose elements are the distinct products ab with $a \in A$, $b \in B$.

The complexes AB and BA may be distinct. If we have $AB = BA$, then we say that A and B are *permutable* or that A and B *commute*. Note that the relation $AB = BA$ does not imply that every element of A commutes with every element of B. The equation $AB = BA$ only means that every product ab can also be written in the form $b_1 a_1$ with suitable elements $a_1 \in A$, $b_1 \in B$, and that, conversely, every product $b_1 a_1$ can be expressed as a product ab. If, in addition, every element of A commutes with every element of B, then A and B are said to commute *elementwise*. If we wish to emphasize that A and

B are permutable in the weaker sense only, then we say that A and B *commute as a whole.*

It is clear that the associative law $(AB)\,C = A(BC)$ holds, because it is satisfied even for the single elements.

The complex of all a^{-1} with $a \in A$ is denoted by A^{-1}. In view of $(ab)^{-1} = b^{-1}a^{-1}$, we have $(AB)^{-1} = B^{-1}A^{-1}$.

Condition (a) is equivalent to $UU \subseteq U$. If U is a subgroup, then we actually have $UU = U$, since U contains the unit element; moreover, 1.1.2 shows that $xU = Ux = U$ for every $x \in U$. By (b), we have $U^{-1} = U$ for every subgroup U. Theorem 1.2.1 may be expressed as follows: A nonempty complex U is a subgroup if and only if $UU^{-1} \subseteq U$. It is clear that, for a subgroup U, we actually have $UU^{-1} = U$.

If U and V are two subgroups, then the product UV need not be a subgroup. The next theorem gives a necessary and sufficient condition for UV to be a subgroup.

1.2.2 *If U and V are subgroups, then UV is a subgroup if and only if U and V are permutable as a whole.*

Indeed, if U and V are permutable, we have in view of $U^{-1} = U$ and $V^{-1} = V$:

$$UV(UV)^{-1} = UVV^{-1}U^{-1} = UVVU = UUVV = UV;$$

hence, UV is a subgroup. Conversely, if UV is a subgroup, then we obtain

$$UV = (UV)^{-1} = V^{-1}U^{-1} = VU.$$

1.3 Cosets

If B is any non-empty complex of the group G, and a an arbitrary element of G, then the products ab, $b \in B$, are all distinct.

Let U be a proper subgroup of G. A complex of the form aU, $a \in G$, is called a *left coset of U*. The element a is said to be a *representative* of that coset. Replacing a by $a_1 = au$, with $u \in U$, we obtain $a_1U = auU = aU$. Conversely, from $a_1U = aU$ it follows that $a_1 \in aU$, since $e \in U$. So, we have $a_1 = au$, with $u \in U$. This shows that *all the elements of aU and only these elements can be used as representatives of aU.*

One of the left cosets of U in G is the subgroup U itself. It is often convenient to take e as a representative.

The following lemma is highly important.

1.3.1 *Two left cosets a_1U and a_2U of the subgroup U are equal or disjoint.*

For, if a_1U and a_2U are not disjoint, then $a_1U \cap a_2U$ contains at least one element x. We have $x = a_1u_1 = a_2u_2$ with $u_1, u_2 \in U$. It follows that $a_1u_1U = a_2u_2U$; hence, $a_1U = a_2U$.

By 1.3.1, we can partition G into left cosets of U:

$$G = aU \cup bU \cup cU \cup \cdots = \bigcup_{x \in L} xU. \tag{1.4}$$

This partition is called the *left coset decomposition of G with respect to the subgroup U*.[1] The complex L of the representatives $a, b, c, \ldots$ in (1.4) is called a *system of left representatives* or a *left transversal* to U in G.

Similarly, we can define *right cosets* Uy of U. What we proved for left cosets carries over to right cosets. In particular, we obtain a decomposition

$$G = \bigcup_{y \in R} Uy$$

of G into right cosets of U. The set R of the right representatives y is said to be a *right transversal* to U in G. Clearly, in an abelian group, we need not distinguish between left and right cosets. As we shall see later, there is also an important class of subgroups of non-abelian groups for which left cosets and right cosets coincide.

If L is a left transversal, then L^{-1} is a right transversal. For, by replacing every element in (1.4) by its inverse, we obtain

$$G = G^{-1} = \bigcup_{x \in L} U^{-1}x^{-1} = \bigcup_{x \in L} Ux^{-1} = \bigcup_{y \in L^{-1}} Uy.$$

This shows, in particular, that *the number*[2] *of left cosets is equal to the number of right cosets*. This number is denoted by $|G:U|$ and is called the *index* of U in G.

If U is the subgroup that consists of the unit element only, then every coset contains a single element, and we have $|G:e| = |G|$.

In a finite group, the index of any subgroup is, of course, finite. Equation (1.4) is a partition into $|G:U|$ cosets, and each coset contains $|U|$ elements. Counting the number of elements on both sides of (1.4), we obtain

1.3.2 Lagrange's Theorem. *For any subgroup U of the finite group G, we have*

$$|G| = |G:U||U| = |G:U||U:e|.$$

In particular, the order as well as the index of any subgroup are divisors of $|G|$.

---

[1] Another notation is

$$G = aU + bU + cU + \cdots = \sum_{x \in L} xU$$

where the signs $+$ and $\sum$ do not refer to any addition but to the union of pairwise disjoint subsets.

[2] If there are infinitely many cosets, then number means the cardinal number. Throughout, we always use the term number in this sense.

By applying 1.3.2 to the case when U is the cyclic subgroup generated by an element of order m, we obtain the following corollary:

1.3.3 *In a finite group G, the order of any element is a divisor of $|G|$.*

An immediate consequence of 1.3.3 is

1.3.4 Fermat's Theorem. *Every element x of a finite group G satisfies the equation $x^{|G|} = e$.*

For, the order m of x is a divisor of $|G|$, say $|G| = mq$. Therefore, we have

$$x^{|G|} = x^{mq} = (x^m)^q = e^q = e.$$

The *exponent* k of a finite group G is the least common multiple of all the orders of elements in G. Thus, we have $x^k = e$ for every $x \in G$. By 1.3.3, k is a divisor of $|G|$. While simple examples show that k may well be a proper divisor of G, we shall prove later that every prime divisor of $|G|$ also divides k (cf. 3.2.2).

Let U and V be two finite subgroups. We shall determine the number of distinct elements in the product UV. Note that we do not assume UV to be a group.

1.3.5 *If U and V are two finite subgroups, then the number of distinct elements in the complex UV is*

$$|U||V||U \cap V|^{-1}.$$

Proof. We decompose U into left cosets and V into right cosets with respect to the intersection $D = U \cap V$:

$$U = u_1 D \cup \cdots \cup u_r D, \quad V = Dv_1 \cup \cdots \cup Dv_s.$$

The union

$$\bigcup_{\substack{i=1,\ldots,r \\ j=1,\ldots,s}} u_i D v_j$$

obviously contains all the elements of UV. It remains to show that all the $|U|\,|V|\,|D|^{-1}$ elements of this union are distinct from one another. If

$$u_i d v_j = u_k d' v_l, \qquad d, d' \in D,$$

then

$$u_k^{-1} u_i d = d' v_l v_j^{-1}.$$

Let x be the element that stands on both sides of the last equation. From $x = u_k^{-1} u_i d$, we conclude that $x \in U$, and $x = d' v_l v_j^{-1}$ shows that $x \in V$. So, we have $x \in D$, which implies that $u_i = u_k$, $v_j = v_l$, $d = d'$.

1.4 Subgroups of Cyclic Groups

In this section we shall determine all the subgroups of cyclic groups. Let $\langle a\rangle$ be a cyclic group of finite or infinite order, and let U be a non-trivial subgroup of $\langle a\rangle$. If a certain power a^n belongs to U, then a^{-n} also belongs to U. This shows that U contains at least one power of a with a positive exponent. Let d be the least positive exponent such that a^d is contained in U. We call d the minimal exponent of U. Now let a^s be an arbitrary element of U. We divide s by d to obtain

$$s = qd + r, \quad 0 \leqq r < d.$$

From $a^s \in U$ and $a^d \in U$, we conclude that $a^s(a^{-d})^q = a^r$ belongs to U. Since d is the minimal exponent of U, this leads to a contradiction unless $r = 0$. Thus, the exponents of all powers of a that belong to U are divisible by d. This means $U = \langle a^d\rangle$. Hence, every subgroup of a cyclic group is cyclic.

In the case of an infinite cyclic group, every subgroup other than e is again an infinite cyclic group.

For any integer n, there is a uniquely determined pair of integers k, j such that

$$n = kd + j, \quad 0 \leqq j \leqq d - 1.$$

This shows that an arbitrary element $a^n = (a^d)^k a^j$ of $\langle a\rangle$ has a unique representation of the form

$$a^n = ua^j \quad \text{with} \quad u \in \langle a^d\rangle, \quad 0 \leqq j \leqq d - 1.$$

Consequently,

$$\langle a\rangle = \langle a^d\rangle \cup \langle a^d\rangle\, a \cup \cdots \cup \langle a^d\rangle\, a^{d-1} \tag{1.5}$$

is the coset decomposition of $\langle a\rangle$ with respect to $\langle a^d\rangle$. Moreover, there is a one-to-one correspondence between the cosets and the residue classes mod d. Note, in particular, that in the case of an infinite cyclic group $\langle a\rangle$ the index

$$|\langle a\rangle : \langle a^d\rangle| = d$$

is always finite.

If $\langle a\rangle$ is an infinite cyclic group and d any given natural number, then there always exists a subgroup whose minimal exponent is d, namely $\langle a^d\rangle$. If $\langle a\rangle$ has the finite order m, then any possible minimal exponent must divide m, since it is the index of a subgroup. Conversely, for any positive divisor d of m, the subgroup $\langle a^d\rangle$ has the minimal exponent d. For, if $m = df$, then $\langle a^d\rangle$ consists of the elements

$$a^0 = e, \quad a^d, a^{2d}, \ldots, a^{(f-1)d}.$$

Since any subgroup is uniquely determined by its minimal exponent and since the minimal exponent coincides with its index, we have the following theorem:

1.4.1 *Every subgroup of a cyclic group is cyclic. There is one and only one subgroup of any possible index. In the case of an infinite cyclic group every natural number occurs as the index of a subgroup. The indices that occur in the case of a finite cyclic group are precisely the positive divisors of its order.*

In a finite group, the order and the index of a subgroup uniquely determine each other. This gives the following corollary for finite cyclic groups:

A finite cyclic group of order m contains one and only one subgroup whose order is a given positive divisor of m.

Every group whose order is a prime number p is cyclic; for every element other than e is of order p and therefore generates the whole group. Moreover, the groups of prime order p contain only the two trivial subgroups. The converse of this proposition is also true:

1.4.2 *If the group G is of order greater than one and contains only trivial subgroups, then the order of G is a prime number.*

Indeed, for every element $\neq e$ of G, we have $\langle a \rangle = G$. Thus, G is cyclic. Our proposition now follows from 1.4.1.

1.5 Subgroups of Finite Index

Let U and V be two subgroups of G such that $V \subseteq U$. We consider the decompositions

$$G = \bigcup_{x \in L} xU, \quad U = \bigcup_{y \in M} yV$$

into left cosets. It is easy to see that the products xy ($x \in L$, $y \in M$) are all distinct and form a left transversal to V in G. For any element a of G can be written in the form $a = x_1 u$, with $x_1 \in L$, $u \in U$; moreover, we have $u = y_1 v$, where $y_1 \in M$, $v \in V$. So, we obtain $a = x_1 y_1 v$. If the intersection $x_1 y_1 V \cap x_2 y_2 V$ contains an element a_0, then we have

$$a_0 = x_1 y_1 v_1 = x_2 y_2 v_2; \qquad v_1, v_2 \in V.$$

Since the left coset of U to which a_0 belongs is unique, we conclude that $x_1 = x_2$. This gives $y_1 v_1 = y_2 v_2$, which implies that $y_1 = y_2$, because M is a left transversal to V in U. Consequently,

$$G = \bigcup_{\substack{x \in L \\ y \in M}} xyV.$$

In the case of a finite index $|G: V|$, it follows that

$$|G : V| = |G : U| |U : V|. \tag{1.6}$$

Now let

$$G = \bigcup_{x \in N} xA \tag{1.7}$$

be the decomposition of G into left cosets of a subgroup A. If B is another subgroup of G, then we have

$$B = G \cap B = \bigcup_{x \in N} (xA \cap B). \tag{1.8}$$

Some of the complexes $xA \cap B$ may be empty, but those that are not empty are certainly disjoint because this is even true for the cosets on the right-hand side of (1.7). By omitting the empty complexes on the right-hand side of (1.8), we obtain a partition of B into disjoint complexes. We shall see that this partition is the decomposition of B into left cosets of $A \cap B$. For, if $xA \cap B$ is not empty, then we can choose a representative b of xA that is in B, i. e. $xA = bA$ with $b \in B$.
Thus, we have

$$xA \cap B = bA \cap B = bA \cap bB = b(A \cap B),$$

which shows that $xA \cap B$ is a left coset of $A \cap B$.

If $|G{:}A|$ is finite, then the number of non-empty complexes on the right-hand side cannot exceed $|G{:}A|$. This gives:

1.5.1 *If A and B are subgroups of G and if the index $|G{:}A|$ is finite, then*

$$|B : A \cap B| \leqq |G : A|.$$

If $|G{:}B|$ is also finite, the last inequality yields

$$|G : B||B : A \cap B| \leqq |G : A||G : B|$$

or, in view of (1.6),

$$|G : A \cap B| \leqq |G : A||G : B|. \tag{1.9}$$

So, we obtain:

1.5.2 Poincaré's Theorem. *If A and B are subgroups of G whose indices are finite, then the index of $A \cap B$ in G is finite and satisfies* (1.9).

By induction, this theorem can be extended to the case of any finite number of subgroups all of whose indices are finite.
1.5.1 can be completed as follows:

1.5.3 *If A and B are subgroups of finite index in G, then the equation*

$$|B : A \cap B| = |G : A| \tag{1.10}$$

holds if and only if $G = AB = BA$.

Proof. If $G = BA$, then every coset of A in (1.7) can be represented by an element of B. Therefore, none of the intersections on the right-hand side of (1.8) is empty, and hence (1.10) holds.

Conversely, (1.10) implies that none of the intersections $xA \cap B$ is empty. Thus, every coset xA has a representative that belongs to B. This means that $G = BA$. By 1.2.2, this implies that $AB = BA$.

In case A and B generate a finite subgroup of G, 1.5.1 and 1.5.3 may be stated as follows:

1.5.4 *Let A and B be subgroups of G such that $\langle A, B\rangle$ is finite. Then*

$$|A||B| \leqq |\langle A, B\rangle||A \cap B|.$$

The equality sign holds if and only if $AB = BA$.

For, from 1.5.1, it follows that

$$\frac{|B|}{|A \cap B|} = |B : A \cap B| \leqq |\langle A, B\rangle : A| = \frac{|\langle A, B\rangle|}{|A|}.$$

As a useful corollary we obtain:

1.5.5 *If A and B are subgroups of G whose indices are relatively prime, then $G = AB = BA$ and*

$$|G : A \cap B| = |G : A||G : B|. \tag{1.11}$$

Proof. By (1.6), we have

$$|G : A \cap B| = |G : A||A : A \cap B| = |G : B||B : A \cap B|. \tag{1.12}$$

Since $|G : A|$ is relatively prime to $|G : B|$, we infer that $|G : A|$ is a divisor of $|B : A \cap B|$. In view of 1.5.1, this implies that

$$|B : A \cap B| = |G : A|.$$

By 1.5.3, this gives $G = AB$. Then, (1.12) implies (1.11).

1.6 Double Modules

In this section, we generalize the coset decomposition of a group with respect to a subgroup.

From 1.2.1, we conclude: If U is a subgroup of G, then so is $a^{-1}Ua$ for any $a \in G$.

Let U and V be two subgroups of G. By a *double coset modulo* (U, V) we mean a complex of the form UaV, where a is an arbitrary element of G. Two such double cosets coincide or are disjoint. Suppose that

$$uav \in UbV$$

for some elements $u \in U$, $v \in V$. Then, we have

$$uav = u_1 b v_1; \quad u_1 \in U, \quad v_1 \in V,$$

hence,

$$UaV = UuavV = Uu_1bv_1V = UbV.$$

Consequently, there exists a partition

$$G = UV \cup UxV \cup UyV \cup \cdots \tag{1.13}$$

of G into pairwise disjoint complexes. We call (1.13) the *decomposition of G with respect to the double module* (U, V). Every double coset modulo (U, V) coincides with one of the double cosets in (1.13).

Clearly, every double coset UaV consists of complete right cosets of U as well as of complete left cosets of V. Under the assumption that both indices $|G: U|$ and $|G: V|$ are finite, we shall determine the number of right cosets of U and the number of left cosets of V that belong to UaV. Any right coset of U contained in UaV has the form Uav with some $v \in V$. If two such cosets coincide, i. e.

$$Uav_1 = Uav_2; \quad v_1, v_2 \in V, \tag{1.14}$$

then we have $av_2v_1^{-1}a^{-1} \in U$ or $v_2v_1^{-1} \in a^{-1}Ua$. Since $v_2v_1^{-1}$ also belongs to V, (1.14) implies that $v_2v_1^{-1}$ belongs to $V \cap a^{-1}Ua = D$, which shows that $Dv_2 = Dv_1$. Conversely, assume that $Dv_2 = Dv_1$. Then, $v_1 = dv_2$ where $d \in D$. Since $d = a^{-1}ua$, with $u \in U$, we obtain

$$Uav_1 = Uadv_2 = Uaa^{-1}uav_2 = Uav_2.$$

This gives the following result: The number of right cosets of U contained in UaV is equal to the number of right cosets of D contained in V, that is, equal to the index $|V:(V \cap a^{-1}Ua)|$.

By similar reasoning, we find that the number of left cosets of V contained in UaV is equal to $|U:(U \cap aVa^{-1})|$.

Exercises

1. Give an example to show that one cannot dispose of the finiteness condition in 1.1.3.

2. Give an example of a quasigroup of three elements that is not a loop. Give an example of a loop of five elements that is not a group.

3. Show that every abelian group in additive notation occurs as the additive group of some ring.

4. Let S be a semigroup with the unit element e. An element u of S is called a unit if there are elements x, y in S such that $xu = uy = e$. Prove that $x = y$. Let U denote the set of all units of S. Prove that U is a subgroup of S and that every subgroup of S containing e is a subgroup of U. Show by an example that S may contain subgroups that do not belong to U.

5. Let $a_1, \ldots, a_n$ be a basis of an n-dimensional vector space A over a field. Find all the bases of A.

6. A subgroup S of a finite-dimensional vector space A over a skew field Λ is said to be admissible if $x \in S$ implies that $\lambda x \in S$ for every $\lambda \in \Lambda$. Prove that every admissible subgroup of A is a vector space over Λ.

7. Determine the elements of finite order and the elements of infinite order of the group of the real numbers x, $0 \leq x < 1$, under addition mod 1.

8. Prove that in every abelian group the elements of finite order form a subgroup. Do the elements of infinite order together with the identity also form a subgroup?

9. Show that an infinite cyclic group contains no minimal subgroup.

10. Show that a group of type p^∞ contains no maximal subgroup.

11. Determine the centre of $GL(n, F)$.

12. Find the normalizer of a diagonal matrix in $GL(n, F)$.

13. Let a_1, a_2, a_3 be a basis of a three-dimensional vector space over the real numbers, and let H be the subgroup generated by a_1, a_2. Consider a_1, a_2, a_3 as vectors in a three-dimensional euclidean space and give a geometric description of H and its cosets.

14. Give examples of infinite groups G_1, G_2, G_3 such that G_1 is torsion-free, G_2 is periodic, and G_3 contains both elements of infinite order and elements of finite order other than e.

15. Prove that every group of exponent 2 is abelian.

16. Determine all groups of order 4 and of order 6.

17. Let n be relatively prime to the exponent of G. Prove that $x^n = y^n$ implies that $x = y$. If G is finite and $a \in G$, show that there exists one and only one $x \in G$ such that $x^n = a$.

18. Prove that every proper subgroup of a group of type p^∞ is finite. Thus, a group of type p^∞ is not finitely generated.

19. Let G be a cyclic group of order m and put $m = hk$, where $h > 1$, $k > 1$. Show that the unique subgroup H of order h can be characterized in either of the following ways:

(a) H consists of all elements x^k, $x \in G$.

(b) H consists of all $x \in G$ such that $x^h = e$.

20. Let H be a subgroup of G and define the relation xRy to mean that $xy^{-1} \in H$. Show that this is an equivalence relation and determine the equivalence classes. Consider in the same way the relation xLy that holds if and only if $y^{-1}x \in H$. Give an example to show that the two relations need not coincide. Find a necessary and sufficient condition on H under which the relations coincide.

CHAPTER 2

HOMOMORPHISMS

2.1 Homomorphic and Isomorphic Mappings

Let G and $\bar{G}$ be two groups. We assume that there exists a mapping σ of G onto $\bar{G}$ that assigns to every $x \in G$ a unique image $x\sigma \in \bar{G}$ such that

$$(xy)\sigma = (x\sigma)(y\sigma) \quad \text{for any two elements} \quad x, y \text{ of } G \tag{2.1}$$

Such a mapping σ is called a *homomorphic mapping* or a *homomorphism* of G onto $\bar{G}$.

As an example, we consider the group $GL(n, F)$ (cf. Example 2 in section 1.1). If we assign to every matrix a of that group its determinant $|a|$, then the multiplication theorem for determinants gives $|ab| = |a||b|$. This shows that the mapping $\sigma\colon a \to |a|$ is a homomorphism of $GL(n, F)$ onto the multiplicative group of all non-zero elements of the field F.

Occasionally, one has to consider a slightly more general situation. Let $\bar{H}$ be an algebraic system with a binary operation, which will be denoted as multiplication. Moreover, let σ be a mapping of the multiplicative group G into $\bar{H}$ for which (2.1) holds. Then, the set of all the images $x\sigma$, $x \in G$, is a certain subset $\bar{G} \subseteq \bar{H}$.[1] We shall prove that $\bar{G}$ is a group. Let $x\sigma$ and $y\sigma$ be two arbitrary elements of $\bar{G}$. Then, x and y are in G and, since G is a group, we have $xy \in G$. This implies that $(xy)\sigma = (x\sigma)(y\sigma)$ is contained in $\bar{G}$. Thus, $\bar{G}$ contains the product of any two of its elements, i. e. the binary operation defined on $\bar{H}$ yields a binary operation on $\bar{G}$. For any three elements $x\sigma$, $y\sigma$, $z\sigma$ of G, we have preimages x, y, z in G. In view of (2.1), the associative law $(xy)z = x(yz)$ in G gives

$$[(x\sigma)(y\sigma)](z\sigma) = (x\sigma)[(y\sigma)(z\sigma)].$$

This proves the associative law for $\bar{G}$. The image $e\sigma$ of the unit element of G has the property $(x\sigma)(e\sigma) = x\sigma$ and is therefore a right unit element of $\bar{G}$. Finally, $xx^{-1} = e$ gives $(x\sigma)\,(x^{-1}\sigma) = e\sigma$, which shows that $x^{-1}\sigma$ is a right

[1] In case $\bar{G} = \bar{H}$, we call σ an *epimorphism* of G onto $\bar{H}$.

inverse of $x\sigma$. Thus, $\bar{G}$ is a group. Since $\bar{G}$ is the image of G under σ, we write $\bar{G} = G\sigma$.

We emphasize once more the two properties of every homomorphism which we have just proved. *Let σ be any homomorphism of G onto $G\sigma$. Then, the image $e\sigma$ of the unit element e of G is the unit element of $G\sigma$. The image $(x^{-1})\sigma$ of x^{-1} is the inverse of $x\sigma$, i. e. $(x^{-1})\sigma = (x\sigma)^{-1}$.*

In the case of an arbitrary homomorphism σ of G onto $G\sigma$, it may happen that distinct elements of G have one and the same image in $G\sigma$. But, if the mapping σ is one-to-one, i. e. if $x \neq y$ implies that $x\sigma \neq y\sigma$, then it is called an *isomorphic* mapping or an *isomorphism* of G onto $G\sigma$.[1] In this case, the inverse mapping σ^{-1} exists and is an isomorphism of $G\sigma$ onto G. A homomorphism that is not one-to-one is called a *proper* homomorphism.

We say that G is isomorphic to $\bar{G}$, and write $G \cong \bar{G}$, if there exists at least one isomorphism σ of G onto $\bar{G}$. As we observed in the last paragraph, the inverse mapping σ^{-1} is then an isomorphism of $\bar{G}$ onto G so that we have $\bar{G} \cong G$. Clearly $G \cong G$, because the identity mapping is an isomorphism. Finally, $G \cong \bar{G}$ and $\bar{G} \cong \tilde{G}$ imply that $G \cong \tilde{G}$, for if σ is an isomorphism of G onto $\bar{G}$ and τ an isomorphism of $\bar{G}$ onto $\tilde{G}$, then the combined mapping $\sigma\tau$ is obviously an isomorphism of G onto $\tilde{G}$. This shows that $\cong$ is an equivalence relation. Groups that belong to the same equivalence class are said to be isomorphic to each other or to belong to the same type. Isomorphic groups may differ from one another in the nature of their elements (e. g. permutations, matrices, numbers), in the operation (e. g. product of permutations, multiplication of matrices, addition or multiplication of numbers), and in the notation, but all the formal laws and relations that do not refer to the particular nature of the group elements or the operation are the same for all groups of the same type. We therefore speak of the type of the infinite cyclic group, of the four-group, and of the symmetric group of degree n or, briefly, of *the* infinite cyclic group, *the* four-group, and *the* symmetric group of degree n. In this book, we are mainly concerned with the common properties of all groups of the same type. The Examples 4 and 5 in section 1.1 are isomorphic groups belonging to the type of the infinite cyclic group. The Examples 6 and 7 belong to the type of the cyclic group of order m.

If G and $\bar{G}$ are isomorphic, then there are, in general, several isomorphisms of G onto $\bar{G}$, possibly infinitely many.

If σ is a homomorphism of G onto $\bar{G} = G\sigma$, then the image $\bar{U} = U\sigma$ of any subgroup U of G is a subgroup of $\bar{G}$. For, let $x\sigma$ and $y\sigma$ be arbitrary ele-

[1] A mapping σ of G into $\bar{H}$ is said to be a *monomorphism* if distinct elements of G have distinct images in $\bar{H}$.

ments of $U\sigma$. Then x and y belong to U, so that $xy^{-1} \in U$. This implies that $(xy^{-1})\sigma = (x\sigma)(y\sigma)^{-1}$ is in $U\sigma$. Thus, $U\sigma$ is a subgroup.

Conversely, if $\bar{U}$ is any subgroup of $\bar{G}$, then the complex U of all the elements x of G with the property $x\sigma \in U$ is a subgroup of G. The proof is obvious: For two arbitrary elements x, y of U, we have $x\sigma, y\sigma \in U$. Since $\bar{U}$ is a subgroup, we conclude that $(x\sigma)(y\sigma)^{-1} = (xy^{-1})\sigma$ belongs to $\bar{U}$. This implies that $xy^{-1} \in U$, hence U is a subgroup.

We now assume that G and $\bar{G}$ are isomorphic and that σ is an isomorphism of G onto $\bar{G}$. In this case, corresponding subgroups U and $U\sigma$ have the same order. Moreover, $|G:U| = |G\sigma:U\sigma|$, because σ carries a right (left) transversal to U in G into a right (left) transversal to $U\sigma$ in $G\sigma$. An element x of G and its image $x\sigma$ have the same order, since either of the two equations $x^n = e$ and $(x\sigma)^n = e\sigma$ implies the other.

By an *endomorphism* α of a group G we mean a homomorphism of G into itself, i. e. a mapping of G into itself such that

$$(xy)\alpha = (x\alpha)(y\alpha) \quad \text{for any two elements} \quad x, y \text{ of } G. \tag{2.2}$$

The complex $G\alpha$ of all the images $x\alpha$, $x \in G$ is a subgroup of G.

Example. In the group S_3 (cf. Example 1 in section 1.1) define $a\alpha = e$, $b\alpha = b$. Then, α is an endomorphism of S_3 onto the subgroup $\langle b \rangle$.

If the endomorphism α is a one-to-one mapping and if the set $G\alpha$ of the images is the whole group G, then α is called an *automorphism* of G. In other words, an automorphism of G is an isomorphism of G onto itself.[1]

Let α and β be two endomorphisms of G. We define the product $\alpha\beta$ as the mapping of G into itself that we obtain by first carrying out α and then β,

$$x(\alpha\beta) = (x\alpha)\beta \qquad (x \in G).$$

We have

$$(xy)(\alpha\beta) = [(xy)\alpha]\,\beta = [(x\alpha)(y\alpha)]\,\beta = [(x\alpha)\,\beta]\,[(y\alpha)\,\beta] = [x(\alpha\beta)]\,[y(\alpha\beta)]\,.$$

Consequently, $\alpha\beta$ is an endomorphism of G. The same argument as in section 1.1, Example 1, shows that the multiplication of endomorphisms is associative. Thus, all the endomorphisms of G form a semigroup under multiplication. The identity mapping is the unit element of this semigroup. If an endomorphism is one-to-one, i. e. if it is an automorphism, then the inverse mapping exists and is obviously an automorphism. Moreover, the product of two automorphisms is an automorphism. This shows that all the automorphisms of G

[1] An isomorphism of a group into itself is called a *meromorphism*. Proper meromorphisms, i. e. meromorphisms that are not automorphisms, can obviously occur only in infinite groups.

form a multiplicative group. This group is called the *automorphism group* of G and will be denoted by $\mathsf{A}(G)$.

2.2 Inner Automorphisms and Conjugate Elements

An important kind of automorphism of a group G is obtained as follows:

For a fixed element a of G, define the mapping τ_a by

$$x\tau_a = a^{-1}xa \quad \text{for all} \quad x \in G.$$

This mapping is certainly one-to-one, since $a^{-1}xa = a^{-1}x_1a$ implies that $x = x_1$. Moreover, it is a mapping of G onto G; for if x is an arbitrary element of G, then we have

$$(axa^{-1})\tau_a = a^{-1}(axa^{-1})\,a = x,$$

so that x is the image of some element of G under τ_a. Finally, for any two elements x, y of G,

$$(xy)\,\tau_a = a^{-1}xya = a^{-1}xaa^{-1}ya = (x\tau_a)(y\tau_a),$$

This shows that τ_a is an automorphism of G. We call τ_a an *inner automorphism* or, in more detail, the inner automorphism induced by a. We also say that $a^{-1}xa$ is obtained from x by *transformation* by a.

We now form the product of two inner automorphisms τ_a, τ_b:

$$x(\tau_a\tau_b) = (x\tau_a)\,\tau_b = b^{-1}(a^{-1}xa)\,b = (ab)^{-1}x(ab) = x\tau_{ab}.$$

Since this holds for every $x \in G$, we have

$$\tau_a\tau_b = \tau_{ab}. \tag{2.3}$$

This gives, in particular, $\tau_a\,\tau_{a^{-1}} = \tau_e$, where τ_e, is the identity mapping of G onto itself. We infer that the inner automorphisms of G form a group. This group will be denoted by $\mathsf{J}(G)$.

Let us assign to any element a of G the inner automorphism τ_a. By (2.3), we obtain a homomorphism of G onto $\mathsf{J}(G)$.

The inner automorphism τ_z is the identity automorphism if and only if z is an element of the centre of G. It follows that the only inner automorphism of an abelian group is the identity. On the other hand, there are groups all of whose automorphisms are inner ones.

All automorphisms of G other than the inner ones are called *outer* automorphisms.

An element x of G is called *conjugate* to y if there exists at least one inner automorphism of G that carries x into y, in other words, if $y = a^{-1}xa$ for at least one $a \in G$. It is easy to see that conjugacy is an equivalence relation. First, every element x is conjugate to itself since $x = e^{-1}xe$. If x is conjugate to y, i. e. $y = a^{-1}xa$, then we have $x = (a^{-1})^{-1}ya^{-1}$, which means that y is conjugate to x. Finally, from $y = a^{-1}xa$ and $z = b^{-1}yb$, we conclude that $z = (ab)^{-1}x(ab)$ so that z is conjugate to x if x is conjugate to y and y conjugate to z. Thus, conjugacy yields a partition of G into pairwise disjoint complexes such that two elements belong to the same complex if and only if they are conjugate. These complexes are called the *conjugacy classes*, or, briefly, the classes of G.

Since all the elements of one class are images of each other under suitable (inner) automorphisms, it follows that they have the same order. Moreover, if x and y are conjugate, then so are x^k and y^k.

The class of x consists of all distinct conjugates of x. We shall determine their number. Let $N = \mathsf{N}(x)$ be the normalizer of x so that $u^{-1}xu = x$ if and only if u belongs to N. We decompose G into right cosets of N:

$$G = N \cup Na \cup Nb \ \ldots$$

Transformation of x by elements of the same coset yields the same element; for if ua is an arbitrary element of Na, then we have

$$(ua)^{-1}x(ua) = a^{-1}u^{-1}xua = a^{-1}xa,$$

where the right-hand side depends on the coset Na, but is independent of the choice of the particular element ua in that coset. Conversely, suppose that $a^{-1}xa = a_1^{-1}xa_1$. This gives

$$x = aa_1^{-1}xa_1a^{-1} = (a_1a^{-1})^{-1}x(a_1a^{-1}),$$

which implies that $a_1a^{-1} \in N$ or $a_1 \in Na$. This shows that there is a one-to-one correspondence between the distinct conjugates of x and the right cosets of N. So we have the following result:

2.2.1 *The number of elements in the class of x is equal to the index $|G:\mathsf{N}(x)|$, where $\mathsf{N}(x)$ is the normalizer of x.*

It follows, in particular, that in the case of a finite group G the number of elements in any class is a divisor of $|G|$.

Obviously, one of the classes consists of the single element e. Moreover, x is the only element in its class if and only if x belongs to the centre of G.

Conjugacy can also be defined for arbitrary complexes, in particular, for subgroups of G. Two complexes K and L are called *conjugate* if there exists at least one element a of G such that $a^{-1}Ka = L$. It is evident that, in this case too, conjugacy is an equivalence relation.

The elements u of G with the property $u^{-1}Ku = K$ form a subgroup $\mathsf{N}(K)$ of G, which is called the *normalizer* of K in G. Note that $u^{-1}Ku = K$ only means that from $x \in K$ it follows that $u^{-1}xu \in K$, where x and $u^{-1}xu$ need not coincide. In other words: $u^{-1}Ku = K$ means that transformation by u leaves K unchanged as a whole, but not necessarily elementwise. The condition $u^{-1}Ku = K$ is equivalent to $Ku = uK$, which means that u commutes with K as a whole, but not necessarily elementwise. In the same way as in the case of a single element we deduce:

2.2.2 *The number of distinct complexes that are conjugate to a given complex K of G is equal to the index $|G:\mathsf{N}(K)|$ of the normalizer $\mathsf{N}(K)$.*

Note that two distinct conjugates $a^{-1}Ka$ and $b^{-1}Kb$ may have a non-empty intersection.

Since transformation by an element a plays an important role, we frequently write

$$a^{-1}xa = x^a, \quad a^{-1}Ka = K^a.$$

By the *centralizer* $\mathsf{C}(K)$ of the complex K we mean the subgroup of all elements of G that are elementwise permutable with K. It is obvious that $\mathsf{C}(K) \subseteq \mathsf{N}(K)$ and

$$\mathsf{C}(K) = \bigcap_{x \in K} \mathsf{N}(x).$$

Occasionally, the following observation proves useful:

2.2.3 *Let H be a proper subgroup of the finite group G. Then, the union of all subgroups conjugate to H is a proper subset of G.*

For, if $|G:H| = r$, then $H \subseteq \mathsf{N}(H)$ gives $|G:\mathsf{N}(H)| \leqq r$. Hence, there are at most r distinct conjugates of H so that their union cannot contain more than $r|H| = |G|$ elements. But the number of distinct elements in that union is less than $|G|$, because the conjugates of H are not disjoint: the unit element, at least, belongs to all of them.

If σ is an arbitrary automorphism of G, then we denote by $K\sigma$ the image of the complex K under σ. Now, $xy = yx$ implies that $(x\sigma)(y\sigma) = (y\sigma)(x\sigma)$ and vice versa. It follows that $\mathsf{N}(K)\sigma$ and $\mathsf{C}(K)\sigma$, respectively, are the normalizer and the centralizer of $K\sigma$. In particular, if $\sigma = \tau_a$ is an inner automorphism, we obtain

$$\mathsf{N}(K^a) = \mathsf{N}(K)^a, \quad \mathsf{C}(K^a) = \mathsf{C}(K)^a.$$

Let ϱ be a homomorphism of G onto $G\varrho$. If K_1 and K_2 are conjugate complexes of G, then the images $K_1\varrho$ and $K_2\varrho$ are conjugate in $G\varrho$. For, $K_2 = a^{-1}K_1a$ implies that

$$K_2\varrho = (a\varrho)^{-1}(K_1\varrho)(a\varrho).$$

A complex (subgroup) of G is called *invariant* or *normal* if every inner automorphism of G maps that complex (subgroup) onto itself. In other words, a complex (subgroup) is invariant if and only if it coincides with all its conjugates. For this reason, invariant complexes (subgroups) are sometimes called *self-conjugate.* It is obvious that a complex (subgroup) is invariant if and only if it consists of complete classes of conjugate elements. In particular, every complex that belongs to the centre is invariant. The great importance of normal subgroups will become evident in the next section.

If a complex K satisfies the condition $a^{-1}Ka \subseteq K$ for every $a \in G$, then K is invariant. For $a^{-1}Ka \subset K$ would imply for $b = a^{-1}$ that $K \subset b^{-1}Kb$, which contradicts our assumption.

To indicate that S is a normal subgroup or a proper normal subgroup of G, we write $S \trianglelefteq G$ or $S \triangleleft G$, respectively.

If $S \trianglelefteq G$, then $Sx = xS$ for every $x \in G$ and vice versa. Thus, every right (left) coset of a normal subgroup is at the same time a left (right) coset.

If S is a normal subgroup of G and H any subgroup such that $S \subseteq H \subseteq G$, then S is a normal subgroup of H. For, $S \trianglelefteq H$ is a consequence of $x^{-1}Sx = S$ for all $x \in H$, and this is even true for all $x \in G$.

In general, however, the relation of being a normal subgroup is not transitive. That is, from $S \trianglelefteq H$ and $H \trianglelefteq G$ it does not follow that $S \trianglelefteq G$.

As an immediate consequence of the definition of an invariant complex we obtain:

2.2.4 *The intersection of any set of invariant complexes (subgroups) is an invariant complex (subgroup).*

Let A be a normal subgroup and U an arbitrary subgroup of G. It follows that $AU = UA$, because $xA = Ax$ even holds for each single element x. Thus, by Theorem 1.2.2, AU is a subgroup of G. If, moreover, U is a normal subgroup, then AU too is a normal subgroup. Indeed, for any $x \in G$ we have $x^{-1}AUx = x^{-1}Axx^{-1}Ux = AU$. This gives:

2.2.5 *If at least one of the subgroups U, V is normal in G, then UV is a subgroup. If both U and V are normal subgroups, then so is UV.*

Let A and B be two normal subgroups of G. The commutator $[a, b]$ of elements $a \in A$, $b \in B$ can be written as follows:

$$[a, b] = a^{-1}b^{-1}ab = a^{-1}(b^{-1}ab) = (a^{-1}b^{-1}a)b.$$

This shows that $[a, b]$ is contained in A as well as in B. In particular, if $A \cap B = e$, then a and b commute. Hence:

2.2.6 *Two normal subgroups whose intersection is the unit element commute elementwise.*

Let $N = \mathsf{N}(H)$ be the normalizer of the subgroup H of G. We partition G into right cosets of N:

$$G = \bigcup_l Nx_l$$

where l runs through a certain index set Λ. We know that the subgroups $x_l^{-1} Hx_l$ $(l \in \Lambda)$ are all the distinct conjugates of H. This set is called a complete class of conjugate subgroups. Transformation by any element of G only permutes the subgroups in that class.

2.2.7 *Suppose that the subgroups $x_l^{-1} Hx_l$ $(l \in \Lambda)$ form a complete class of conjugate subgroups. Then,*

$$D = \bigcap_{l \in \Lambda} x_l^{-1} Hx_l \quad \textit{and} \quad \langle x_l^{-1} Hx_l,\ l \in \Lambda \rangle$$

are normal subgroups of G.

Of course, it may happen that $D = e$.

If, in particular, the index $|G{:}H|$ is finite, then we infer from $H \subseteq \mathsf{N}(H)$ that $|G{:}\,\mathsf{N}(H)|$ is also finite. Consequently, there exist only finitely many conjugates of H. Thus, by Theorem 1.5.2, the index $|G{:}D|$ is finite. This gives:

2.2.8 *If G contains a subgroup of finite index, then G also contains a normal subgroup of finite index.*

An invariant complex C obviously generates a normal subgroup $\langle C \rangle$.

2.2.9 *Let C be a finite invariant complex whose elements all have finite order. Then, the normal subgroup generated by C is finite.*

Proof. Suppose that C contains n elements, and let m denote the least common multiple of their orders. Every element of $\langle C \rangle$ can be represented, possibly in several ways, as a product of elements of C. It is sufficient to show that, for every $x \in \langle C \rangle$, there is at least one representation that contains not more than $n(m - 1)$ factors. Suppose that

$$x = c_1 c_2 \cdots c_s, \quad c_i \in C \quad (i = 1, 2, \ldots, s)$$

is such a representation and that $s > n(m - 1)$. Then, at least one element of C, say a, occurs at least m times among the c_i. Let c_k be the first element in the sequence $c_1, c_2, \ldots, c_s$ that is equal to a. Putting

$$a^{-1} c_j a = c_j', \quad j = 1, \ldots, k - 1,$$

we obtain

$$x = ac_1' \cdots c_{k-1}' c_{k+1} \cdots c_s.$$

Note that $c'_1, \ldots, c'_{k-1}$ belong to C, because C is invariant. We then apply the same process to the first element of the sequence $c_{k+1}, \ldots, c_s$ that is equal to a. After m steps we arrive at a representation of the form

$$x = a^m \bar{c}_1 \cdots \bar{c}_{s-m}, \quad \bar{c}_i \in C.$$

Since $a^m = e$, this is a representation of x as a product of $s - m$ elements of C. If $s - m > n(m - 1)$, we proceed in the same way. After a finite number of steps, we obtain a representation of x as a product of at most $n(m - 1)$ elements of C. This completes the proof.

The image of a commutator $[x, y] = x^{-1}y^{-1}xy$ under an arbitrary endomorphism σ is

$$[x, y]\sigma = (x\sigma)^{-1}(y\sigma)^{-1}(x\sigma)(y\sigma) = [x\sigma, y\sigma], \tag{2.4}$$

which is again a commutator. This shows that the complex of all commutators of G is not only normal but is even carried into itself by any endomorphism. It follows, in particular, that the commutator subgroup G' of G is normal.

2.3 The Homomorphism Theorem

Let σ be a homomorphism of the group G onto the group $\bar{G}$. By K we denote the complex of all those elements of G whose image under σ is the unit element $\bar{e}$ of $\bar{G}$. For arbitrary elements u, v of K, we have $u\sigma = v\sigma = \bar{e}$ which gives $(uv^{-1})\sigma = (u\sigma)(v\sigma)^{-1} = \bar{e}\bar{e} = \bar{e}$, and hence uv^{-1} belongs to K. This shows that K is a subgroup of G. For an arbitrary element x of G and any $u \in K$, we have

$$(x^{-1}ux)\sigma = (x\sigma)^{-1}(u\sigma)(x\sigma) = (x\sigma)^{-1}\bar{e}(x\sigma) = \bar{e}.$$

Thus, $u \in K$ implies that $x^{-1}ux \in K$ for every $x \in G$ so that K is a normal subgroup of G.

The normal subgroup K is called the *kernel* of the homomorphism σ.

The homomorphism σ can now be described as follows: Let

$$G = K \cup Ka \cup Kb \cup \cdots$$

be the partition of G into cosets of the kernel K. (Since K is a normal subgroup, we need not distinguish between right and left cosets.) Every element of a coset Ka has the form ua with $u \in K$. Applying σ to this element, we find

$$(ua)\sigma = (u\sigma)(a\sigma) = \bar{e}(a\sigma) = a\sigma.$$

This shows that σ carries all the elements of a coset of K into one and the same element of $\bar{G}$. Conversely, if two elements of G have the same image under σ, then they belong to the same coset of K. For $x\sigma = y\sigma$ implies that $(xy^{-1})\sigma = (x\sigma)(y\sigma)^{-1} = \bar{e}$, which means that $xy^{-1} \in K$ or $x \in Ky$. Consequently, there is a one-to-one correspondence between the cosets of K and the elements of $\bar{G}$.

Let us mention two extreme cases. If $K = G$, then we have the trivial homomorphism of G onto the group $\bar{e}$ of order 1. On the other hand, if K consists of the unit element only, then any two distinct elements of G have distinct images under σ. In this case, σ is a one-to-one mapping, i. e. σ is an isomorphism of G onto $\bar{G}$.

So far, our starting point was a given homomorphism of G onto some group. We saw that this homomorphism has a normal subgroup of G as its kernel. We now ask whether, conversely, every normal subgroup of G is the kernel of a suitable homomorphism of G. The answer is in the affirmative. Let K be any normal subgroup of G. We shall construct a group $\bar{G}$ and a homomorphism of G onto $\bar{G}$ whose kernel is K.

First, we construct the group $\bar{G}$. Let

$$G = K \cup Ka \cup Kb \cup \cdots$$

be the decomposition of G into cosets of K. We prove that the cosets of K form a group with respect to multiplication. Since K is a normal subgroup, we have

$$(Ka)(Kb) = KaKb = KKab = Kab = Kc,$$

where c is the representative of the coset that contains the element ab. This shows that the product of two cosets of K is a coset of K. The coset K plays the role of the unit element, since

$$(Ka)K = KaK = KKa = Ka.$$

Finally, if Kd is the coset that contains a^{-1}, then we see that $KaKd = K$; for, as we have proved, $KaKd$ is a coset of K, and since aa^{-1} is contained in that coset it must be the coset K. We conclude that the cosets of the normal subgroup K form a group under multiplication. This group is called the *factor group* of G with respect to K and is denoted by G/K.

A mapping of G onto G/K can be defined in a most obvious way: to every element of G we assign the coset of K in which it is contained. From $x \in Ka$ and $y \in Kb$, it follows that $xy \in KaKb$. This shows that our mapping is a homomorphism of G onto G/K. It is called the *natural homomorphism* of G onto G/K. The unit element of G/K is the coset K, and the natural homomorphism maps an element of G onto K if and only if it is contained in K, in other words, K is the kernel of the natural homomorphism of G onto G/K.

Summarizing our results, we obtain one of the most important theorems of group theory:

2.3.1 Homomorphism Theorem. *To any homomorphism of a group G onto a group $\bar{G}$ there corresponds, as its kernel, a normal subgroup K of G consisting of all elements of G that are mapped onto the unit element of $\bar{G}$. The group $\bar{G}$ is isomorphic to the factor group G/K. Conversely, if K is any normal subgroup of G, then the natural homomorphism of G onto the factor group G/K has K as its kernel.*

The kernel of the example at the beginning of section 2.1 consists of all matrices of $GL(n, F)$ whose determinant is equal to the unit element of F. The decomposition (1.5) shows that every factor group of a cyclic group is cyclic; for the factor group $\langle a \rangle / \langle a^d \rangle$ is generated by the coset $\langle a^d \rangle a$.

Equation (2.3) shows that we obtain a homomorphism of G onto the group $\mathsf{J}(G)$ of the inner automorphisms of G if we assign to an arbitrary element a of G the inner automorphism τ_a induced by a. We observed that τ_z is the identity automorphism if and only if z belongs to the centre $\mathsf{Z}(G)$ of G. Thus, $G/\mathsf{Z}(G) \cong \mathsf{J}(G)$.

We remark that the factor group G/Z of a group G with respect to its centre cannot be cyclic unless $G = Z$. For, suppose that G/Z is cyclic, generated by Za. Then, we have $G = \langle Z, a \rangle$, and hence G is abelian, i. e. $G = Z$.

Let G' be the subgroup of G generated by all commutators $[x, y]$ of elements x, y of G. In section 2.2, we observed that G' is a normal subgroup of G. The factor group G/G' is abelian, for we have

$$xG'yG' = xyG' = yx[x, y]G' = yxG' = yG'xG'.$$

Now, let T be a normal subgroup of G such that G/T is abelian. Then, we have

$$T = Tx^{-1}Ty^{-1}TxTy = Tx^{-1}y^{-1}xy.$$

Consequently, T contains all commutators of G, and hence $G' \subseteq T$.

2.3.2 *The factor group G/G' of a group G with respect to its commutator subgroup G' is abelian. If T is any normal subgroup of G such that G/T is abelian, then G' is contained in T.*

Every group contains two trivial normal subgroups, the group itself and the unit subgroup. A group that contains no other normal subgroups is called *simple*. All groups whose order is a prime number are obviously simple. In an abelian group, every subgroup is normal. Thus, by Theorem 1.4.2, the only simple abelian groups are the cyclic groups of prime order. Examples of non-abelian simple groups will be given in section 7.4 and section 13.7.

If two elements x and y of G belong to the same coset of the normal subgroup K, we write

$$x \equiv y \pmod K \tag{2.5}$$

(x congruent to y modulo K).
It is evident that this is an equivalence relation. Moreover, (2.5) remains true if we multiply both sides on the right or on the left by any element of G, i. e. (2.5) implies that

$$xz \equiv yz \quad \text{and} \quad zx \equiv zy \pmod K$$

for any $z \in G$. For, (2.5) is equivalent to $Kx = Ky$ so that we have $Kxz = Kyz$, which means that $xz \equiv yz \pmod K$. Since K is a normal subgroup, (2.5) is also equivalent to $xK = yK$, which implies that $zxK = zyK$, i. e. $zx \equiv zy \pmod K$. From (2.5) and

$$x_1 \equiv y_1 \pmod K$$

we obtain

$$xx_1 \equiv yy_1 \pmod K. \tag{2.6}$$

For, from $Kx = Ky$ and $Kx_1 = Ky_1$ it follows that

$$Kxx_1 = KxKx_1 = KyKy_1 = Kyy_1,$$

which is (2.6)

2.4 The Isomorphism Theorems

Let K be a normal subgroup of the group G. The natural homomorphism of G onto $\bar{G} = G/K$ maps any subgroup H of G onto a subgroup $\bar{H}$ of $\bar{G}$. The kernel of the homomorphism of H onto $\bar{H}$ is obviously $H \cap K$. Thus, by Theorem 2.3.1, $H \cap K$ is a normal subgroup of H, and we have

$$H/(H \cap K) \cong \bar{H}. \tag{2.7}$$

We now determine the complex H^* of *all* those elements of G whose images are contained in $\bar{H}$. From section 2.1, we know that H^* is a subgroup of G; moreover, it is clear that H is contained in H^*. In general, however, H^* is larger than H; for if a is an arbitrary element of H, then the whole coset aK is contained in H^*. On the other hand, H^* contains only those cosets of K that can be represented by an element of H. This shows that $H^* = HK$, and by the definition of H^*, we have

$$H^*/K = HK/K = \bar{H}. \tag{2.8}$$

Combining (2.7) and (2.8), we obtain:

2.4.1 I. Isomorphism Theorem. *If K is a normal subgroup and H an arbitrary subgroup of G, then $H \cap K$ is a normal subgroup of H and*

$$HK/K \cong H/(H \cap K).$$

We now restrict our attention to the subgroups H of G that contain K. Our arguments show that there is a one-to-one correspondence between these subgroups and the subgroups of G/K; for in this case, we have $H^* = HK = H$. The following theorem describes this correspondence in detail.

2.4.2 II. Isomorphism Theorem. *Let K be a normal subgroup of G. There is a one-to-one correspondence between the subgroups H of G that contain K and the subgroups of G/K defined by*

$$H \leftrightarrow H/K \qquad (K \subseteq H \subseteq G).$$

Two subgroups H_1 and H_2 are conjugate in G if and only if H_1/K and H_2/K are conjugate in G/K. In particular, H is a normal subgroup of G if and only if H/K is a normal subgroup of G/K. In this case, the respective factor groups are isomorphic, i. e.

$$G/H \cong (G/K)/(H/K).$$

Proof. Clearly, to every subgroup H such that $K \subseteq H \subseteq G$ there corresponds a unique subgroup H/K of G/K. Conversely, let $\bar{H}$ be an arbitrary subgroup of G/K. Let H denote the subgroup of all those elements of G whose images under the natural homomorphism of G onto G/K belong to $\bar{H}$. Then, we have $K \subseteq H$, and H consists of complete cosets of K. The group $\bar{H}$ consists of all cosets of K that are contained in H, hence $\bar{H} = H/K$.

If H_1 and H_2 are conjugate in G, then their homomorphic images H_1/K and H_2/K are conjugate in G/K, because conjugacy is preserved under every homomorphism.

The subgroup H_1/K of G/K consists of cosets of K. Taking a representative of each one of these cosets, we obtain a complex R_1 such that $H_1 = KR_1$. A subgroup of G/K that is conjugate to H_1/K has the form

$$H_2/K = (Ka)^{-1}(H_1/K)(Ka).$$

Thus, we have

$$H_2 = (Ka)^{-1}KR_1(Ka) = a^{-1}KR_1a = a^{-1}H_1a.$$

This shows that conjugate subgroups of G (containing K) correspond to conjugate subgroups of G/K and vice versa.

Since a subgroup is normal if and only if it coincides with all its conjugates, it follows that H/K is a normal subgroup of G/K if and only if H is a normal subgroup of G.

Finally, suppose that H/K is a normal subgroup of G/K. We first perform the natural homomorphism of G onto G/K and then the natural homomorphism of G/K onto $(G/K)/(H/K)$. This yields a homomorphism of G onto $(G/K)/(H/K)$.

It is easily checked that H is the kernel of this homomorphism. Thus, we have

$$G/H \cong (G/K)/(H/K).$$

This completes the proof.

As an application of the Isomorphism Theorems we prove a theorem that will be useful later.

2.4.3 Zassenhaus' Lemma. *Let U and V be two subgroups of the group G, and let U_0 and V_0 be normal subgroups of U and V, respectively. Then,*

$$U_0(U \cap V_0) \quad \textit{is a normal subgroup of} \quad U_0(U \cap V),$$

$$V_0(V \cap U_0) \quad \textit{is a normal subgroup of} \quad V_0(V \cap U),$$

and we have

$$U_0(U \cap V)/U_0(U \cap V_0) \cong V_0(V \cap U)/V_0(V \cap U_0).$$

Proof. Since U_0 is a normal subgroup of U, we have $uU_0 = U_0u$ for every $u \in U$. Thus, U_0 commutes with $U \cap V_0$ and $U \cap V$. By symmetry, V_0 commutes with $V \cap U_0$ and $V \cap U$. This shows that the products that occur in our theorem are groups.

By Theorem 2.4.1, we obtain

$$U_0(U \cap V)/U_0 = (U \cap V)/(U_0 \cap U \cap V) = (U \cap V)/(V \cap U_0). \qquad (2.9)$$

By symmetry, $(V \cap U_0) \trianglelefteq (U \cap V)$ implies that $(U \cap V_0) \trianglelefteq (U \cap V)$, and this gives

$$(V \cap U_0)(U \cap V_0) \trianglelefteq U \cap V.$$

By Theorem 2.4.2, to this normal subgroup of $U \cap V$ there corresponds a unique normal subgroup of $(U \cap V)/(V \cap U_0)$, namely

$$(V \cap U_0)(U \cap V_0)/(V \cap U_0). \qquad (2.10)$$

Under the isomorphism (2.9), the normal subgroup (2.10) of the right-hand side and the normal subgroup

$$U_0(V \cap U)(U \cap V_0)/U_0 = U_0(U \cap V_0)/U_0$$

of the left-hand side correspond to each other. Since the factor groups with respect to corresponding normal subgroups on both sides of (2.9) are isomorphic, we have

$$U_0(U \cap V)/U_0(U \cap V_0) \cong (U \cap V)/(V \cap U_0)(U \cap V_0). \qquad (2.11)$$

Since the right-hand side of (2.11) is symmetric with respect to U_0, U and V_0, V, we conclude

$$V_0(V \cap U)/V_0(V \cap U_0) \cong (V \cap U)(U \cap V_0)(V \cap U_0). \qquad (2.12)$$

Combining (2.11) and (2.12), we obtain the theorem.

Clearly, every subgroup is contained in its normalizer. The following theorem gives an example of subgroups that coincide with their normalizers.

2.4.4 *If H is a subgroup of the finite group G for which $|H|$ and $|G{:}H|$ are relatively prime, then the normalizer of H is its own normalizer in G.*

Proof. Let N be the normalizer of H. We have to prove that $x^{-1}Nx = N$ implies that $x^{-1}Hx = H$, i. e. $x \in N$. Suppose that $x^{-1}Hx = H_1 \neq H$. Then, H as well as H_1 are normal subgroups of N. By Theorem 2.4.1, we have $HH_1/H_1 \cong H/(H \cap H_1)$. This shows that every prime divisor of $|HH_1|$ divides $|H| = |H_1|$.

On the other hand, $|HH_1{:}H|$ is greater than 1 and divides $|G{:}H|$. Owing to the condition of our theorem, this implies that $|HH_1{:}H|$, and hence $|HH_1|$, has at least one prime divisor that does not divide $|H|$. This is a contradiction, and hence the theorem is proved.

2.5 Factor Groups of Free Groups

Let

$$B^1 = \{x_l^1\} \quad \text{and} \quad B^{-1} = \{x_l^{-1}\}$$

be two non-empty sets of symbols x_l^1 and x_l^{-1}.[1] We assume that B^1 and B^{-1} are disjoint and that in both sets l ranges over one and the same index set Λ. The elements of the union

$$B = B^1 \cup B^{-1}$$

are called letters. Two letters x_l^1 and x_l^{-1} are said to be *associate* if they have the same index $l \in \Lambda$. By a *word*, we mean any finite sequence

$$u = u_1^{\varepsilon_1} u_2^{\varepsilon_2} \cdots u_n^{\varepsilon_n} \quad (u_i^{\varepsilon_i} \in B)$$

of letters $u_i^{\varepsilon_i}$. The superscripts ε_i are 1 or -1 according as $u_i^{\varepsilon_i}$ belongs to B^1 or B^{-1}. This means that each u_j^1 is some element x_l^1 of B^1 and each u_k^{-1} some x_m^{-1} of B^{-1}, where one and the same element of B^1 or B^{-1} may occur several times in u. Among the words, we include the *empty word* which contains no letters. The word u is called *reduced* if it contains no pair of adjacent letters that are associate, i. e. if it contains no pair of the form

$$u_i^{\varepsilon_i} u_{i+1}^{\varepsilon_{i+1}} = x_l^1 x_l^{-1} \quad \text{or} \quad x_l^{-1} x_l^1 \quad (i = 1, \ldots, n-1).$$

[1] The exponent -1 of the elements of B^{-1} should not suggest that they are inverses of the elements of B. We do not assume that the elements of B^1 and B^{-1} belong to a group, so that neither multiplication nor inverses are defined as yet. For the time being, we regard 1 and -1 as superscripts by which we distinguish the two sets.

If k, l, m are distinct indices, then, for instance, $x_k^1 x_k^1 x_l^{-1} x_m^1 x_l^1$ is reduced, but $x_k^1 x_l^{-1} x_l^1$ and $x_k^{-1} x_l^{-1} x_m^{-1} x_m^1 x_l^1 x_k^1$ are not reduced. If a word is not reduced, then we can apply the following reduction process: First delete all pairs of adjacent associate letters. By these cancellations new pairs of adjacent associate letters may arise. If this happens, then again delete these new pairs. Continue this process as long as there are adjacent associate letters. This reduction process yields a reduced word, possibly the empty word. In our first example, we obtain x_k^1, in the second example, the empty word. The reduction process is not uniquely determined but, as we shall see later, all possible ways of carrying out the successive cancellations yield a unique reduced word.

Let F denote the set of all reduced words. In F, we define a binary operation as follows: Let

$$a = a_1^{\varepsilon_1} a_2^{\varepsilon_2} \cdots a_r^{\varepsilon_r} \qquad (a_i^{\varepsilon_i} \in B),$$

$$b = b_1^{\eta_1} b_2^{\eta_2} \cdots b_s^{\eta_s} \qquad (b_j^{\eta_j} \in B)$$

be two words of F. By juxtaposition we form the word

$$a_1^{\varepsilon_1} a_2^{\varepsilon_2} \cdots a_r^{\varepsilon_r} b_1^{\eta_1} b_2^{\eta_2} \cdots b_s^{\eta_s}. \tag{2.13}$$

This word need not be reduced. Since the words of F are reduced, there is no pair of adjacent associate letters among the letters $a_i^{\varepsilon_i}$ alone or among the letters $b_j^{\eta_j}$ alone. Thus, the word (2.13) is not reduced if and only if $a_r^{\varepsilon_r}$ and $b_1^{\eta_1}$ are associate. In that case, we cancel the two letters. Then, $a_{r-1}^{\varepsilon_{r-1}}$ and $b_2^{\eta_2}$ are adjacent. If they are associate, we cancel them too. We continue this procedure until all possible cancellations are carried out. Contrary to the general reduction process, the process just described is unique. Consequently, (2.13) yields a unique reduced word which is denoted by ab. Of course, it may happen that ab is the empty word.

We shall prove that F is a group with respect to this binary operation.

If e denotes the empty word, then we have

$$ae = ea = a$$

for every $a \in F$. Thus, e plays the role of the unit element. For a given word a of F, let a^{-1} denote the word that is obtained by replacing all the letters of a by their associates and by reversing the order. Clearly, a^{-1} is reduced, and we have

$$aa^{-1} = a^{-1} a = e.$$

It remains to prove that the associative law is satisfied. For an arbitrary word

$$c = c_1^{\delta_1} c_2^{\delta_2} \cdots c_t^{\delta_t} \qquad (c_k^{\delta_k} \in B)$$

we have to show that

$$(ab)\,c = a(bc). \tag{2.14}$$

We shall proceed by induction on the number of letters in the word b. Clearly, (2.14) holds if b is the empty word. Next, we assume that b consists of a single letter, $b = b_1^{\eta_1}$ say. We consider four cases.

(a) $a_r^{\varepsilon_r}$ and $b_1^{\eta_1}$ not associate; $b_1^{\eta_1}$ and $c_1^{\delta_1}$ not associate. Then, no cancellations are possible, and hence (2.14) holds.

(b) $a_r^{\varepsilon_r}$ and $b_1^{\eta_1}$ associate; $b_1^{\eta_1}$ and $c_1^{\delta_1}$ not associate. Then, we have $ab = a_1^{\varepsilon_1} \cdots a_{r-1}^{\varepsilon_{r-1}}$ and $(ab)\,c$ coincides with

$$a_1^{\varepsilon_1} \cdots a_{r-1}^{\varepsilon_{r-1}}\, c_1^{\delta_1} \cdots c_t^{\delta_t} \tag{2.15}$$

or is obtained from this word by a unique sequence of cancellations. On the other hand, to obtain $a(bc)$ we have to apply the unique cancellation procedure to

$$a_1^{\varepsilon_1} \cdots a_{r-1}^{\varepsilon_{r-1}}\, a_r^{\varepsilon_r}\, b_1^{\eta_1}\, c_1^{\delta_1} \cdots c_t^{\delta_t}.$$

The first possible cancellation consists in deleting $a_r^{\varepsilon_r}\, b_1^{\eta_1}$. This gives

$$a_1^{\varepsilon_1} \cdots a_{r-1}^{\varepsilon_{r-1}}\, c_1^{\delta_1} \cdots c_t^{\delta_t}.$$

The remaining cancellations, if any, are the same as in (2.25). This gives $(ab)\,c = a\,(bc)$.

(c) $a_r^{\varepsilon_r}$ and $b_1^{\eta_1}$ not associate; $b_1^{\eta_1}$ and $c_1^{\delta_1}$ associate. This case is analogous to (b).

(d) $a_r^{\varepsilon_r}$ and $b_1^{\eta_1}$ associate; $b_1^{\eta_1}$ and $c_1^{\delta_1}$ associate. In this case, we have $a_r^{\varepsilon_r} = c_1^{\delta_1}$. The product $(ab)\,c$ coincides with

$$a_1^{\varepsilon_1} \cdots a_{r-1}^{\varepsilon_{r-1}}\, c_1^{\delta_1} \cdots c_t^{\delta_t}$$

or is obtained from this word by cancellations. $a(bc)$ is equal to the word

$$a_1^{\varepsilon_1} \cdots a_r^{\varepsilon_r}\, c_2^{\delta_2} \cdots c_t^{\delta_t}$$

or is obtained from it by cancellations. Owing to $a_r^{\varepsilon_r} = c_1^{\delta_1}$, this yields the same reduced word.

Thus, (2.14) is proved in case b consists of a single letter. Suppose now that the associative law holds if the middle factor contains $s - 1$ letters. We write

$$b = b_1^{\eta_1} \cdots b_{s-1}^{\eta_{s-1}}\, b_s^{\eta_s} = (b_1^{\eta_1} \cdots b_{s-1}^{\eta_{s-1}})\, b_s^{\eta_s} = b_0\, b_s^{\eta_s}.$$

Since b_0 consists of only $s - 1$ letters, we have

$$\begin{aligned}(ab)c &= [a(b_0\, b_s^{\eta_s})]\, c = [(ab_0)\, b_s^{\eta_s}]\, c \\ &= (ab_0)\,(b_s^{\eta_s} c) = a\,[b_0(b_s^{\eta_s} c)] \\ &= a\,[(b_0\, b_s^{\eta_s})\, c] = a(bc).\end{aligned}$$

This completes the proof of the associative law.

It follows that F is a group.

Every single letter is a reduced word, and every reduced word is the product of its letters. Thus, the letters of the set B^1 generate F. We call F the *free group* generated by B^1, and B^1 is referred to as a *system of free generators*. By the rank of F, we mean the cardinal number of B^1. To simplify the notation, we now omit the superscript 1 of the letters in B^1. For any natural number k, we denote the product of k letters x_l of B^1 by x_l^k and the product of k letters x_l^{-1} of B^{-1} by x_l^{-k}.

As we mentioned above, for a given word there are, in general, several distinct ways of carrying out the cancellations that lead to a reduced word. The different ways to reduce the word

$$u_1^{\varepsilon_1} u_2^{\varepsilon_2} \cdots u_r^{\varepsilon_r} \qquad (u_i^{\varepsilon_i} \in B)$$

can be described by different systems of brackets and therefore correspond to different ways of computing the product of $u_1^{\varepsilon_1}, \ldots, u_r^{\varepsilon_r}$ in F. Owing to the associative law in F, all systems of brackets yield the same final result. Hence, to any given word there corresponds a unique reduced word.

The following theorem gives an indication of the importance of free groups.

2.5.1 *Every group is isomorphic to a factor group of a free group.*

Proof. Let G be an arbitrary group and denote by A a system of generators of G. (We may take $A = G$, for instance.) We denote the elements of A by a_l, where l ranges over some index set Λ. We now take a set X of symbols x_l $(l \in \Lambda)$ and denote by F the free group generated by X. To any reduced word

$$x_{l_1}^{\varepsilon_1} x_{l_2}^{\varepsilon_2} \cdots x_{l_n}^{\varepsilon_n} \qquad (\varepsilon_i = \pm 1;\ l_i \in \Lambda)$$

of F we assign the element

$$a_{l_1}^{\varepsilon_1} a_{l_2}^{\varepsilon_2} \cdots a_{l_n}^{\varepsilon_n}$$

of F. It is evident that this defines a homomorphism of F onto G. By Theorem 2.3.1, F contains a normal subgroup R such that

$$F/R \cong G.$$

This completes the proof.

If G is finitely generated, by n elements say, then G is isomorphic to a factor group of a free group with n free generators.

The normal subgroup R of F consists of all elements

$$x_{m_1}^{\eta_1} x_{m_2}^{\eta_2} \cdots a_{m_r}^{\eta_r}$$

of F such that

$$a_{m_1}^{\eta_1} a_{m_2}^{\eta_2} \cdots a_{m_r}^{\eta_r} = e. \tag{2.16}$$

We call this equation a *relation* between the elements of A. Thus, to every word of R there corresponds a relation in G. Let R_0 be a system of words of R such that $\langle R_0 \rangle = R$. The relations that correspond to the elements of R_0 are called a system of *defining relations* of G. Since R_0 generates the whole group R, every relation in G can be obtained as a consequence of the defining relations.

Conversely, suppose that we are given a set A of elements a_l $(l \in \Lambda)$ and some symbol e. Let Σ be an arbitrary set of expressions of the form (2.16). Then, there is always a group G with the following properties: (1) G possesses a system of generators that has the same cardinal as Λ. (2) If the generators of G are denoted by a_l $(l \in \Lambda)$ and the unit element by e, then Σ is a system of defining relations of G. In view of our previous arguments, it is easy to construct G: Take the free group F generated by x_l $(l \in \Lambda)$ and determine the normal subgroup R of F generated by the words

$$x_{m_1}^{\eta_1} x_{m_2}^{\eta_2} \cdots x_{m_r}^{\eta_r},$$

which correspond to the left-hand sides of the given relations. Then, F/R is generated by the cosets $a_l = x_l R$ with Σ as a system of defining relations.

2.5.2 *An arbitrary set of generators together with an arbitrary system of relations define one and, up to isomorphism, only one group.*

The following theorem shows the effect of adding new relations to the given ones.

2.5.3 Dyck's Theorem. *Suppose that two groups G_1 and G_2 have the same system of generators and that each defining relation of G_1 occurs among the defining relations of G_2. Then, G_2 is isomorphic to a factorgroup of G_1.*

Proof. In a suitable free group F, the defining relations of G_1 and G_2 determine normal subgroups R_1 and R_2, respectively, such that

$$G_1 \cong F/R_1, \quad G_2 \cong F/R_2.$$

Since every defining relation of G_1 occurs among the defining relations of G_2, we have $R_1 \subseteq R_2$. To the normal subgroups R_2/R_1 of F/R_1, there corresponds a normal subgroup T of G_1 and we obtain

$$G_1/T \cong (F/R_1)/(R_2/R_1) \cong F/R_2 \cong G_2,$$

which proves the theorem.

If a group is given by generators and defining relations, then one of the main problems is to decide whether two words in the generators represent one and the same element of the group or, in other words, whether two words can be transformed into each other by applying the given relations. This problem is called the *word problem*. P. S. Novikov [54] has proved that even in the case of a finite number of generators and defining relations the word problem is undecidable in the following sense: There is not always an algorithm by which the word problem can be decided in a finite number of steps. For several special classes of groups, however, the word problem is decidable. In many applications, it is natural to exhibit a group by generators and defining relations (cf. [12]).

2.6 Operators

By an *operator domain* Ω of the group G we mean a set of symbols τ, ω, ... with the following properties:

(a) To every $\omega \in \Omega$ and every $x \in G$, there corresponds a unique element $x\omega$ of G.

(b) For arbitrary elements x, y of G and ω of Ω, we have

$$(x\,\omega)(y\,\omega) = (xy)\,\omega\,.$$

In view of (b), every operator $\omega \in \Omega$ induces an endomorphism of G. But we distinguish between an operator and the corresponding endomorphism and, in particular, we do not exclude the possibility that two distinct operators define the same endomorphism of G. This distinction is one of the reasons for the great importance of groups with operators. For instance, when we are given two groups with the same operator domain, then two distinct operators may induce distinct endomorphisms in one of the groups, but the same endomorphism in the other. Or, the operator domain may consist of the elements of some given algebraic system (e. g. a ring) so that equality of operators is defined qua elements of that system; but this does not imply that distinct operators induce distinct endomorphisms. However, many of the operator domains we consider in the sequel simply consist of given sets of endomorphisms or automorphisms.

A subgroup U of G is called *Ω-admissible* if every operator $\omega \in \Omega$ maps U into itself, i. e.

$$U\omega \subseteq U \quad \text{for every} \quad \omega \in \Omega\,.$$

Here, of course, $U\omega$ means the group of all $u\omega$ with $u \in U$. If it is clear from the context to which operator domain we refer, then we simply speak of admissible subgroups.

It is obvious that the intersection of an arbitrary set of Ω-admissible subgroups is an Ω-admissible subgroup.

There are special names for subgroups that are admissible with respect to some important operator domains:

As we know, a *normal* subgroup is one that is admissible with respect to all inner automorphisms.

Subgroups that are admissible with respect to all automorphisms of G are called *characteristic* subgroups of G.

The centre is a characteristic subgroup. Indeed, $zx = xz$ for every $x \in G$ implies that $(z\alpha)(x\alpha) = (x\alpha)(z\alpha)$ for any automorphism α of G. Since the elements $x\alpha$, $x \in G$, range over the whole group G, it follows that $z\alpha$ lies in the centre.

Any automorphism α maps a characteristic subgroup onto itself, not only into itself; for $U\alpha \subset U$ would imply that $U \subset U\alpha^{-1}$, which is impossible if U is characteristic.

Subgroups of G that are admissible with respect to all endomorphisms of G are called *fully invariant*. Equation (2.4) shows that the commutator subgroup is fully invariant. Every characteristic subgroup is normal, but the converse need not be true. Every fully invariant subgroup is characteristic, but there are characteristic subgroups which are not fully invariant.

In section 2.2 we remarked that, in general, the relation of being a normal subgroup is not transitive. However, as regards characteristic and fully invariant subgroups, we have:

2.6.1 *If U is a characteristic (fully invariant) subgroup of V and V a characteristic (fully invariant) subgroup of G, then U is a characteristic (fully invariant) subgroup of G.*

For any automorphism (endomorphism) of G induces an automorphism (endomorphism) of V, and every automorphism (endomorphism) of V maps U onto (into) itself.

In the case of normal subgroups, we have only a weaker form of transitivity:

2.6.2 *If U is a characteristic subgroup of V and V a normal subgroup of G, then U is a normal subgroup of G.*

Indeed, every inner automorphism of G induces an automorphism of V; this need not be an inner automorphism of V, but it maps U onto itself since U is characteristic in V.

Let G be a group and Ω an arbitrary operator domain of G. An automorphism (endomorphism) α of G is called an *Ω-operator automorphism* or, briefly, an *Ω-automorphism (Ω-endomorphism)* if

$$(x\omega)\alpha = (x\alpha)\omega \quad \text{for every} \quad x \in G \quad \text{and every} \quad \omega \in \Omega.$$

This equation means that the mapping α is permutable with every mapping induced by an operator.

An analogous concept arises in connection with homomorphisms of one group onto another. Let G and $\bar{G}$ be two groups with the same operator domain Ω. A homomorphism σ of G onto $\bar{G}$ is called an *Ω-homomorphism* if

$$(x\,\omega)\,\sigma = (x\,\sigma)\,\omega \quad \text{for every} \quad x \in G \quad \text{and every} \quad \omega \in \Omega. \tag{2.17}$$

Thus, an Ω-homomorphism is a homomorphism which commutes with every endomorphism induced by an operator or, in other words, if an element x of G is mapped onto $x\sigma$, then, for every $\omega \in \Omega$, the element $x\omega$ is mapped onto $(x\sigma)\,\omega$. A special case of an Ω-homomorphism is an *Ω-isomorphism* between two groups.

We shall now prove that the Homomorphism Theorem 2.3.1 and the two Isomorphism Theorems 2.4.1 and 2.4.2 remain valid if we consider only admissible subgroups and admissible normal subgroups and if the term homomorphism (isomorphism) is replaced by operator homomorphism (operator isomorphism). Instead of repeating our previous proofs, we shall only supplement them with a few remarks.

Suppose that the groups G and $\bar{G}$ have the same operator domain Ω and that σ is an Ω-homomorphism of G onto $\bar{G}$. Let K be the kernel of σ. We prove that K is admissible with respect to Ω. For, K consists of all elements u of G such that $u\sigma$ is the unit element $\bar{e}$ of $\bar{G}$; by (2.17), $u\sigma = \bar{e}$ gives

$$(u\,\omega)\,\sigma = (u\,\sigma)\,\omega = \bar{e}\,\omega = \bar{e},$$

hence $u\omega \in K$.

By Theorem 2.3.1, the factor group G/K is isomorphic to $\bar{G}$ in the ordinary sense (i. e. disregarding Ω). It remains to show that Ω can be regarded as an operator domain of G/K and that G/K and $\bar{G}$ are even Ω-isomorphic.

First, we have to define how the operators act on the cosets of K. All the elements of the coset Ka have the form ua with $u \in K$. By applying an operator ω, we obtain

$$(ua)\,\omega = (u\,\omega)\,(a\,\omega) = u_1(a\,\omega)$$

where u_1 belongs to K, since K is admissible. Thus, ω carries all the elements of Ka into $K(a\omega)$. Therefore, we define

$$(Ka)\,\omega = K(a\,\omega).$$

Then, we obtain

$$(KaKb)\,\omega = K(ab)\,\omega = K(a\,\omega)(b\,\omega) = K(a\,\omega)\,K(b\,\omega) = [(Ka)\,\omega][(Kb)\,\omega].$$

Consequently, Ω is an operator domain of G/K.

It is now easy to prove that G/K and $\bar{G}$ are Ω-isomorphic. Under the ordinary isomorphism between G/K and $\bar{G}$, the cosets Kx and the elements $x\sigma$ of $\bar{G}$ correspond to each other. An operator ω carries Kx into $K(x\omega)$ and $x\sigma$ into

$(x\sigma)\,\omega = (x\omega)\,\sigma$. But under the ordinary isomorphism between G/K and $\bar{G}$, $K(x\omega)$ and $(x\omega)\,\sigma$ also correspond to each other. This shows that G/K and $\bar{G}$ are Ω-isomorphic.

This completes the proof of our generalized Homomorphism Theorem, which can be stated as follows:

2.6.3 *To any operator homomorphism of a group G onto a group $\bar{G}$ there belongs as its kernel an admissible normal subgroup K of G. The kernel consists of all elements of G that are mapped onto the unit element of $\bar{G}$. The group $\bar{G}$ is operator isomorphic to the factor group G/K. Conversely, if K is an admissible normal subgroup of G, then the natural homomorphism of G onto G/K is an operator homomorphism.*

The proof of the Isomorphism Theorems may now be left as an exercise for the reader. Finally, as is readily seen, Zassenhaus' Lemma can also be extended to groups with operators.

2.7 Series of Subgroups

Let G be an arbitrary group. By a *subnormal series* of G we mean a finite sequence.

$$G = G_0 \supseteq G_1 \supseteq G_2 \supseteq \cdots \supseteq G_{r-1} \supseteq G_r = e \tag{2.18}$$

of subgroups G_i of G beginning with G and ending with the unit subgroup, such that every G_i is a normal subgroup of G_{i-1}. If every term is a proper subgroup of the preceding one, then we speak of a subnormal series *without repetitions*. Every subgroup of G that occurs in some subnormal series of G is called a *subnormal subgroup*. Clearly, every normal subgroup is subnormal, but the converse is not true, since a normal subgroup of a normal subgroup of G need not be normal in G.

We call r the length of the subnormal series, and the factor groups

$$G_0/G_1,\ G_1/G_2,\ \ldots,\ G_{r-1}/e$$

its factors.

If an operator domain Ω of G is given, then we consider only those subnormal series whose terms are admissible with respect to Ω. To emphasize the existence of an operator domain Ω, we sometimes speak of Ω-subnormal series.

If Ω consists of all inner automorphisms of G, then the admissible subgroups are the normal subgroups, and the Ω-subnormal series are called *normal series*. Thus, (2.18) is a normal series if and only if each G_i is a normal subgroup not only of G_{i-1}, but of G.

If Ω consists of all automorphisms of G, then the G_i in (2.18) are characteristic subgroups of G, and we speak of a *characteristic series*. Finally, if Ω is the set of all endomorphisms of G, then (2.18) is called a *fully invariant series* since all G_i are fully invariant subgroups of G.[1]

First, we exhibit a relation between a subnormal series of a group and those of its subgroups.

2.7.1 *Let (2.18) be a subnormal series of G. Every subgroup A of G possesses a subnormal series whose factors are isomorphic to subgroups of distinct factors of the series (2.18).*

Proof. We write $A_i = A \cap G_i$ $(i = 0, 1, \ldots, r)$. Applying Theorem 2.4.3 with $U = A$, $U_0 = e$, $V = G_{i-1}$, $V_0 = G_i$ we find that A_i is a normal subgroup of A_{i-1} and that

$$A_{i-1}/A_i \cong G_i A_{i-1}/G_i.$$

We have $G_{i-1} \supseteq G_i A_{i-1} \supseteq G_i$, which shows that A_{i-1}/A_i is isomorphic to a subgroup of G_{i-1}/G_i. Thus, the subnormal series

$$A = A_0 \supseteq A_1 \supseteq A_2 \supseteq \cdots \supseteq A_{r-1} \supseteq A_r = e$$

has the required property.

Two subnormal series of G are called *isomorphic* if they have the same length and if the factors of the first series, taken in a suitable order, are isomorphic to the factors of the second series. This means that the subnormal series

$$G = \bar{G}_0 \supseteq \bar{G}_1 \supseteq \bar{G}_2 \supseteq \cdots \supseteq \bar{G}_{t-1} \supseteq \bar{G}_t = e$$

is isomorphic to the subnormal series (2.18) if and only if $t = r$ and

$$G_{i-1}/G_i \cong \bar{G}_{k_i-1}/\bar{G}_{k_i} \qquad (i = 1, \ldots, r)$$

where $k_1, \ldots, k_r$ is a suitable permutation of $1, \ldots, r$.

Two isomorphic subnormal series of G with repetitions obviously yield two isomorphic subnormal series without repetitions if we remove the repetitions by cancelling corresponding terms in both series.

By a *refinement* of the subnormal series (2.18), we understand (2.18) itself and every subnormal series of G that is obtained from (2.18) by inserting additional terms. *A proper refinement* of (2.18) is one that contains at least one new term, distinct from all the G_i.

We now prove the basic theorem on refinements.

---

[1] The terminology is not uniform. Sometimes subnormal series as defined above (i. e. with an empty set of operators) are called normal series. However, we shall systematically denote the series according to the special properties of their terms, i. e. subnormal series if all terms are subnormal subgroups, normal series if all terms are normal subgroups, etc.

2.7.2 Schreier's Refinement Theorem. *Any two Ω-subnormal series of the group G have refinements that are Ω-isomorphic.*

Proof. Let

$$G = G_0 \supseteq G_1 \supseteq \cdots \supseteq G_r = e,$$
$$G = H_0 \supseteq H_1 \supseteq \cdots \supseteq H_s = e$$

be two Ω-subnormal series of G. We construct the following refined series: Between G_{i-1} and G_i we insert

$$G_{i,k} = G_i(G_{i-1} \cap H_k) \qquad (k = 1, \ldots, s-1),$$

and between H_{k-1} and H_k we insert

$$H_{i,k} = H_k(H_{k-1} \cap G_i) \qquad (i = 1, \ldots, r-1).$$

By Theorem 2.4.3, $G_{i,k}$ is an admissible normal subgroup of $G_{i,k-1}$, and $H_{i,k}$ is an admissible normal subgroup of $H_{i-1,k}$ so that we obtain Ω-subnormal series. The same theorem tells us that

$$G_i(G_{i-1} \cap H_{k-1})/G_i(G_{i-1} \cap H_k) \cong H_k(H_{k-1} \cap G_{i-1})/H_k(H_{k-1} \cap G_i)$$

or

$$G_{i,k-1}/G_{i,k} \cong H_{i-1,k}/H_{i,k} \qquad (i = 1, \ldots, r;\ k = 1, \ldots, s).$$

Therefore, the refinements are Ω-isomorphic.

An Ω-subnormal series without repetitions that has no proper refinement is called an *Ω-composition series*. It must not be assumed that every group possesses a composition series, but this is certainly true in the case of finite groups. The existence of an Ω-composition series of a given group may also depend on the operator domain Ω.

An immediate consequence of Theorem 2.7.2 is the:

2.7.3 Jordan-Hölder Theorem. *Any two Ω-composition series of a given group are Ω-isomorphic.*

For, by Theorem 2.7.2, any two Ω-composition series have isomorphic refinements, while, on the other hand, they possess no proper refinements so that they are isomorphic without any insertion.

In the above definition of isomorphism between subnormal series, we disregarded the order of the factors. The following example shows that this is necessary for the validity of Theorem 2.7.3. The cyclic group $\langle a \rangle$ of order 6 has the two composition series

$$\langle a \rangle \supset \langle a^2 \rangle \supset e, \quad \langle a \rangle \supset \langle a^3 \rangle \supset e.$$

If we speak of a composition series without referring to any operator domain, then we tacitly assume that the operator domain is empty. Thus,

$$G = G_0 \supset G_1 \supset G_2 \supset \cdots \supset G_{l-1} \supset G_l = e$$

is a composition series if and only if each G_i is a *maximal normal* subgroup of G_{i-1}, i. e. G_i is a normal subgroup of G_{i-1}, and there is no normal subgroup G_i^* of G_{i-1} such that $G_{i-1} \supset G_i^* \supset G_i$. By Theorem 2.4.2, this means that the factor groups G_{i-1}/G_i are simple. Consequently, we have:

2.7.4 *A subnormal series without repetitions is a composition series if and only if its factors are simple groups.*

The factors of a composition series of G are called the *composition factors* of G; their orders are the *composition indices.*

If a group possesses a composition series, then every subnormal series can be refined to a composition series. Let N be a normal subgroup of G. If G possesses a composition series, then we can refine the subnormal series $G \supset N \supset e$ to obtain a composition series. This shows that in a group with composition series there always exists a composition series containing a given normal subgroup.

If the operator domain Ω consists of all inner automorphisms of G, then the Ω-composition series are called *principal series* or *chief series.*

A subnormal series

$$G = H_0 \supset H_1 \supset H_2 \supset \cdots \supset H_{m-1} \supset H_m = e \tag{2.19}$$

is a chief series if and only if each H_i is a normal subgroup of G such that there is no normal subgroup N_i of G with $H_{i-1} \supset N_i \supset H_i$. The infinite cyclic group, for instance, has no chief series.

The factors of a chief series of G are called the *chief factors* of G; their orders are the *chief indices*. Theorem 2.7.3 tells us that there are isomorphisms between the factors of two chief series of G that commute with all inner automorphisms of G regarded as automorphisms of the chief factors.

In a group having chief series, every normal series can be refined to a chief series. In particular, there always exists a chief series that contains a given normal subgroup.

A group G having chief series has no proper homomorphism onto itself. For, such a homomorphism implies the existence of a normal subgroup $T \neq e$ such that $G \cong G/T$. This contradicts Theorem 2.7.3, since G possesses a chief series containing T. For the same reason, a group with chief series is not isomorphic to a proper normal subgroup.

A group G is called *characteristically simple* if it does not contain any characteristic subgroup other than G and e.

2.7.5 *The chief factors of G are characteristically simple.*

Proof. Suppose that the factor H_{i-1}/H_i of the chief series (2.19) contains the non-trivial characteristic subgroup C/H_i. Then, C/H_i is a characteristic subgroup of the normal subgroup H_{i-1}/H_i of G/H_i. Thus, by Theorem 2.6.2, C/H_i is a normal subgroup of G/H_i. Consequently, C is a normal subgroup of G such that $H_{i-1} \supset C \supset H_i$. But, this is a contradiction, since (2.19) is a chief series.

The following theorem will be used later.

2.7.6 *Let A and B be normal subgroups of G such that the factor groups G/A and G/B possess chief series. Then, $G/(A \cap B)$, too, possesses a chief series, and the factors of this chief series are isomorphic to chief factors of G/A or G/B.*

Proof. Without loss of generality, we may assume that $A \cap B = e$. Then, A and B commute elementwise, and every subgroup of AB that contains B has the form A^*B with some subgroup A^* of A. Let

$$G/A \supset G_1/A \supset \cdots \supset G_{r-1}/A \supset A/A$$

be a chief series of G/A. Since G/B possesses a chief series, there is a chief series of G/B that contains the normal subgroup AB/B, say

$$G/B \supset \cdots \supset AB/B \supset A_1B/B \supset \cdots \supset A_{s-1}B/B \supset B/B.$$

Since A_iB and A are normal subgroups of G, the same holds for $A_iB \cap A = A_i$. In the series

$$G \supset \cdots \supset AB \supset A_1B \supset \cdots \supset A_{s-1}B \supset B,$$

we form the intersections of AB and the subsequent terms with A. This gives

$$G \supset \cdots \supset A \supset A_1 \supset \cdots \supset A_{s-1} \supset A \cap B = e.$$

In this series, we replace the part between G and A by

$$G \supset G_1 \supset \cdots \supset G_{r-1} \supset A$$

to obtain a chief series of G with the required property.

2.8 Soluble Groups

A group is called *soluble*[1] if it possesses a subnormal series whose factors are abelian. Chapter 11 is devoted to a more detailed study of soluble groups. Here, we shall deal only with some simple properties of this class of groups. The importance of solubility in the theory of finite groups can be judged from the following extremely deep theorem, which was conjectured for decades, but was proved only quite recently.

[1] This name refers to the fact that an algebraic equation is 'soluble by radicals' if and only if its Galois group is soluble. Finite soluble groups were first considered in this context.

2.8.1 Feit–Thompson Theorem. *Every finite group of odd order is soluble.*

The proof of this theorem is by far beyond the scope of this book (cf. [17]).

Clearly, every abelian group is soluble.

Any refinement of a subnormal series with abelian factors has abelian factors. Thus, Theorem 1.4.2 gives:

2.8.2 *A finite group is soluble if and only if its composition indices are prime numbers.*

For arbitrary groups we have:

2.8.3 *If G contains a normal subgroup A such that A and G/A are soluble, then G is soluble.*

Proof. Since G/A is soluble, there exists a subnormal series

$$G/A = G_0/A \supset G_1/A \supset \cdots \supset G_{r-1}/A \supset G_r/A = A/A$$

with abelian factors

$$(G_{i-1}/A)/(G_i/A) \cong G_{i-1}/G_i.$$

Moreover, A has a subnormal series

$$A \supset A_1 \supset \cdots \supset A_{s-1} \supset A_s = e$$

with abelian factors A_{k-1}/A_k. Then,

$$G = G_0 \supset G_1 \supset \cdots \supset G_{r-1} \supset A \supset A_1 \supset \cdots \supset A_{s-1} \supset A_s = e$$

is a subnormal series of G whose factors are abelian. This completes the proof.

The *higher commutator subgroups* of an arbitrary group are defined as follows

$$G^{(0)} = G,$$

$$G^{(i)} = \text{commutator subgroup of } G^{(i-1)} \qquad (i = 1, 2, \ldots).$$

The subgroup $G^{(i)}$ is also called the i-th *derived group* of G. For small values of the superscript, we write $G^{(1)} = G'$, $G^{(2)} = G''$, etc. The fact that the commutator subgroup is fully invariant and the transitivity of this property shows that all the $G^{(i)}$ are fully invariant subgroups of G. The sequence

$$G = G^{(0)} \supseteq G^{(1)} \supseteq G^{(2)} \supseteq \cdots$$

is called the *commutator chain*.

2.8.4 *A group G is soluble if and only if its commutator chain terminates in the unit subgroup after a finite number of steps.*

Proof. Suppose that G is soluble and that

$$G = G_0 \supset G_1 \supset \cdots \supset G_{l-1} \supset G_l = e$$

is a subnormal series with abelian factors. We shall prove that

$$G^{(i)} \subseteq G_i \qquad (i = 1, \ldots, l) \tag{2.20}$$

which implies that $G^{(k)} = e$ for some $k \leqq l$. Since G/G_1 is abelian, Theorem 2.3.2 shows that $G^{(1)} \subseteq G_1$. Thus, (2.20) is true for $i = 1$. Suppose that $G^{(i-1)} \subseteq G_{i-1}$. Since G_{i-1}/G_i is abelian, we can apply Theorem 2.3.2 to obtain $G'_{i-1} \subseteq G_i$, where G'_{i-1} is the commutator subgroup of G_{i-1}. From $G^{(i-1)} \subseteq G_{i-1}$, it is obvious that $G^{(i)} \subseteq G'_{i-1}$. This gives

$$G^{(i)} \subseteq G'_{i-1} \subseteq G_i$$

and proves (2.20).

Conversely, if $G^{(k)} = e$ for some k, then

$$G \supset G^{(1)} \supset G^{(2)} \supset \cdots \supset G^{(k)} = e$$

is a subnormal series (in fact, even a fully invariant series) whose factors are abelian. Thus, G is soluble.

Soluble groups are also called metabelian. More precisely, if $G^{(k-1)} \neq e$ and $G^{(k)} = e$, then G is said to be *k-step metabelian.*

2.8.5 *If G is k-step metabelian, then every factor group G/K and every subgroup H of G is at most k-step metabelian.*

Proof. The commutator subgroup of G/K is generated by the cosets

$$a^{-1}Kb^{-1}KaKbK = a^{-1}b^{-1}abK \; (a, b \in G).$$

This gives

$$(G/K)' = G'K/K$$

and, by induction,

$$(G/K)^{(i)} = G^{(i)}K/K. \qquad (i = 1, 2, \ldots)$$

From $G^{(k)} = e$, it follows that $(G/K)^{(k)} = K/K$. Moreover, $H \subseteq G$ obviously yields $H^{(i)} \subseteq G^{(i)}$ $(i = 1, 2, \ldots)$. Hence, $G^{(k)} = e$ implies that $H^{(k)} = e$.

Exercises

1. Prove that the mapping $x \to x^{-1}$ for every $x \in G$ is an automorphism of G if and only if G is abelian.
2. Prove that the mapping $x \to x^2$ for every $x \in G$ is an endomorphism of G if and only if G is abelian. Show that this mapping is an automorphism if G is abelian and of odd order.
3. Determine the automorphism groups of the cyclic groups.
4. Find the automorphism group of an n-dimensional vector space over a field.
5. Determine the automorphism group of the four-group.
6. Prove that the symmetric group S_3 has only inner automorphisms and is isomorphic to its automorphism group.
7. Prove that the endomorphisms of a group form a semigroup. Find the endomorphism semigroups of the four-group, a cyclic group, and an n-dimensional vector space over a field.

8. Find two non-isomorphic groups whose automorphism groups are of order 2.

9. Find all the isomorphisms of a cyclic group of order m onto another cyclic group of the same order.

10. Let $f(x)$ and $g(x)$ be rational functions of a variable x. Define the product fg to mean $fg(x) = f(g(x))$.
Show that under this operation the functions $f(x) = 1/x$ and $g(x) = 1/(1-x)$ generate a group isomorphic to S_3. Find all the isomorphisms of this group onto S_3.

11. Let the operator domain Ω of the four-group consist of all automorphisms. Show that there is no Ω-automorphism of the four-group other than the identity.

12. Let e, a, b, ab be the elements of the four-group and let the operator domain Ω consist of the single operator ω defined as follows: $a\omega = b$, $b\omega = ab$. Determine the Ω-automorphisms of the four-group.

13. Prove that a subgroup U is normal if every right coset of U is also a left coset.

14. Consider the symmetric group S_4 to show that the property of being a normal subgroup is not transitive.

15. Characterize the normalizer of a subgroup U of the group G in terms of the decomposition of G with respect to the double module (U, U).

16. Find all the homomorphisms of the four-group onto a group of order 2.

17. Show that every subgroup of index 2 in an arbitrary group is normal.

18. Prove that a subgroup of G is normal in G if it contains the commutator subgroup G'.

19. Let C denote the multiplicative group of all non-zero complex numbers. Find endomorphisms of C whose kernels are
(a) the subgroup of all complex numbers with modulus 1,
(b) the subgroup of all positive real numbers.
Determine the subgroup of the elements of finite order of C.

20. Show that the multiplicative group of all complex numbers with modulus 1 is isomorphic to the group of all real numbers x, $0 \leqq x < 1$, under addition mod 1.

21. Prove that every subgroup of a cyclic group is fully invariant.

22. Determine the normalizer and the centralizer of the subgroup of all diagonal matrices in $GL(n, F)$.

23. Give an example of a normal subgroup that is not characteristic.

24. Show that the permutations

$$a = \begin{pmatrix} 1 & 2 & 3 & 4 & 5 & 6 \\ 2 & 3 & 4 & 1 & 5 & 6 \end{pmatrix}, \quad b = \begin{pmatrix} 1 & 2 & 3 & 4 & 5 & 6 \\ 1 & 4 & 3 & 2 & 5 & 6 \end{pmatrix}, \quad c = \begin{pmatrix} 1 & 2 & 3 & 4 & 5 & 6 \\ 1 & 2 & 3 & 4 & 6 & 5 \end{pmatrix}$$

satisfy the relations
$a^4 = e,\ b^2 = c^2 = e,\ ab = ba^{-1},\ ac = ca,\ bc = cb$
and generate a subgroup U of order 16 of S_6. Prove that the centre Z of U is generated by a^2 and c and that the mapping $a \rightarrow b$, $b \rightarrow b$, $c \rightarrow b$ is an endomorphism of U. Show that this mapping does not carry Z into itself, so that Z is not fully invariant.

25. Determine all normal series of a cyclic group of order 24.

26. Find all composition series and all chief series of the symmetric group S_4.

27. Determine the commutator chain of S_4.

28. Show that $GL(n, F) = UD$, where U is the normal subgroup of all matrices of determinant 1 and D is the subgroup of all diagonal matrices. Verify that $GL(n, F)/U \cong D/U \cap D \cong F^*$, where F^* denotes the multiplicative group of all non-zero elements of the field F.

29. Represent the four-group as a factor group of a free group.

CHAPTER 3

Sylow Subgroups of Finite Groups

3.1 Permutational Representations of Finite Groups

A homomorphism of a group G into a group Γ is sometimes called a *representation of G by Γ*. In chapter 13, we shall deal with representations by $GL(n, F)$. In case Γ is the symmetric group of degree m, we speak of a *permutational representation* of degree m of G In chapter 7, we shall determine all permutational representations of finite groups. Here, we prove a result that will be used in the next section.

A permutational representation of degree m of the group G assigns to every $x \in G$ a permutation of some set Ω containing m elements. For any element α of Ω, let αx denote the image of α under the permutation of Ω that corresponds to x. Since the mapping of G onto the permutations is a homomorphism, we have

$$\alpha(xy) = (\alpha x)\, y \qquad (\alpha \in \Omega;\; x, y \in G).$$

We say that two elements α and β are connected under G, and write $\alpha \sim \beta$, if there exists at least one $x \in G$ such that $\alpha x = \beta$. It is easily seen that this relation is an equivalence relation. The permutation that corresponds to the unit element e of G is the identity permutation of Ω, i. e. $\alpha e = \alpha$ for every $\alpha \in \Omega$. Hence, we have $\alpha \sim \alpha$ and our relation is reflexive. From $\alpha \sim \beta$, i. e. $\alpha x = \beta$, we obtain $\alpha = \beta x^{-1}$ or $\beta \sim \alpha$. Our relation is therefore symmetric. Finally, if $\alpha \sim \beta$ and $\beta \sim \gamma$, then we have $\alpha x = \beta$ and $\beta y = \gamma$ for some $x, y \in G$. It follows that $\alpha(xy) = (\alpha x)\, y = \beta y = \gamma$, i. e. $\alpha \sim \gamma$. Thus, our relation is transitive. We conclude that G defines a partition of Ω into mutually disjoint subsets such that two elements of Ω belong to the same subset if and only if they are connected under G. These subsets are called the *orbits* of the permutational representation.

Our aim is to determine the number of elements in a given orbit T. Let τ be an arbitrary element of T, and denote by G_τ the complex of all elements u of G such that $\tau u = \tau$. It is obvious that G_τ is a subgroup of G. We call G_τ the *stabilizer subgroup* of τ. Let

$$G = G_\tau \cup G_\tau a \cup G_\tau b \cup \cdots$$

be the decomposition of G into right cosets of G_τ. All permutations that correspond to the elements of a coset $G_\tau c$ carry τ in one and the same element of Ω; for, if u is an arbitrary element of G_τ, then we have $\tau(uc) = (\tau u)c = \tau c$. Conversely, $\tau x = \tau y$ implies that $\tau(xy^{-1}) = \tau$ and hence that $xy^{-1} \in G_\tau$, i. e. $x \in G_\tau y$. Thus, the number of elements in the orbit T is equal to the number of cosets of G_τ.

3.1.1 *Let T be an orbit of a permutational representation of the group G. Then, the number of elements in T is equal to the index $|G:G_\tau|$, where G_τ is the stabilizer subgroup of an arbitrary element τ of T.*

We need the following corollary.

3.1.2 *If G is a finite group, then the number of elements in every orbit of a permutational representation of G is a divisor of $|G|$.*

3.2 Sylow's Theorem

Throughout this section we assume that G is a finite group.

The converse of Lagrange's Theorem 1.3.2 is not true, i. e. if d is a given divisor of $|G|$, then G does not necessarily contain a subgroup of order d. A basic result of the theory of finite groups is the following theorem, which asserts that the converse of Lagrange's Theorem holds if we confine ourselves to subgroups of prime power order.

3.2.1 *If p^t is any prime power that divides the order of the group G, then G contains at least one subgroup of order p^t.*

Proof. We denote the order of G by g and put $g = p^t r$. (Note that r may be divisible by p.) Let $K_1, K_2, \ldots, K_m$ be all the complexes of G containing p^t elements each. Their number m is given by

$$m = \binom{g}{p^t} = \frac{p^t r(p^t r - 1) \cdots (p^t r - k) \cdots (p^t r - p^t + 1)}{p^t \cdot 1 \cdots k \cdots (p^t - 1)}.$$

Let p^s be the highest power of p that divides r, where the case $s = 0$ is not excluded. It is easily seen that m is not divisible by p^{s+1}; for in the representation above of m as a fraction numerator and denominator of each pair $(p^t r - k)$, k of factors are divisible by the same power of p.

To any element x of G, we assign the permutation

$$\begin{pmatrix} K_1 & K_2 & \cdots K_m \\ K_1 x & K_2 x & \cdots K_m x \end{pmatrix}$$

of the complexes $K_1, K_2, \ldots, K_m$. Since $K_i(xy) = (K_ix)y$, we obtain a permutational representation of G. We consider its orbits. Since m is not divisible by p^{s+1}, there is at least one orbit T such that the number of complexes contained in T is not divisible by p^{s+1}. Let l denote the number of complexes in T. By renumbering the complexes, if necessary, we may assume that K_1 belongs to T. Let H be the stabilizer subgroup of K_1 and put $|H| = h$. By Theorem 3.1.1, we have $g = hl$. The divisor l of g is not divisible by p^{s+1}, which implies that l divides r; hence, in particular, $l \leqq r$. If a is any element of K_1, then $K_1 = K_1H$ shows that K_1 contains at least the h distinct elements au, where u ranges over all the elements of H. This shows that $h \leqq p^t$. So, we obtain the inequalities

$$g = hl \leqq hr \leqq p^t r = g,$$

which imply that $h = p^t$. Consequently, H is a subgroup of order p^t.

In particular, if p^n is the highest power of p that divides the order of G, then G contains at least one subgroup of order p^n. Every subgroup of order p^n of G is called a *Sylow p-subgroup*.

A special case of Theorem 3.2.1 is:

3.2.2 Cauchy's Theorem. *If p is a prime divisor of the order of G, then G contains at least one element of order p.*

The next theorem relates the Sylow p-subgroups of G to those of normal subgroups and factor groups of G.

3.2.3 *If P is a Sylow p-subgroup and T a normal subgroup of G, then*

PT/T *is a Sylow p-subgroup of G/T, and*

$P \cap T$ *is a Sylow p-subgroup of T.*

Proof. A subgroup is a Sylow p-subgroup if and only if its order is a power of p and its index is prime to p. From

$$PT/T \cong P/(P \cap T) \tag{3.1}$$

we infer that the order of PT/T is a power of p. On the other hand,

$$|(G/T):(PT/T)| = |G:PT|$$

is a divisor of $|G:P|$ and therefore prime to p. This shows that PT/T is a Sylow p-subgroup of G/T.

Since $|PT:T| = |P:(P \cap T)|$, which follows from (3.1), the equation

$$|PT:T||T:(P \cap T)| = |PT:P||P:(P \cap T)|$$

gives $|T:(P \cap T)| = |PT:P|$. Since $|PT:P|$ divides $|G:P|$, we conclude that $|T:(P \cap T)|$ is prime to p. Clearly, $|P \cap T|$ is a power of p, and hence $P \cap T$ is a Sylow p-subgroup of T. This completes the proof.

If the Sylow p-subgroup P of G is normal in G, then P is the only Sylow p-subgroup of G. For, suppose that P^* is another Sylow p-subgroup. Then, we have

$$PP^*/P \cong P^*/(P \cap P^*)$$

which shows that $|PP^*|$ is a power of p. On the other hand, $|P|$ is the highest power of p that divides $|G|$. So, we obtain $|PP^*| = P$ and hence $P^* = P$, a contradiction.

This gives:

3.2.4 *A Sylow p-subgroup P of G is the only Sylow p-subgroup of its normalizer* $\mathsf{N}(P)$.

The main facts on Sylow p-subgroups are summarized in:

3.2.5 Sylow's Theorem.

(a) *Let p^n be the highest power of the prime number p that divides the order of the group G. Then, G contains at least one Sylow p-subgroup, i. e. a subgroup of order p^n.*

(b) *Any two Sylow p-subgroups of G are conjugate.*

(c) *Every subgroup of G whose order is a power of p is contained in a Sylow p-subgroup.*

(d) *If r denotes the number of Sylow p-subgroups in G, then $r \equiv 1$ (mod p).*

Proof. (a) has been proved above. Let $P_1, P_2, \ldots, P_r$ be all the Sylow p-subgroups of G. To any element x of P_1, we assign the permutation

$$\begin{pmatrix} P_1 & P_2 & \cdots & P_r \\ x^{-1}P_1x & x^{-1}P_2x & & x^{-1}P_rx \end{pmatrix}$$

to obtain a permutational representation of degree r of P_1. All these permutations leave P_1 fixed. Moreover, P_1 is the only Sylow p-subgroup that remains fixed under all these permutations. For, suppose that we have $x^{-1}P_kx = P_k$ for all $x \in P_1$ and some $k \neq 1$, then P_1 would be contained in the normalizer of P_k; this contradicts Theorem 3.2.4. We now consider the orbits of our permutational representation of P_1. One of the orbits consists of the single Sylow p-subgroup P_1, whereas all the other orbits contain more than one P_i. By Theorem 3.1.2, the numbers of Sylow p-subgroups in the latter orbits are divisors of $|P_1|$, so they are powers of p with positive exponents. So, we have $r \equiv 1$ (mod p), which proves (d).

b

Suppose that $P_1, \ldots, P_s$ are conjugate in G. For the proof of (b), we have to show that $s < r$ leads to a contradiction. Transformation by elements of P_1 permutes $P_1, \ldots, P_s$. As above, we conclude that $s \equiv 1 \pmod p$. In case $s < r$, there is another system, $P_{s+1}, \ldots, P_{s+t}$ say, of Sylow p-subgroups that are conjugate in G. Transformation by elements of P_1 yields a permutational representation of P_1 of degree t. From our arguments above, we conclude that the numbers of Sylow p-subgroups in the orbits of this representation are powers of p with positive exponents. This gives $t \equiv 0 \pmod p$. On the other hand, we can transform $P_{s+1}, \ldots, P_{s+t}$ by elements of P_{s+1}. In the same way as above, we then obtain $t \equiv 1 \pmod p$. Thus, the assumption $s < r$ leads to a contradiction. This proves (b).

Finally, let U be a subgroup of G whose order is a power of p. If we transform $P_1, \ldots, P_r$ by the elements of U, then we obtain a permutational representation of U of degree r. Since $|U|$ is a power of p, the numbers of Sylow p-subgroups in the orbits of this representation are powers of p. From $r \equiv 1 \pmod p$, we conclude that at least one orbit consists of a single Sylow p-subgroup, P_k say. Then, we have $u^{-1} P_k u = P_k$ for every $u \in U$. This shows that UP_k is a subgroup of G whose order is a power of p. But, since $|P_k|$ is the highest power of p dividing $|G|$, we have $|UP_k| = |P_k|$ and hence $U \subseteq P_k$. This completes the proof of (c).

3.3 Further Theorems on Sylow Subgroups

By a *finite p-group (p-subgroup)* we understand a group (subgroup) whose order is a power of the prime number p.

3.3.1 *If U is a p-subgroup but not a Sylow p-subgroup of G, then the normalizer $\mathsf{N}(U)$ of U in G contains U as a proper subgroup.*

Proof. If p does not divide $|G: \mathsf{N}(U)|$, then $\mathsf{N}(U)$ contains a Sylow p-subgroup of G, which in its turn contains U properly.

Now assume that $|G: \mathsf{N}(U)| = pk$. There are pk distinct conjugates of U. By transforming them by elements of U, we obtain a permutational representation of U of degree pk. The numbers of conjugates of U in the orbits are powers of p. Since one orbit consists of U only and since the total number of conjugates is divisible by p, there is at least one orbit other than U that consists of a single conjugate. (The number of orbits consisting of a single conjugate is divisible by p.) Let $U_1 = a^{-1}Ua$ be such an orbit. Then, U is contained in the normalizer $\mathsf{N}(U_1)$ of U_1, and hence $\mathsf{N}(U_1)$ contains U_1 properly. It follows that $\mathsf{N}(U) = a\mathsf{N}(U_1)a^{-1}$ contains $U = a U_1 a^{-1}$ properly.

3.3.2 *If the subgroup H of G contains the normalizer $\mathsf{N}(P)$ of a Sylow p-subgroup P of G, then H is its own normalizer in G.*

Proof. We have to show that $a^{-1}Ha = H$ implies that $a \in H$. From $a^{-1}Ha = H$, we conclude that P and $a^{-1}Pa$ are two Sylow p-subgroups of H. By Theorem 3.2.5, there is an element b of H such that $b^{-1}a^{-1}Pab = P$. Consequently, we have $ab \in \mathsf{N}(P)$ and hence $ab \in H$. Since $b \in H$, this gives $a \in H$.

3.3.3 *If the normal subgroup T of G contains a Sylow p-subgroup P of G, then $G = \mathsf{N}(P)\, T$.*

Proof. The Sylow p-subgroups of T are also Sylow p-subgroups of G. If x is an arbitrary element of G, then P and $x^{-1}Px$ are Sylow p-subgroups of T. By Theorem 3.2.5, we can find an element y of T such that $y^{-1}Py = x^{-1}Px$. So, we have $(xy^{-1})\; P(xy^{-1}) = P$, which means that $xy^{-1} \in \mathsf{N}(P)$. This shows that an arbitrary element x of G can be represented as a product of an element of T and an element of $\mathsf{N}(P)$.

3.3.4 *Let H be a subgroup of G and let P_1^* and P_2^* be two distinct Sylow p-subgroups of H. Then P_1^* and P_2^* are not contained in the same Sylow p-subgroup of G.*

Proof. Assume that P_1^* and P_2^* are contained in the Sylow p-subgroup P of G. Then, $\langle P_1^*, P_2^*\rangle$ is contained in P, and hence $\langle P_1^*, P_2^*\rangle$ is a p-subgroup. But, since $\langle P_1^*, P_2^*\rangle$ is contained in H and since P_1^* is a Sylow p-subgroup of H, it follows that $\langle P_1^*, P_2^*\rangle$ and P_1^* have the same order. This implies that $P_1^* = P_2^*$, which contradicts the hypothesis.

3.3.5 Burnside's Theorem. *Let K and L be two complexes of G satisfying the following conditions:*

(a) *K and L are conjugate in G,*

(b) *$y^{-1}Ky = K$, $y^{-1}Ly = L$ for every element y of a Sylow p-subgroup P of G.*

Then, there exists an element u of $\mathsf{N}(P)$ such that $u^{-1}Ku = L$.

Proof. From (b), we know that $P \subseteq \mathsf{N}(K)$ and $P \subseteq \mathsf{N}(L)$. By (a), we have $x^{-1}Kx = L$ for some $x \in G$ and hence $x^{-1}\mathsf{N}(K)x = \mathsf{N}(L)$. This gives $x^{-1}Px \subseteq \mathsf{N}(L)$. Thus, P and $x^{-1}Px$ are two Sylow p-subgroups of G contained in $\mathsf{N}(L)$. Of course, P and $x^{-1}Px$ are also Sylow p-subgroups of $\mathsf{N}(L)$. By Theorem 3.2.5, we have $P = z^{-1}x^{-1}Pxz$ for some $z \in \mathsf{N}(L)$. This means that $u = xz \in \mathsf{N}(P)$, and one readily checks that u has the required property:

$$u^{-1}Ku = z^{-1}x^{-1}Kxz = z^{-1}Lz = L.$$

3.4 Simple Properties of p-Groups

A *p-group* is a periodic group in which the order of every element is a power of the prime number p. By 3.2.2, a finite group is a p-group if and only if its order is a power of p. Finite and infinite p-groups have been thoroughly studied and we refer to [29] and [41]. In chapter 10, we shall deal with finite p-groups. Here, we derive only some of their simplest properties.

3.4.1 *If a p-group G contains a subgroup U of finite index, then* $|G:U|$ *is a power of p.*

Proof. By Theorem 2.2.8, G contains a normal subgroup D such that $D \subseteq U$ and $|G:D|$ is finite. The factor group G/D is a finite p-group so that $|G:D|$ is a power of p. Hence, $|G:U|$, as a divisor of $|G:D|$, is a power of p.

The next theorem is basic for the theory of finite p-groups.

3.4.2 *Every finite p-group possesses a non-trivial centre.*

Proof. Let G be a group of order p^n. Denote by $K_1, K_2, \ldots, K_t$ the classes of conjugate elements of G. By Theorem 2.2.1, we have $|K_i| = p^{n_i}$, where $n_i \geqq 0$ $(i = 1, \ldots t)$. One of the classes, K_1 say, consists of the unit element only so that $n_1 = 0$. Counting the elements in each class, we obtain

$$p^n = 1 + p^{n_2} + \cdots + p^{n_t}.$$

Since the right-hand side is divisible by p, we conclude that not all the exponents $n_2, \ldots, n_t$ are positive. Thus, K_1 is not the only class that consists of a single element and this means that e is not the only element in the centre.

There are infinite p-groups with trivial centre (cf. [41], vol II, p. 270–271). But, an argument similar to that in the proof of the last theorem can be used to give the following result:

3.4.3 *If the p-group G has a class of conjugate elements other than e that contains a finite number of elements, then G possesses a non-trivial centre.*

Proof. Let K be a class of G that contains finitely many elements. Then, by Theorem 2.2.9, the normal subgroup $\langle K \rangle$ is finite. Consequently, every element of $\langle K \rangle$ belongs to a finite class of conjugate elements. Thus, the normalizers of all elements of $\langle K \rangle$ have finite indices in G and, by Theorem 3.4.1, these indices are powers of p. Suppose that $\langle K \rangle$ consists of r classes of conjugate elements of G. The numbers of elements in these classes are powers $p^{k_1}, p^{k_2}, \ldots, p^{k_r}$ where $k_i \geqq 0$. Since e occurs among these classes, we have $p^{k_1} = 1$, say. We obtain

$$p^m = 1 + p^{k_2} + \cdots + p^{k_r}$$

where $p^m = |\langle K \rangle|$. As in the proof of Theorem 3.4.2, it now follows that G has a non-trivial centre.

In a finite p-group G, we put $Z^0(G) = e$ and define subgroups $Z^i(G)$ as follows

$$Z^i(G)/Z^{i-1}(G) = \text{ centre of } G/Z^{i-1}(G) \qquad (i = 1, 2, \ldots).$$

Since each factor group $G/Z^{i-1}(G)$ is a p-group, Theorem 3.4.2 shows that $Z^i(G)$ is larger than $Z^{i-1}(G)$. We finally arrive at $Z^m(G) = G$ for some natural number m. The series

$$e = Z^0(G) \subset Z^1(G) \subset Z^2(G) \subset \cdots \subset Z^{m-1}(G) \subset Z^m(G) = G \tag{3.2}$$

is called the *upper central series* of G. Since the centre of any group is a characteristic subgroup, it is easily seen that all the terms $Z^i(G)$ are characteristic subgroups of G so that (3.2) is a characteristic series. All the factor groups $Z^i(G)/Z^{i-1}(G)$ are abelian. Hence:

3.4.4 *Every finite p-group is soluble.*
The characteristic series (3.2) can be refined to a composition series. Since G is soluble, all composition indices are equal to p. From the fact that $Z^i(G)/Z^{i-1}(G)$ is the centre of $G/Z^{i-1}(G)$, we conclude that all subgroups $U/Z^{i-1}(G)$ of $Z^i(G)/Z^{i-1}(G)$ are normal in $G/Z^{i-1}(G)$. Hence, all subgroups U of G such that $Z^{i-1}(G) \subseteq U \subseteq Z^i(G)$ are normal subgroups of G. This shows that every refinement of (3.2) is a normal series. If we refine (3.2) to a composition series, we actually obtain a chief series. So, we have proved:

3.4.5 *The chief indices of a finite p-group are equal to p.*

Let U be a proper subgroup of the finite p-group G. Suppose that U contains $Z^i(G)$ but not $Z^{i+1}(G)$. The factor group $U/Z^i(G)$ commutes elementwise with the centre $Z^{i+1}(G)/Z^i(G)$ of $G/Z^i(G)$. Thus, $Z^{i+1}(G)/Z^i(G)$ is contained in the normalizer of $U/Z^i(G)$ in $G/Z^i(G)$, and hence $Z^{i+1}(G)$ is contained in the normalizer of U in G. This gives the following result:

3.4.6 *The normalizer of any proper subgroup U of a finite p-group contains U properly.*

From the Theorems 3.4.4 and 3.4.6, we obtain, in particular:

3.4.7 *Every maximal subgroup of a finite p-group is normal and of index p.*

Theorem 3.4.5 shows that every finite p-group contains normal subgroups of order p. As to these normal subgroups, we have the following theorem:

3.4.8 *Every normal subgroup of order p of a finite p-group G is contained in the centre of G.*

Proof. If $\langle a \rangle$ is a normal subgroup of order p in G, then all the conjugates of a belong to the set $\{a, a^2, \ldots, a^{p-1}\}$. The number of distinct conjugates of a is some power of p. Since this number cannot exceed $p - 1$, it must be equal to 1. Thus, a is equal to all its conjugates, which means that a belongs to the centre.

In section 2.3, we remarked that the factor group of a nonabelian group with respect to its centre is not cyclic. This gives:

3.4.9 *The index of the centre of a non-abelian finite p-group is divisible by* p^2.

From this, we obtain:

3.4.10 *Every group of order* p^2 *is abelian. Every normal subgroup of index* p^2 *in an arbitrary group contains the commutator subgroup. The centre of a non-abelian group of order* p^3 *coincides with the commutator subgroup and is of order* p.

Exercises

1. The elements a, b subject to the relations $a^2 = b^3 = (ab)^3 = e$ generate a group of order 12. Show that this group contains no subgroup of order 6
2. Find all Sylow subgroups of the symmetric group S_4.
3. Determine the normalizers of the Sylow subgroups of S_4.
4. Prove that the exponent of a finite group G is divisible by every prime factor of $|G|$.
5. Let T be a normal subgroup of the finite group G such that $|G:T|$ is not divisible by the prime divisor p of $|G|$. Show that T contains all Sylow p-subgroups of G.
6. Show that a normal Sylow p-subgroup of a finite group is fully invariant.
7. Let G be a group of order pq, where p and q are prime numbers and $p < q$. Prove that G contains a single, and hence normal, Sylow q-subgroup.

CHAPTER 4

DIRECT PRODUCTS

4.1 Definition and Basic Properties

If a group G contains two proper subgroups P and Q such that $G = PQ$ and $P \cap Q = e$, then G is said to be factorizable.

In this chapter, we shall deal with a special but important kind of factorization. We emphasize that all groups to be studied in this chapter may have an operator domain. In this case, subgroup (normal subgroup) is to mean admissible subgroup (admissible normal subgroup) and all homomorphisms are supposed to be operator homomorphisms.

A factorization of a group G into the product of its subgroups $G_1, G_2, \ldots, G_n$ is called a *direct decomposition* if the following conditions are satisfied:

(a) *Every element x of G has a unique representation*

$$x = x_1 x_2 \cdots x_n, \quad x_i \in G_i \qquad (i = 1, 2, \ldots, n). \tag{4.1}$$

(b) *Any two distinct subgroups G_i, G_j commute elementwise.*

Direct decomposition is denoted by

$$G = G_1 \times G_2 \times \cdots \times G_n. \tag{4.2}$$

The subgroups G_i are called *direct factors* of G, and G is said to be the *direct product* of its subgroups $G_1, G_2, \ldots, G_n$.

In the case of a finite group, we obviously have

$$|G| = |G_1||G_2| \cdots |G_n|$$

To exclude trivial cases, we always assume that direct factors are nontrivial subgroups. A group is called *directly decomposable* if it can be represented as the direct product of proper subgroups.

By (b), the direct factors may be arranged arbitrarily.

The element x_i in the representation (4.1) is called the *G_i-component* of x. To indicate that the representation (4.1) corresponds to the direct decomposition (4.2), we also write

$$x = x_1 \times x_2 \times \cdots \times x_n$$

instead of (4.1).

If

$$y = y_1 \times y_2 \times \cdots \times y_n$$

is another element of G, then (b) yields the following rule for the product

$$xy = x_1y_1 \times x_2y_2 \times \cdots \times x_ny_n$$

where x_iy_i is the G_i-component of xy. To put it briefly, multiplication is carried out componentwise. It follows that

$$y^{-1} = y_1^{-1} \times y_2^{-1} \times \cdots \times y_n^{-1}$$

and

$$y^{-1}xy = y_1^{-1}x_1y_1 \times y_2^{-1}x_2y_2 \times \cdots \times y_n^{-1}x_ny_n. \tag{4.3}$$

If x is an element of G_i for which $x = x_i$, $x_j = e$ for $j \neq i$, then (4.3) gives

$$y^{-1}x_iy = y_i^{-1}x_i y_i, \tag{4.4}$$

which shows that $y^{-1}x_iy$ belongs to G_i. So, we have:

(c) *Every direct factor is a normal subgroup.*

It is readily seen that

$$G/G_i \cong G_1 \times \cdots \times G_{i-1} \times G_{i+1} \times \cdots \times G_n.$$

As another easy consequence, we obtain the following property of a direct product:

(d) $G_i \cap \langle G_1, \ldots, G_{i-1}, G_{i+1}, \ldots, G_n \rangle = e, \qquad (i = 1, 2, \ldots, n).$

For, if x_i is an element of this intersection, then (b) gives

$$x_i = x_1 \cdots x_{i-1} x_{i+1} \cdots x_n \quad x_j \in G_j,$$

and using (a) we conclude that $x_j = e$ for $j = 1, 2, \ldots, n$.

Conversely, (c) and (d) imply that G is the direct product of $G_1, G_2, \ldots, G_n$. We may even replace (d) by a weaker condition.

4.1.1 *Let $G_1, G_2, \ldots, G_n$ be normal subgroups of G such that $G = \langle G_1, G_2, \ldots, G_n \rangle$ and*

$$G_i \cap \langle G_1, \ldots, G_{i-1} \rangle = e \qquad (i = 2, \ldots, n). \tag{4.5}$$

Then, G is the direct product of $G_1, G_2, \ldots, G_n$.

Proof. By (4.5), we have $G_j \cap G_i = e$ if $j \neq i$. From Theorem 2.2.6, we conclude that G_j and G_i commute elementwise. Thus, (b) is satisfied.

Since $G_1, G_2, \ldots, G_n$ are normal subgroups, we obtain

$$G = \langle G_1, G_2, \ldots, G_n \rangle = G_1 G_2 \cdots G_n.$$

Hence, every $x \in G$ has at least one representation (4.1). Suppose that

$$x = x_1 x_2 \cdots x_n = x_1' x_2' \cdots x_n'$$

are two such representations. Then, there is some integer i, $1 \leq i \leq n$, such that $x_i \neq x_i'$, but $x_{i+1} = x_{i+1}', \ldots, x_n = x_n'$. This gives

$$x_i x_i'^{-1} = (x_1^{-1} x_1')(x_2^{-1} x_2') \cdots (x_{i-1}^{-1} x_{i-1}'),$$

which shows that $x_i x_i'^{-1} \neq e$ is contained in G_i as well as in $\langle G_i, \ldots, G_{i-1} \rangle$. Since this contradicts (4.5), we infer that the representation (4.1) is unique. Thus, (a) is satisfied, and our proof is complete.

The concept of a direct product can also be used to construct new groups from given ones. Let A and B be two groups whose unit elements are denoted by e_A and e_B, respectively. We take all pairs (a, b), $a \in A$, $b \in B$, and define multiplication as follows

$$(a_1, b_1)(a_2, b_2) = (a_1 a_2, b_1 b_2).$$

It is easily verified that these pairs form a group G with respect to this multiplication. The unit element is (e_A, e_B). Moreover, the elements (a, e_B) form a subgroup A^*, that is obviously isomorphic to A, and the pairs (e_A, b) form a subgroup B^* isomorphic to B. It is evident that $G = A^* \times B^*$. In view of the isomorphisms $A \cong A^*$ and $B \cong B^*$, the group G we have constructed is called the direct product of A and B. In the same way, one can construct the direct product of any finite number of given groups. The method may also be used to construct the direct product of infinitely many groups.

The next theorem is an immediate consequence of the fact that multiplication in a direct product is carried out componentwise.

4.1.2 *The centre of a direct product*

$$G = G_1 \times G_2 \times \cdots \times G_n$$

is the direct product of the centres of the direct factors. The commutator subgroup of G is the direct product of the commutator subgroups G_i'. Any normal subgroup of a direct factor G_i is a normal subgroup of G.

To simplify our next arguments, we confine ourselves to a direct product $G = G_1 \times G_2$ of two factors.

If U is a subgroup of G, then the G_i-components of the elements of U obviously form a subgroup U_i of G_i. The subgroup of G generated by U_1 and U_2 is the direct product $U_1 \times U_2$. It is evident that

$$U \subseteq U_1 \times U_2. \tag{4.6}$$

In general, however, U is a proper subgroup of $U_1 \times U_2$. For example, let $G_1 = \langle a \rangle$, $G_2 = \langle b \rangle$ be two cyclic groups of the same order m. The subgroup $U = \langle ab \rangle$ is a cyclic subgroup of order m of $\langle a \rangle \times \langle b \rangle$, and we have $U_1 = \langle a \rangle$, $U_2 = \langle b \rangle$, but only the elements of the form $a^k b^k$ belong to U.

Clearly, $U = U_1 \times U_2$ implies that $U \cap G_1 = U_1$. We prove the converse:

4.1.3 *If* $U_1 = U \cap G_1$, *then* $U_2 = U \cap G_2$ *and* $U = U_1 \times U_2$. *In particular,* $G_1 \subseteq U$ *implies that* $U = G_1 \times (U \cap G_2)$.

Proof. Let $u = u_1 \times u_2$ be any element of U. From $U_1 = U \cap G_1$, we obtain $u_1 \in U \cap G_1$ and hence $u_1 \in U$. It follows that $u_2 = u_1^{-1} u$ belongs to U. Therefore, $U_1 \subseteq U$ and $U_2 \subseteq U$, so that (4.6) gives $U = U_1 \times U_2$. And, also, $U_2 = U \cap G_2$.

The next result is the converse of the last proposition in Theorem 4.1.2.

4.1.4 *If U is a normal subgroup of G, then so are U_1 and U_2.*

Proof. For any element $u = u_1 \times u_2$ of U and an arbitrary element $x = x_1 \times x_2$

$$x^{-1} u x = x_1^{-1} u_1 x_1 \times x_2^{-1} u_2 x_2,$$

and since U is a normal subgroup, $x^{-1} u x$ belongs to U. Consequently, $x_1^{-1} u_1 x_1$ belongs to U_1. If x ranges over all elements of G, then x_1 ranges over all elements of G_1. Thus, U_1 is a normal subgroup of G_1. It follows from Theorem 4.1.2 that U_1 is normal in G. A similar argument applies to U_2.

4.2 Completely Reducible Groups

A group is called *completely reducible* if it is the direct product of simple groups. One should bear in mind that an operator domain may be given. In this case, a group is said to be simple if it contains no admissible normal subgroups other than the unit element and the whole group. There may exist other normal subgroups, but no admissible ones.

Every completely reducible group possesses a chief series. For let

$$G = G_1 \times G_2 \times \cdots \times G_n$$

be a direct decomposition into simple factors and put

$$P_i = G_1 \times \cdots \times G_i \qquad (i = 1, 2, \ldots, n).$$

Then,

$$G = P_n \supset P_{n-1} \supset \cdots \supset P_2 \supset P_1 \supset e$$

is a chief series, for the factors are, respectively, isomorphic to $G_n, G_{n-1}, \ldots, G_2, G_1$, which are simple groups.

We observe that the simple direct factors of a completely reducible group are isomorphic to its chief factors. Then, Theorem 2.7.3 shows that the simple direct factors are unique in the following sense: If

$$G = G_1 \times \cdots \times G_n = H_1 \times \cdots \times H_n$$

are two direct decompositions of the completely reducible group G into simple factors, then $n = m$ and $G_i \cong H_i$ $(i = 1, \ldots, n)$ if the order of the factors is properly chosen.

By a *minimal normal subgroup* F of a group G, we mean a non-trivial normal subgroup that contains no normal subgroup of G other than e. There may be non-trivial normal subgroups of F, but they cannot be normal in G. A minimal normal subgroup of G is also called a *foot* of G. By Theorem 2.6.2, every minimal normal subgroup is characteristically simple.

The infinite cyclic group is an example of a group that contains no minimal normal subgroup. However, every group that possesses a chief series contains at least one minimal normal subgroup. For, otherwise, we could find an infinite sequence of normal subgroups every term of which is properly contained in the preceding one; by Theorem 2.7.3, this contradicts the existence of a chief series.

The following theorem gives us information about the structure of characteristically simple groups.

4.2.1 *If the group G is characteristically simple and possesses a chief series, then G is simple or a direct product of isomorphic simple groups.*

Proof. Since G possesses a chief series, there is at least one minimal normal subgroup F. Any automorphism α of G carries F into a minimal normal subgroup $F\alpha$. Let U denote the subgroup generated by the images $F\alpha$ under all automorphisms of G. Since every automorphism of G only permutes the images $F\alpha$, it maps U onto itself, i. e. U is a characteristic subgroup of G. As G is characterically simple, we have $U = G$.

The intersection of an $F\alpha$ with any normal subgroup T of G is e or $F\alpha$; for, otherwise, $F\alpha \cap T$ would be a nontrivial subgroup of $F\alpha$, which is normal in G; but this is impossible, since $F\alpha$ is minimal. In case $F\alpha \cap T = e$, we have $T(F\alpha) = T \times F\alpha$.

We now proceed as follows: For the identity automorphism α_1 of G, we have $F\alpha_1 = F$. If there exists some $F\alpha_2$ such that $F\alpha_2 \neq F\alpha_1$, then we get $F\alpha_1 F\alpha_2 = F\alpha_1 \times F\alpha_2$. If there is an $F\alpha_3$ that is not contained in $F\alpha_1 \times F\alpha_2$, then we can form the direct product $F\alpha_1 \times F\alpha_2 \times F\alpha_3$. We continue in the same way. The general step is as follows: If we have formed the direct product $F\alpha_1 \times F\alpha_2 \times \cdots \times F\alpha_k$ and if there is some $F\alpha$ not contained in that direct product, then we form

$$F\alpha_1 \times F\alpha_2 \times \cdots \times F\alpha_k \times F\alpha.$$

Since G has a chief series, this process terminates after a finite number of steps, i. e. we arrive at a direct product

$$F\alpha_1 \times F\alpha_2 \times \cdots \times F\alpha_r$$

that contains all the images $F\alpha$. Thus,

$$G = F\alpha_1 \times F\alpha_2 \times \cdots \times F\alpha_r.$$

It remains to show that the direct factors $F\alpha_1$ are simple. Suppose that $F\alpha_1$ contains a non-trivial noımal subgroup D. By Theorem 4.1.2, D is a normal subgroup of G, but this contradicts the fact that $F\alpha_1$ is a minimal normal subgroup.

Thus, Theorem 4.2.1 is proved. In view of Theorem 2.7.5, we can apply Theorem 4.2.1 to obtain:

4.2.2 *The chief factors of a finite group G are simple or direct products of isomorphic simple groups. If G is soluble, then the chief factors are cyclic groups of prime order or direct products of cyclic groups of the same prime order.*

The next theorem, on completely reducible groups, is frequently applied in group theory and other branches of algebra.

4.2.3 *Every normal subgroup H of a completely reducible group G is a direct factor of G, i. e. there exists a direct decomposition $G = H \times B$.*

Proof. Let

$$G = G_1 \times \cdots \times G_n$$

be a direct decomposition into simple factors G_i. Clearly,

$$G = HG = HG_1 \cdots G_n. \tag{4.7}$$

We shall show that suitable factors G_i on the right-hand side can be cancelled such that the product of the remaining factors is direct. Since G_1 is simple and H is normal in G, the intersection $G_1 \cap H$ coincides with G_1 or is e. In the first case, the factor G_1 on the right-hand side of (4.7) is redundant and can be cancelled. In the second case, we have $HG_1 = H \times G_1$. In the same way, we deal with all the factors G_i: Suppose that among $G_1, \ldots, G_{i-1}$ precisely $G_{k_1}, \ldots, G_{k_r}$ have not been cancelled so that we have

$$HG_1 \cdots G_{i-1} = H \times G_{k_1} \times \cdots \times G_{k_r}.$$

The intersection $G_i \cap HG_i \ldots G_{i-1}$ is G_i or e. In the first case, G_i is cancelled, in the second case, we have

$$HG_1 \cdots G_i = H \times G_{k_1} \times \cdots \times G_{k_r} \times G_i.$$

Eventually, we obtain

$$G = HG_1 \cdots G_n = H \times G_{k_1} \times \cdots \times G_{k_t} = H \times B.$$

Incidentally, we have proved that the complementary direct factor B can be chosen as the direct product of suitable G_i.

In general, even a finite group has several minimal normal subgroups. It is useful to study the set of all minimal normal subgroups.

Let G be a group that possesses a chief series. The normal subgroup S of G generated by all minimal normal subgroups of G is called the *socle* of G.

If G contains a single minimal normal subgroup F_1, then $S = F_1$. In case F_1 is not the only minimal normal subgroup, we proceed as follows: Let F_2 be another minimal normal subgroup. Then, we have $F_1 \cap F_2 = e$ and hence $F_1F_2 = F_1 \times F_2$. If there is a minimal normal subgroup F_3 that is not contained in $F_1 \times F_2$, then we obtain $F_1F_2F_3 = F_1 \times F_2 \times F_3$. Since G possesses a chief series, this procedure must terminate after a finite number of steps. Eventually, we arrive at a representation

$$S = F_1 \times F_2 \times \cdots \times F_r \tag{4.8}$$

of the socle as a direct product of suitable minimal normal subgroups.

There may exist minimal normal subgroups that do not occur as a direct factor in (4.8). Suppose that this is the case and let A be one of them. Then, every $a \in A$ has a unique representation

$$a = a_1 \times a_2 \times \cdots \times a_r, \quad a_i \in F_i.$$

The F_i-components of all elements of A form a subgroup A_i of F_i. By Theorem 4.1.4, A_i is a normal subgroup of G. Since F_i is minimal, we have either $A_i = F_i$ or $A_i = e$. Using a suitable notation, we may assume that

$$A_j = F_j \quad \text{for} \quad j = 1, \ldots, s,$$
$$A_k = e \quad \text{for} \quad k = s+1, \ldots, r.$$

We now prove that every element of F_j ($j = 1, \ldots s$) occurs as the F_j-component of a single element of A. It is clearly sufficient to prove this for $j = 1$. Suppose that a and b are two distinct elements of A with the same F_1-components. Then, the F_1-component of ab^{-1} is e, and hence ab^{-1} is contained in $F_2 \times \cdots \times F_r$. This shows that $A \cap (F_2 \times \cdots \times F_r)$ is not the unit subgroup and, as A is minimal, this intersection coincides with A. This would mean that all the elements of A have the F_1-component e, which contradicts $A_1 = F_1$.

We conclude that the mapping

$$a \to a_j \qquad (a \in A)$$

is an isomorphism of A onto F_j $(j = 1, \ldots, s)$. Thus, all the minimal normal subgroups $F_1, \ldots, F_s$ are isomorphic to each other, and isomorphisms between them are provided by A. Such an isomorphism of F_1 onto F_j $(j = 2, \ldots, s)$ can be described as follows: To any element x_1 of F_1, there corresponds the (unique) element of F_j that occurs as the F_j-component of the (unique) element of A whose F_1-component is x_1.

Let σ_{12} denote this isomorphism of F_1 onto F_2. Thus, for $a \in A$ and

$$a = a_1 \times a_2 \times \cdots \times a_s \times e \times \cdots \times e$$

we have

$$a_1 \sigma_{12} = a_2 . \tag{4.9}$$

For an arbitrary element x of G,

$$x^{-1} a x = x^{-1} a_1 x \times x^{-1} a_2 x \times \cdots$$

is contained in A, which implies that

$$(x^{-1} a_1 x)\, \sigma_{12} = x^{-1} a_2 x = x^{-1} (a_1 \sigma_{12})\, x . \tag{4.10}$$

The last equation shows that the isomorphism σ_{12} commutes with all the inner automorphisms of G. In other words: If the operator domain Ω consists of all the inner automorphisms of G, then σ_{12} is an Ω-isomorphism in the sense of section 2.6. Such isomorphisms are called *normal isomorphisms*.

If two minimal normal subgroups are normally isomorphic, then they are abelian. Indeed, if the element x in (4.10) belongs to F_1, then it commutes with the element $a_1\sigma_{12}$ of F_2 so that we obtain

$$(x \sigma_{12})^{-1} (a_1 \sigma_{12}) (x \sigma_{12}) = a_1 \sigma_{12} .$$

Since the elements $a_1\sigma_{12}$ range over the whole group F_2, this shows that $x\sigma_{12}$ belongs to the centre of F_2. But, every element of F_2 is of the form $x\sigma_{12}$ for some $x \in F_1$, hence F_2 is abelian.

4.3 Normal Endomorphisms and Direct Decompositions

In the next two sections, we adopt a slightly modified notation, which might be more suggestive. If σ is an endomorphism of the group G, then the image of an element x of G under σ will be denoted by x^σ. Similarly, K^σ denotes the image of the complex K. Powers of group elements will not occur, so that there is no danger of confusion.

Let ε denote the identity automorphism of G, and 0 the null endomorphism, i. e. the endomorphism that carries every element of G into e.

An endomorphism α is called *idempotent* if $\alpha^2 = \alpha$. Trivial examples of idempotent endomorphisms are ε and 0. We shall see that certain idempotent endomorphisms of G other than ε and 0 are closely connected with direct decompositions of G. An endomorphism $\sigma \neq 0$ is called *nilpotent* if $\sigma^n = 0$ for some natural number n.

We now turn to the definition of addition of two endomorphisms. Let α and β be two endomorphisms of G. The sum $\alpha + \beta$ is not defined unless the images G^α and G^β commute elementwise. If this condition is satisfied, then we define the mapping $\alpha + \beta$ as follows:

$$x^{\alpha+\beta} = x^\alpha x^\beta \qquad (x \in G).$$

It is easily verified that $\alpha + \beta$ is an endomorphism:

$$(xy)^{\alpha+\beta} = (xy)^\alpha (xy)^\beta = x^\alpha y^\alpha x^\beta y^\beta = x^\alpha x^\beta y^\alpha y^\beta = x^{\alpha+\beta} y^{\alpha+\beta}.$$

Since G^α and G^β commute elementwise, we have

$$\alpha + \beta = \beta + \alpha.$$

One readily proves that the associative law

$$(\alpha + \beta) + \gamma = \alpha + (\beta + \gamma)$$

holds, provided that the sums of any two of the endomorphisms α, β, γ exist. Let α and β be two endomorphisms such that $\alpha + \beta$ is defined, and let γ denote an arbitrary endomorphism of G. Then, $\alpha\gamma + \beta\gamma$ exists and

$$(\alpha + \beta)\gamma = \alpha\gamma + \beta\gamma;$$

moreover, $\gamma\alpha + \gamma\beta$ exists and

$$\gamma(\alpha + \beta) = \gamma\alpha + \gamma\beta.$$

For, if G^α and G^β commute elementwise, then the same holds for $(G^\alpha)^\gamma = G^{\alpha\gamma}$ and $(G^\beta)^\gamma = G^{\beta\gamma}$ so that $\alpha\gamma + \beta\gamma$ exists. For any $x \in G$, we have

$$x^{(\alpha+\beta)\gamma} = (x^\alpha x^\beta)^\gamma = x^{\alpha\gamma} x^{\beta\gamma} = x^{\alpha\gamma+\beta\gamma}.$$

The proof of the second distributive law is just as straightforward.

Since $G^0 = e$, the sum of the null endomorphism and any other endomorphism α exists, and we have $\alpha + 0 = \alpha$. If α and α' are two endomorphisms such that $\alpha + \alpha'$ exists and $\alpha + \alpha' = 0$, then we write $\alpha' = -\alpha$. Instead of $\beta + (-\alpha)$, we write $\beta - \alpha$.

An endomorphism of G is called *normal* if it commutes with every inner automorphism of G. In other words: The normal endomorphisms of G are the Ω-endomorphisms, where the operator domain Ω consists of all the inner automorphisms of G. This means that α is normal if and only if

$$(y^{-1}xy)^\alpha = y^{-1}x^\alpha y$$

for any two elements x, y of G. It is obvious that the product of two normal endomorphisms is normal. Clearly, ε and 0 are normal. If α and β are normal and if $\alpha + \beta$ exists, then $\alpha + \beta$ is normal; for

$$(y^{-1}xy)^{\alpha+\beta} = (y^{-1}xy)^{\alpha}(y^{-1}xy)^{\beta} = y^{-1}x^{\alpha}yy^{-1}x^{\beta}y = y^{-1}x^{\alpha}x^{\beta}y = y^{-1}x^{\alpha+\beta}y.$$

The following property of normal endomorphisms accounts for their name:

4.3.1 *Every normal endomorphism α of G carries any normal subgroup H of G into a normal subgroup H^{α}.*

Indeed, for arbitrary elements $x \in G$ and $y \in H$

$$x^{-1}y^{\alpha}x = (x^{-1}yx)^{\alpha}$$

belongs to H^{α}, which means that $x^{-1}H^{\alpha}x = H^{\alpha}$.

The next theorem gives a characterization of normal automorphisms.

4.3.2 *An automorphism α of G is normal if and only if for every element x of G, the product $x^{\alpha}x^{-1}$ lies in the centre of G.*

Proof. If α is normal, then we have

$$x^{-1}y^{\alpha}x = (x^{-1}yx)^{\alpha} = (x^{\alpha})^{-1}y^{\alpha}(x^{\alpha}),$$

which gives

$$x^{\alpha}x^{-1}y^{\alpha} = y^{\alpha}x^{\alpha}x^{-1},$$

so that $x^{\alpha}x^{-1}$ belongs to the centre of G^{α}. But, since α is an automorphism, we have $G^{\alpha} = G$; hence $x^{\alpha}x^{-1}$ lies in the centre of G. Conversely, suppose that $x^{\alpha}x^{-1}$ is contained in the centre of G. Then, we have, for any $y \in G$,

$$x^{\alpha}x^{-1}y^{\alpha} = y^{\alpha}x^{\alpha}x^{-1},$$

which implies that

$$x^{-1}y^{\alpha}x = (x^{\alpha})^{-1}y^{\alpha}(x^{\alpha}) = (x^{-1}yx)^{\alpha},$$

so that α is normal.

On account of Theorem 4.3.2, normal automorphisms are also called *central automorphisms*.

Two subgroups U and V of G are called *normally isomorphic* if there exists a normal automorphism of G that maps U onto V.

4.3.3 *If*

$$G = A \times B = A \times C$$

are two direct decompositions, then B and C are normally isomorphic.

Proof. Since both B and C are isomorphic to G/A, there exists an isomorphism γ of B onto C such that $Ab = Ab^{\gamma}$ for every $b \in B$. Thus, $b^{\gamma}b^{-1}$ lies in A, and hence it commutes with every element of B. Moreover, $b^{\gamma}b^{-1}$ commutes with every element of A, since b^γ as an element of C and b^{-1} as an element of B have this property. This shows that $b^\gamma b^{-1}$ belongs to the centre of $G = A \times B$. Hence, it follows from Theorem 4.3.2 that the mapping

$$ab \to ab^\gamma \qquad (a \in A,\, b \in B)$$

is a normal automorphism of G that maps B onto C.

We consider a direct decomposition

$$G = G_1 \times G_2 \times \cdots \times G_n. \tag{4.11}$$

Let γ_i denote the mapping that assigns to an arbitrary element

$$x = x_1 \times x_2 \times \cdots \times x_n$$

its G_i-component x_i, i. e.

$$x^{\gamma_i} = x_i \qquad (i = 1, 2, \ldots, n).$$

Obviously, γ_i is an endomorphism of G that maps G onto G_i. Since every element of G_i coincides with its G_i-component, we have $\gamma_i^2 = \gamma_i$. The mappings γ_i are often called the *projections* of G onto the direct factors G_i.

For $i \neq j$, the images $G^{\gamma_i} = G_i$ and $G^{\gamma_j} = G_j$ commute elementwise so that the sum $\gamma_i + \gamma_j$ is defined. We have

$$\varepsilon = \gamma_1 + \gamma_2 + \cdots + \gamma_n. \tag{4.12}$$

Since the G_j-component of every element of G_i is e, we obtain

$$\gamma_i \gamma_j = 0 \quad \text{for} \quad i \neq j.$$

Accordingly, $\gamma_1, \gamma_2, \ldots, \gamma_n$ are said to be *pairwise orthogonal*. Equation (4.4) can be written as follows:

$$y^{-1} x^{\gamma_i} y = (y^{-1} x y)^{\gamma_i}.$$

This shows that γ_i is a normal endomorphism. Summarizing the previous remarks, we obtain:

4.3.4 *To the direct decomposition (4.11) of G, there belongs a decomposition (4.12) of the identity automorphism ε into a sum of pairwise orthogonal idempotent normal endomorphisms.*

The close connection between direct decompositions and orthogonal idempotent normal endomorphisms is made evident by the fact that the converse of the last theorem is also true.

4.3.5 *Let (4.12) be a decomposition of the identity automorphism ε of G into a sum of pairwise orthogonal idempotent normal endomorphisms. Then, G has a direct decomposition (4.11), where $G_i = G^{\gamma_i}$ $(i = 1, 2, \ldots, n)$.*

Proof. As the endomorphisms γ_i are normal, the images $G^{\gamma_i} = G_i$ are normal subgroups of G by Theorem 4.3.1. From $\gamma_i^2 = \gamma_i$, we see that γ_i induces the identity automorphism in G_i. Since $\gamma_j \gamma_i = 0$ for $i \neq j$, the image of any G_j, $j \neq i$, under γ_i is e. Let x be an arbitrary element of the intersection

$$D_i = G_i \cap \langle G_1, \ldots, G_{i-1} \rangle.$$

From $x \in G_i$, we conclude that $x^{\gamma_i} = x$, but $x \in \langle G_1, \ldots, G_{i-1} \rangle$ gives $x^{\gamma_i} = e$. So, we obtain $D_i = e$. It follows that

$$G_1 G_2 \cdots G_n = G_1 \times G_2 \times \cdots \times G_n .$$

By (4.12), we have for an arbitrary element y of G

$$y = y^{\varepsilon} = y^{\gamma_1} y^{\gamma_2} \cdots y^{\gamma_n}.$$

This shows that $G_1 G_2 \ldots G_n = G$. Hence, G admits the direct decomposition (4.11).

In the next section, we shall prove an important uniqueness theorem for direct decompositions. Here, we derive three lemmas, which will be used in the proof.

Lemma 1. Let α be a homomorphism of the group B onto the group A. Suppose that α induces an isomorphism of some normal subgroup B^* of B onto A. Then, B has a direct decomposition $B = B^* \times B_0$, where B_0 is the kernel of α.
Proof. We have $B/B_0 \cong A \cong B^*$; hence $B = B^* B_0$. Clearly, $B^* \cap B_0 = e$, so that $B = B^* \times B_0$.

Lemma 2. Let G be a group with a chief series. If there exists a normal endomorphism σ of G that is neither an automorphism nor nilpotent, then G is directly decomposable.

Proof. Since σ is normal,

$$G \supset G^{\sigma} \supseteq G^{\sigma^2} \supseteq G^{\sigma^3} \supseteq \cdots$$

is a decreasing sequence of normal subgroups of G. As G has a chief series, not all the terms of this series are distinct. Consequently, there is a natural number n such that $G^{\sigma^n} = G^{\sigma^{n+1}}$. Hence, $G^{\sigma^n} = G^{\sigma^{2n}}$. Put $\sigma^n = \tau$ and $G^{\tau} = H$. Then, H is a normal subgroup of G and $H \neq e$, because σ is not nilpotent. Moreover, $H \neq G$, because G^{σ} is a proper normal subgroup of G; for, otherwise, σ would be a proper homomorphism of G onto itself, contradicting the existence of a chief series. We have $G^{\tau^2} = G^{\tau}$; hence $H^{\tau} = H$. It follows that τ is an automorphism of H; for H, being a normal subgroup of G, has a chief series, so that there is no proper homomorphism of H onto itself. Thus, τ is a homomorphism of G onto H that induces an isomorphism of H onto itself. Hence, by Lemma 1, G is directly decomposable.

Lemma 3. Suppose that G is directly indecomposable and has a chief series. Let $\sigma_1, \ldots, \sigma_r$ be normal endomorphisms of G for which the sums $\sigma_i + \sigma_j$, $i \neq j$, are defined. If $\sigma_1 + \cdots + \sigma_r$ is an automorphism of G, then at least one σ_i is an automorphism of G.

Proof. It is obviously sufficient to prove this for $r = 2$. Put $(\sigma_1 + \sigma_2)^{-1} = \sigma$. Then, we have $\sigma(\sigma_1 + \sigma_2) = \varepsilon$, hence

$$\sigma\sigma_1 + \sigma\sigma_2 = \varepsilon. \tag{4.13}$$

Suppose that both σ_1 and σ_2 are proper endomorphisms of G. This implies that $\sigma\sigma_1$ and $\sigma\sigma_2$ are proper endomorphisms. Since G is directly indecomposable, it follows from Lemma 2 that $\sigma\sigma_1$ and $\sigma\sigma_2$ are nilpotent, say

$$(\sigma\sigma_1)^n = (\sigma\sigma_2)^n = 0. \tag{4.14}$$

By (4.13), $\sigma\sigma_1$ and $\sigma\sigma_2$ commute with each other. So, we arrive at the contradiction

$$\varepsilon = \varepsilon^{2n} = (\sigma\sigma_1 + \sigma\sigma_2)^{2n} = \sum_{k=0}^{2n} \binom{2n}{k} (\sigma\sigma_1)^{2n-k} (\sigma\sigma_2)^k = 0;$$

for (4.14) shows that each term of the sum is 0.

4.4 The Remak–Schmidt Theorem

A group G having a chief series is directly indecomposable or it is a direct product of directly indecomposable factors. For, if every direct decomposition of G contained at least one factor that is again directly decomposable, then we would obtain an infinite strictly decreasing sequence of normal subgroups, which contradicts the existence of a chief series.

We now turn to the basic uniqueness theorem for direct decompositions.

4.4.1 Remak–Schmidt Theorem. *Suppose that the group G has a chief series and let*

$$\text{(a)} \qquad G = A_1 \times A_2 \times \cdots \times A_m,$$
$$\text{(b)} \qquad G = B_1 \times B_2 \times \cdots \times B_n$$

be two direct decompositions into directly indecomposable factors. Then, $m = n$ and there exists a normal automorphism ω of G such that with a suitable numbering

$$A_i^\omega = B_i \qquad (i = 1, 2, \ldots, n).$$

Moreover, any factor A_i may be replaced by $B_i = A_i^\omega$, i. e.

$$\text{(c)} \qquad G = B_1 \times \cdots \times B_k \times A_{k+1} \times \cdots \times A_n.$$

Proof. Let

$$\varepsilon = \alpha_1 + \alpha_2 + \cdots + \alpha_m, \quad \varepsilon = \beta_1 + \beta_2 + \cdots + \beta_n$$

be the decompositions of the identity automorphism ε of G into sums of pairwise orthogonal idempotent normal endomorphisms belonging to the direct decompositions (a) and (b), respectively, in the sense of the Theorems 4.3.4 and 4.3.5. Then,

$$\alpha_1 = \varepsilon\alpha_1 = (\beta_1 + \cdots + \beta_n)\,\alpha_1 = \beta_1\alpha_1 + \cdots + \beta_n\alpha_1 .$$

Since α_1 is an automorphism of A_1 and A_1 is directly indecomposable, we can apply Lemma 3 of section 4.3 to show that at least one of the endomorphisms $\beta_1\alpha_1, \ldots, \beta_n\alpha_1$ is an automorphism of A_1. We may assume that the numbering in (b) is such that $\beta_1\alpha_1$ is an automorphism of A_1.

We now consider the mapping $\beta_1\alpha_1$ more closely. β_1 induces a homomorphism of A_1 onto some normal subgroup B_1^* of B_1. Then, α_1 induces a homomorphism of B_1^* onto A_1. Since the product $\beta_1\,\alpha_1$ is an automorphism of A_1, i. e. a one-to-one mapping, we conclude that β_1 induces an isomorphism of A_1 onto B_1^* and α_1 induces an isomorphism of B_1^* onto A_1. Since B_1 is directly indecomposable, it follows from Lemma 1 of section 4.3 that $B_1^* = B_1$. Hence, $A_1^{\beta_1} = B_1$ and $B_1^{\alpha_1} = A_1$, where these mappings are isomorphisms.

It is easily seen that $B_1 = A_1^{\beta_1}$ is elementwise permutable with every A_k, $k > 1$; for, if a_1 and a_k are arbitrary elements of A_1 and A_k, $k > 1$, respectively, then

$$a_k^{-1}\, a_1^{\beta_1} a_k = (a_k^{-1}\, a_1 a_k)^{\beta_1} = a_1^{\beta_1},$$

because β_1 is normal and a_1 commutes with a_k. This shows that the sum

$$\omega_1 = \alpha_1\beta_1 + \alpha_2 + \cdots + \alpha_m$$

exists. We consider the elements y of G for which $y^{\omega_1} = e$. In particular, the A_1-component of y^{ω_1} is the unit element so that $y^{\omega_1\alpha_1} = e$. Since $\alpha_i\,\alpha_1 = 0$ for $i = 2, \ldots, n$, this implies that $y^{\alpha_1\beta_1\alpha_1} = e$. Now, $\beta_1\alpha_1$ is an automorphism of A_1, so that the last equation gives $y^{\alpha_1} = e$. Thus, the components of y^{ω_1} with respect to the direct decomposition (a) are

$$e = y^{\omega_1} = e \times y^{\alpha_2} \times \cdots \times y^{\alpha_m},$$

whence $y^{\alpha_2} = \cdots = y^{\alpha_m} = e$. Consequently, $y^{\omega_1} = e$ implies that $y = e$. This shows that ω_1 is a normal automorphism of G. In particular, ω_1 maps A_1 onto B_1. By applying ω_1 to the direct decomposition (a), we obtain

$$G = B_1 \times A_2 \times \cdots \times A_m .$$

Now, compare this direct decomposition with (b). Theorem 4.3.3 shows that $A_2 \times \cdots \times A_m$ and $B_2 \times \cdots \times B_n$ are normally isomorphic. Thus, $A_2 \times \cdots \times A_m$

contains normal subgroups $\bar{B}_2, \ldots, \bar{B}_n$, normally isomorphic to $B_2, \ldots, B_n$, respectively, such that

$$A_2 \times \cdots \times A_m = \bar{B}_2 \times \cdots \times \bar{B}_n. \tag{4.15}$$

After these preliminaries, it is easy to prove our theorem by induction on m. For $m = 1$, the assertion is obviously true, because G is directly indecomposable. Suppose that the theorem holds for every group having a direct decomposition into fewer than m directly indecomposable factors. Then, (4.15) shows that $m = n$. Moreover, it follows that there is a normal automorphism of $A_2 \times \cdots \times A_n$ that maps A_j onto $\bar{B}_j$ $(j = 2, \ldots, n)$, provided that the $\bar{B}_j$ are suitably numbered. Consequently, there also exists a normal automorphism $\bar{\omega}$ of G that maps $A_2, \ldots, A_n$ onto $B_2, \ldots, B_n$, respectively. Finally, $\omega = \omega_1 \bar{\omega}$ is a normal automorphism of G such that $A_i^\omega = B_i$, $i = 1, \ldots, n$, as required. It is easy to show that

$$\omega = \alpha_1\beta_1 + \alpha_2\beta_2 + \cdots + \alpha_n\beta_n.$$

The direct decomposition (c) is obtained by applying the normal automorphism

$$\alpha_1\beta_1 + \cdots + \alpha_k\beta_k + \alpha_{k+1} + \cdots + \alpha_n$$

to (a).

This completes the proof.

4.5 Direct Products of Infinitely Many Factors

The decomposition of a group into the direct product of subgroups can also be defined in the case of infinitely many factors:

The group G is said to be the direct product of its subgroups G_λ, where λ runs through some index set, if the following conditions are satisfied:

(a) G is generated by the subgroups G_λ.

(b) Any subgroup of G that is generated by a finite number of the subgroups G_λ is the direct product of these subgroups.

We write $G = \prod_\lambda^\times G_\lambda$.

In the case of a finite number of subgroups G_λ, this definition obviously coincides with our previous definition.

From (a) and (b), it follows that every element x of G has a unique representation as a product of finitely many factors belonging to distinct subgroups G_λ, i. e.

$$x = x_{\lambda_1} x_{\lambda_2} \cdots x_{\lambda_r} \qquad (x_{\lambda_i} \in G_{\lambda_i}).$$

We simplify the notation by writing

$$x = \prod_{\lambda} x_{\lambda}$$

and call x_{λ} the G_{λ}-component of x, where only finitely many components are distinct from e. The product of x and

$$y = \prod_{\lambda} y_{\lambda}$$

is

$$xy = \prod_{\lambda} (x_{\lambda} y_{\lambda}).$$

In the same way as in the case of finitely many factors, we find that every G_{λ} is a normal subgroup of G and that, for every index λ,

$$G_{\lambda} \cap \langle G_{\mu}, \mu \neq \lambda \rangle = e. \tag{4.16}$$

Theorems 4.1.1 and 4.1.2 also remain valid for infinitely many factors, with (4.5) replaced by (4.16).

If an arbitrary set G_{λ} of groups is given, then we can use the process described in section 4.1 to construct a group G that contains, for every index λ, a subgroup G_{λ}^{*} isomorphic to G_{λ} and is the direct product of its subgroups G_{λ}^{*}. Let G be the set of the formal expressions of the form

$$x = \prod_{\lambda} x_{\lambda} \qquad (x_{\lambda} \in G_{\lambda}) \tag{4.17}$$

where only finitely many of the elements x_{λ} are distinct from the unit element e_{λ} of the corresponding G_{λ}. In G, we define a binary operation as follows:

$$xy = (\prod_{\lambda} x_{\lambda})(\prod_{\lambda} y_{\lambda}) = \prod_{\lambda} (x_{\lambda} y_{\lambda}). \tag{4.18}$$

It is readily seen that G is a group with respect to this operation. For a fixed μ, the expressions (4.17) in which $x_{\lambda} = e_{\lambda}$ for all $\lambda \neq \mu$ obviously form a subgroup G_{μ}^{*} of G. We leave it as an exercise for the reader to show that G is the direct product of the subgroups G_{λ}^{*}. As in the case of finitely many factors, we call G the direct product of the groups G_{λ}.

The group G just constructed is more accurately called the *restricted direct product* of the groups G_{λ}. In contrast, the so-called *unrestricted direct product* or *cartesian product* $\bar{G}$ of the groups G_{λ} is defined as follows: G consists of all formal expressions (4.17), where now infinitely many factors may be distinct from the unit elements of the corresponding groups G_{λ}. Again, it is easily proved that $\bar{G}$ is a group with respect to the binary operation (4.18). The restricted direct product G is a subgroup of the cartesian product $\bar{G}$, namely the subgroup of $\bar{G}$ that is generated by the subgroups G_{λ}.

Exercises

1. Find all direct decompositions of the four-group.

2. Represent the multiplicative group of all non-zero complex numbers as a direct product.

3. Let A and B be normal subgroups of G. Show that $G/(A \cap B)$ is isomorphic to a subgroup of $(G/A) \times (G/B)$.

4. Suppose that $G = AB$, where A and B are normal subgroups of G. Prove that $G/(A \cap B) \cong (A/A \cap B) \times (B/A \cap B)$.

5. Let N be a non-trivial normal subgroup of a direct product $A \times B$. Prove that $N \cap A = N \cap B = e$ implies that N is abelian.

6. Show that a direct product of isomorphic simple groups is characteristically simple.

7. All the endomorphisms of an abelian group form a ring. Determine the endomorphism rings of a cyclic group and of a vector space over a field.

8. Let G be a cyclic group whose order m admits a proper factorization $m = m_1 m_2$ with $(m_1, m_2) = 1$. Find a decomposition of the identity automorphism of G into the sum of two mutually orthogonal idempotent endomorphisms.

9. Give a specific formulation of Theorem 4.2.3 in case G is a vector space over a field F, regarded as an additive group with the operator domain F.

10. Illustrate Theorems 4.3.4 and 4.3.5 in case G is a vector space over a field.

11. Use the results of section 4.3 to prove the following theorem: Let E be an n-square matrix of rank r with elements in a field F such that $E^2 = E$. Then, E is similar to a matrix $A = [a_{ik}]$, where $a_{ii} = 1$ for $i = 1, \ldots, r$ and $a_{ik} = 0$ otherwise.

12. Suppose that the group G possesses an idempotent endomorphism other than the identity and the null endomorphism. Prove that G contains a normal subgroup N and a subgroup H such that $G = NH$, $N \cap H = e$. Give an example.

CHAPTER 5

ABELIAN GROUPS

5.1 Periodic, Torsion-Free, and Mixed Abelian Groups

In this chapter, we denote the group elements by lower case Greek letters and adopt the additive notation for the group operation. Integers are denoted by lower case Roman letters.

For instance, an element ξ of a group is of order m if m is the least positive integer such that $m\xi = 0$. Since the groups are abelian, we have

$$k(\xi + \eta) = k\xi + k\eta$$

for any two group elements ξ, η and every integer k. Instead of direct products, we now speak of *direct sums*. If a group G is the direct sum of its subgroups $G_1, G_2, \ldots, G_n$, we write

$$G = G_1 \oplus G_2 \oplus \cdots \oplus G_n.$$

Corresponding to this direct decomposition every element of G has a unique representation

$$\xi = \xi_1 + \xi_2 + \cdots + \xi_n, \quad \xi_i \in G \qquad (i = 1, 2, \ldots, n).$$

An abelian group is said to be *torsion-free* if every element other than 0 has infinite order. (This name is derived from certain abelian groups that occur in topology.) If all the elements have finite order, then we speak of a *periodic* group. By a *primary* abelian group we mean an abelian p-group, i. e. a periodic abelian group in which the order of every element is a power of the prime number p. Finally, a *mixed* abelian group is one that is neither torsion-free nor periodic.

Let ξ and η be two elements of finite order of an abelian group, ξ of order m, η of order n, say. Then the order of $\xi + \eta$ is a divisor of the least common multiple k of m and n; for, $m\xi = n\eta = 0$ and $k = m_1 m = n_1 n$ together imply that

$$k(\xi + \eta) = k\xi + k\eta = m_1 m\xi + n_1 n\eta = 0.$$

If m and n are relatively prime, then $\xi + \eta$ is of order mn. For, if$(m, n) = 1$, we have

$$\langle \xi \rangle \cap \langle \eta \rangle = 0;$$

if d is a divisor of mn such that

$$d(\xi + \eta) = d\xi + d\eta = 0,$$

then $d\xi = -d\eta$ is contained in $\langle \xi \rangle \cap \langle \eta \rangle$, which shows that $d\xi = 0$ and hence $d\eta = 0$. Thus, d is divisible by m as well as by n so that $d = mn$. A similar result is easily obtained for sums of more than two elements.

Let F denote the set of all elements of finite order in a mixed abelian group G. By our previous remark, if $\xi \in F$ and $\eta \in F$, then $\xi - \eta \in F$, so that F is a subgroup of G. We call F the *maximal periodic subgroup* or the *torsion subgroup* of G. The factor group G/F is torsion-free. For, if $F + \alpha$ is an element of finite order in G/F, $n(F + \alpha) = 0$ say, then we have $n\alpha \in F$ and therefore $\alpha \in F$.

A similar argument yields the following result. For a given prime number p, the set G_p of all elements of G whose order is a power of p forms a subgroup of G. Of course, G_p may be the null subgroup.

Let ξ be an element of order m of the periodic abelian group G. Let

$$m = p_1^{k_1} p_2^{k_2} \cdots p_r^{k_r}$$

be the decomposition of m into powers of distinct prime numbers. Then, the quotients

$$m_i = \frac{m}{p_i^{k_i}} \qquad (i = 1, 2, \ldots, r)$$

satisfy the condition $(m_1, m_2, \ldots, m_r) = 1$. Hence, there are integers a_1, $a_2, \ldots, a_r$ such that

$$a_1 m_1 + a_2 m_2 + \cdots + a_r m_r = 1.$$

This gives

$$\xi = a_1 m_1 \xi + a_2 m_2 \xi + \cdots + a_r m_r \xi.$$

The element $a_i m_i \xi$ is obviously contained in the primary subgroup G_{p_i} of G $(i = 1, 2, \ldots, r)$. It follows from Theorem 4.1.1 that $\langle G_{p_1}, G_{p_2}, \ldots, G_{p_r} \rangle = G_{p_1} \oplus G_{p_2} \oplus \cdots \oplus G_{p_r}$; for, the conditions (4.5), namely

$$G_{p_i} \cap \langle G_{p_1}, \ldots, G_{p_{i-1}} \rangle = 0 \qquad (i = 2, \ldots, r),$$

are satisfied, since the order of every element of G_{p_i} is a power of p_i, whereas the orders of all elements of $\langle G_{p_1}, \ldots, G_{p_{i-1}} \rangle$ have no prime divisors other than $p_1, \ldots, p_{i-1}$.

This shows, moreover, that G is generated by its primary subgroups G_p, where p ranges over all prime numbers. If G is infinite, then infinitely many of the subgroups G_p may be non-trivial. In accordance with the definition in section 4.5, we now have:

5.1.1 *Every periodic abelian group has a unique decomposition into the restricted direct sum of primary subgroups.*

In particular, a finite abelian group is the direct sum of its Sylow p-subgroups. This result also follows from Theorem 3.2.5.

As another special case, we find that every finite non-primary cyclic group is the direct sum of primary cyclic groups whose orders are powers of distinct primes.

A cyclic group of prime power order, however, is directly indecomposable. For, if $\langle\alpha\rangle$ is of order p^n, then the non-trivial subgroups of $\langle\alpha\rangle$ are $\langle p\alpha\rangle, \ldots, \langle p^{n-1}\alpha\rangle$. Since each is contained in the preceding one, there is no pair of non-trivial subgroups that intersect trivially, whereas a direct decomposition of $\langle\alpha\rangle$ implies the existence of such a pair.

An infinite cyclic group $\langle\alpha\rangle$ is also directly indecomposable, because any two non-trivial subgroups have a non-trivial intersection, since $rs\alpha \in \langle r\alpha\rangle \cap \langle s\alpha\rangle$.

Let $\langle\alpha_i\rangle$ be a finite cyclic group of order n_i $(i = 1, 2, \ldots, k)$ and suppose that $(n_i, n_j) = 1$ for $i \neq j$. Then, the direct sum

$$\langle\alpha_1\rangle \oplus \langle\alpha_2\rangle \oplus \cdots \oplus \langle\alpha_k\rangle$$

is cyclic. For, the direct sum has the order $n = n_1 n_2 \ldots n_k$ and contains the element $\alpha_1 + \alpha_2 + \cdots + \alpha_k$ of order n.

The theory of abelian groups is very extensive and by no means elementary. We refer the reader to [19] and [38]. Here we shall confine ourselves to a narrow but important class of abelian groups, namely to the finitely generated ones. Owing to the fact that finitely generated abelian groups are direct sums of cyclic groups, we arrive at a complete classification of these groups by means of integral parameters.

5.2 Free Abelian Groups

In section 2.5, we proved that every group is isomorphic to a factor group of a free group. Of course, this holds, in particular, for abelian groups. It turns out, however, that all finitely generated abelian groups are also isomorphic to factor groups of so-called free abelian groups. Since free abelian groups, their subgroups and factor groups can be described in an effective manner, they form a useful tool for the study of finitely generated abelian groups.

A direct sum

$$M = \langle \xi_1 \rangle \oplus \langle \xi_2 \rangle \oplus \cdots \oplus \langle \xi_m \rangle$$

of m infinite cyclic groups $\langle \xi_i \rangle$ is called a *free abelian group of rank m*. Among the free abelian groups we include the null group.

Every system $\eta_1, \eta_2, \ldots$ of elements of M such that

$$M = \langle \eta_1 \rangle \oplus \langle \eta_2 \rangle \oplus \cdots$$

is called a *basis* of M. Thus, $\xi_1, \xi_2, \ldots, \xi_m$ form a basis.

5.2.1 *Every basis of M consists of m elements.*

Proof. Let $2M$ denote the complex of all elements ϱ of M such that $\varrho = 2\zeta$ for some $\zeta \in M$. If $\varrho_1 = 2\zeta_1, \zeta_1 \in M$, then $\varrho - \varrho_1 = 2(\zeta - \zeta_1)$ so that $\varrho - \varrho_1 \in 2M$. Consequently, $2M$ is a subgroup of M. Note that $2M$ is defined without reference to a basis of M. Clearly,

$$a_1 \xi_1 + a_2 \xi_2 + \cdots + a_m \xi_m$$

belongs to $2M$ if and only if all the integers a_i are even. Thus, the 2^m elements

$$c_1 \xi_1 + c_2 \xi_2 + \cdots + c_m \xi_m \quad \text{with} \quad c_i = 0 \text{ or } 1$$

form a transversal to $2M$ in M. This shows that $|M:2M| = 2^m$. If we repeat this argument for another basis of M, we must arrive at the same index 2^m, because the definition of $2M$ does not depend on the choice of the basis. Hence, every basis consists of m elements.

As an immediate consequence of (a) in section 4.1, we obtain:

A system of m elements $\eta_1, \eta_2, \ldots, \eta_m$ of M is a basis if and only if every element of M has a unique representation in the form

$$a_1 \eta_1 + a_2 \eta_2 + \cdots + a_m \eta_m$$

with integral coefficients $a_1, a_2, \ldots, a_m$. We call such an expression a *linear combination* of $\eta_1, \eta_2, \ldots, \eta_m$. In particular, the elements of a basis are *linearly independent*; this means that

$$b_1 \eta_1 + b_2 \eta_2 + \cdots + b_m \eta_m = 0$$

for integers b_i implies that $b_1 = b_2 = \cdots = b_m = 0$; for, 0 has only a single representation as a linear combination of $\eta_1, \eta_2, \ldots, \eta_m$.

It is obvious that by an arbitrary permutation of the elements of a basis we again obtain a basis. If $\eta_1, \eta_2, \ldots, \eta_m$ form a basis and if $\zeta_1, \zeta_2, \ldots, \zeta_m$ are defined as follows

$$\begin{aligned} \zeta_1 &= \eta_1 + c\eta_k \qquad && (k \neq 1,\ c \text{ an arbitrary integer}) \\ \zeta_i &= \eta_i && (i = 2, \ldots, m), \end{aligned} \tag{5.1}$$

then $\zeta_1, \zeta_2, \ldots, \zeta_m$ also form a basis. For, since

$$\eta_1 = \zeta_1 - c\zeta_k,$$

every linear combination of $\eta_1, \eta_2, \ldots, \eta_m$, i. e. every element of M, can also be expressed as a linear combination of $\zeta_1, \zeta_2, \ldots, \zeta_m$. Moreover, the representation of the elements of M as linear combinations of $\zeta_1, \zeta_2, \ldots, \zeta_m$ is unique. For,

$$b_1\zeta_1 + b_2\zeta_2 + \cdots + b_m\zeta_m = b_1'\zeta_1 + b_2'\zeta_2 + \cdots + b_m'\zeta_m$$

implies that

$$0 = (b_1 - b_1')\,\zeta_1 + (b_2 - b_2')\,\zeta_2 + \cdots + (b_m - b_m')\,\zeta_m$$

$$= \sum_{i \neq k} (b_i - b_i')\,\eta_i + [(b_k - b_k') + c(b_1 - b_1')]\,\eta_k.$$

Since $\eta_1, \eta_2, \ldots, \eta_m$ are linearly independent, it follows that $b_i = b_i'$ for $i = 1, 2, \ldots, m$.

By repeated application of the basis transformation (5.1), we obtain the following result: Let $\eta_1, \eta_2, \ldots, \eta_m$ be a basis and put

$$\eta_1' = \eta_1 + c_2\eta_2 + \cdots + c_m\eta_m,$$

$$\eta_i' = \eta_i \quad \text{for} \quad i = 2, \ldots, m$$

where $c_2, \ldots, c_m$ are arbitrary integers; then, $\eta_1', \eta_2', \ldots, \eta_m'$ form a basis.

We now prove the main theorem on the subgroups of free abelian groups.

5.2.2 *Every subgroup L of the free abelian group M of rank m is a free abelian group whose rank l does not exceed m. A basis $\lambda_1, \ldots, \lambda_l$ of L and a basis $\eta_1, \ldots, \eta_m$ of M can be chosen such that*

$$\lambda_i = e_i\eta_i \qquad (i = 1, \ldots, l) \tag{5.2}$$

where $e_1, \ldots, e_l$ are natural numbers and e_{i+1} is divisible by e_i for $i = 1, \ldots, l - 1$.

Proof. For $m = 1$, there are no divisibility conditions so that our theorem only states that every subgroup of an infinite cyclic group is itself an infinite cyclic group.

We now assume that our theorem holds for every free abelian group of rank $m - 1$.

Let $\xi_1, \ldots, \xi_m$ be any basis of M. We consider the set of all coefficients that occur in the representations of the elements of L as linear combinations of $\xi_1, \ldots, \xi_m$. This is a certain set of integers and contains a unique least positive integer, which we call the minimal coefficient belonging to the basis $\xi_1, \ldots, \xi_m$. In general, to different bases there belong different minimal coefficients. We now choose a basis such that the corresponding minimal coefficient is as small as possible. Let us assume that $\xi_1, \ldots, \xi_m$ is a basis with this property.

Moreover, we may assume that $\xi_1, \ldots, \xi_m$ are numbered in such a way that the minimal coefficient occurs as a coefficient of ξ_1. If the minimal coefficient belonging to $\xi_1, \ldots, \xi_m$ is denoted by e_1, then there is an element λ_1 of L whose representation as a linear combination of $\xi_1, \ldots, \xi_m$ has the form

$$\lambda_1 = e_1 \xi_1 + a_2 \xi_2 + \cdots + a_m \xi_m . \tag{5.3}$$

It turns out that the integers $a_2, \ldots, a_m$ are divisible by e_1. For suppose the contrary: That for some index j

$$a_j = q e_1 + r, \qquad 0 < r < e_1 .$$

If we replace ξ_1 by $\xi_1' = \xi_1 + q\xi_j$, then $\xi_1', \xi_2, \ldots, \xi_m$ again form a basis and λ_1 has the following representation by the new basis:

$$\lambda_1 = e_1 \xi_1' + \cdots + a_{j-1} \xi_{j-1} + r \xi_j + a_{j+1} \xi_{j+1} + \cdots + a_m \xi_m .$$

The minimal coefficient belonging to the basis $\xi_1', \xi_2, \ldots, \xi_m$ is therefore at most equal to r and hence less than e_1. This contradicts the choice of $\xi_1, \xi_2, \ldots, \xi_m$. Consequently, we have $a_i = q_i e_1$ for $i = 2, \ldots, m$. Putting

$$\eta_1 = \xi_1 + q_2 \xi_2 + \cdots + q_m \xi_m ,$$

we obtain

$$\lambda_1 = e_1 \eta_1 . \tag{5.4}$$

Since $\eta_1, \xi_2, \ldots, \xi_m$ form a basis of M, we have

$$M = \langle \eta_1 \rangle \oplus \langle \xi_2 \rangle \oplus \cdots \oplus \langle \xi_m \rangle = \langle \eta_1 \rangle \oplus M_1 ,$$

where

$$M_1 = \langle \xi_2 \rangle \oplus \cdots \oplus \langle \xi_m \rangle .$$

We now show that $\langle \lambda_1 \rangle$ is a direct summand of L, namely

$$L = \langle \lambda_1 \rangle + L_1 , \tag{5.5}$$

where

$$L_1 = L \cap M_1 .$$

If

$$\lambda = b_1 \xi_1 + b_2 \xi_2 + \cdots + b_m \xi_m$$

is an arbitrary element of L, then b_1 is divisible by e_1. For, writing

$$b_1 = q_1 e_1 + r_1 , \qquad 0 \leqq r_1 < e_1 ,$$

we obtain

$$\lambda - q_1 \lambda_1 = r_1 \xi_1 + \cdots$$

which contradicts the assumption that e_1 is the minimal coefficient unless $r_1 = 0$. It follows that $\lambda - q_1 \lambda_1$ is contained in L_1. Consequently, L is generated by $\langle \lambda_1 \rangle$ and L_1. To complete the proof of (5.5), it remains to show that

$$\langle \lambda_1 \rangle \cap L_1 = 0 . \tag{5.6}$$

If we use the basis $\xi_1, \ldots, \xi_m$, then (5.3) shows that 0 is the only element of $\langle\lambda_1\rangle$ for which the coefficient of ξ_1 vanishes, whereas L_1 consists only of such elements. This proves (5.6), so that (5.5) holds.

If $L_1 = 0$, then $L = \langle\lambda_1\rangle$, and by (5.4) our theorem is proved.

If $L_1 \neq 0$, we apply the inductive assumption to the subgroup L_1 of M_1 to obtain that L_1 is a free abelian group. If $l - 1$ denotes the rank of L_1, then the inductive assumption also gives $l - 1 \leqq m - 1$ and

$$L_1 = \langle\lambda_2\rangle \oplus \cdots \oplus \langle\lambda_l\rangle,$$

$$M_1 = \langle\eta_2\rangle \oplus \cdots \oplus \langle\eta_m\rangle,$$

where

$$\lambda_i = e_i \eta_i \qquad (i = 2, \ldots, l)$$

and the natural numbers e_i satisfy the conditions

$$e_{i+1} \equiv 0 \pmod{e_i} \quad \text{for} \quad i = 2, \ldots, l-1.$$

So, we have

$$L = \langle\lambda_1\rangle \oplus \langle\lambda_2\rangle \oplus \cdots \oplus \langle\lambda_l\rangle,$$

$$M = \langle\eta_1\rangle \oplus \langle\eta_2\rangle \oplus \cdots \oplus \langle\eta_m\rangle.$$

It remains to prove that e_1 divides e_2. Let

$$e_2 = \bar{q} e_1 + \bar{r}, \qquad 0 \leqq \bar{r} < e_1.$$

The elements $\eta_1' = \eta_1 - \bar{q}\eta_2, \eta_2, \ldots, \eta_m$ form a basis of M. We have

$$\lambda_2 - \lambda_1 = e_2\eta_2 - e_1\eta_1 = -e_1\eta_1' + \bar{r}\eta_2.$$

Since $\lambda_2 - \lambda_1$ belongs to L, this contradicts the choice of the basis $\xi_1, \xi_2, \ldots, \xi_m$ unless $\bar{r} = 0$.

This completes the proof.

It can be proved that the natural numbers $e_1, \ldots, e_l$ are uniquely determined by L (cf. for instance [44]). This fact could be used to obtain a uniqueness theorem for direct decompositions of abelian groups. We prefer, however, to establish that uniqueness in the next section in a different way.

If the particular bases $\eta_1, \ldots, \eta_m$ and $\lambda_1, \ldots, \lambda_l$ are replaced by arbitrary bases $\zeta_1, \ldots, \zeta_m$ of M and $\omega_1, \ldots, \omega_l$ of L, then the equations connecting the two bases are

$$\omega_i = \sum_{j=1}^{m} a_{ij}\zeta_j \qquad (i = 1, \ldots, l)$$

instead of (5.2). The integers a_{ij} form an $l \times m$ matrix A. The natural numbers e_i are called the elementary divisors of A. There is an algorithm to compute the elementary divisors, which yields a constructive method of determining the bases $\eta_1, \ldots, \eta_m$ and $\lambda_1, \ldots, \lambda_l$ if L is given by any system of generators (cf. for instance [44]).

Theorem 5.2.2 enables us to describe the structure of the factor group M/L.

To begin with, we remark that some of the natural numbers e_i or even all of them may be equal to 1. Suppose that $e_i = \cdots = e_h = 1$, $e_{h+1} > 1$. Then, among $\eta_1, \ldots, \eta_m$, precisely $\eta_1, \ldots, \eta_h$ belong to L. The remaining elements

$$\eta_{h+1}, \ldots, \eta_l, \eta_{l+1}, \ldots, \eta_m$$

are now denoted in the same order by

$$\zeta_1, \ldots, \zeta_s, \zeta_{s+1}, \ldots, \zeta_n$$

where $s = l - h$, $n = m - h$. Then, the cosets of L in M can be represented by integral linear combinations

$$\alpha = a_1\zeta_1 + \cdots + a_s\zeta_s + a_{s+1}\zeta_{s+1} + \cdots + a_n\zeta_n.$$

By Theorem 5.2.2, α belongs to L if and only if

$$a_i \equiv 0 \pmod{e_{h+i}} \quad \text{for} \quad i = 1, \ldots, s \tag{5.7}$$

and

$$a_j = 0 \quad \text{for} \quad j = s+1, \ldots, n. \tag{5.8}$$

For, if (5.7) and (5.8) are satisfied, $a_i = g_i e_{h+i}$ $(i = 1, \ldots, s)$ say, then we have

$$\alpha = \sum_{i=1}^{s} a_i\zeta_i = \sum_{i=1}^{s} g_i e_{h+i}\eta_{h+i} = \sum_{i=1}^{s} g_i\lambda_{h+i} \in L.$$

Conversely, if α is contained in L, then α can be represented as a linear combination of $\lambda_1, \ldots, \lambda_l$. This shows immediately that (5.7) and (5.8) are satisfied.

Consequently, the elements

$$a_1\zeta_1 + \cdots + a_s\zeta_s + a_{s+1}\zeta_{s+1} + \cdots + a_n\zeta_n, \tag{5.9}$$

whose first s coefficients are subject to the conditions

$$0 \leqq a_i < e_{h+i} \qquad (i = 1, \ldots, s),$$

whereas $a_{s+1}, \ldots, a_n$ are arbitrary integers, form a transversal to L in M. We now assign to each coset of L the sequence

$$(a_1, \ldots, a_s, a_{s+1}, \ldots, a_n)$$

of the coefficients of its representative (5.9). Then, this correspondence between the cosets and the sequences is one-to-one and to the sum of two cosets there corresponds the sum of the sequences, where addition of the sequences is carried out componentwise and such that, for $i = 1, \ldots, s$, the sum $a_i + a'_i$ of two i-th components has to be replaced by $a_i + a'_i - e_{h+i}$ if $a_i + a'_i \geqq e_{h+i}$.

This shows that the structure of M/L can be described as follows:

5.2.3 *The factor group M/L is the direct sum of n cyclic groups. The first s direct summands have finite orders, namely $e_{h+1}, \ldots, e_{h+s}$, whereas the remaining direct summands are infinite cyclic groups.*

In particular, we may have $s = 0$, i. e. $e_1 = \cdots = e_l = 1$. Then, all the direct summands are infinite cyclic groups so that M/L is a free abelian group. If, on the other hand, $s = n$ and hence $l = m$, then M/L is the direct sum of finite cyclic groups.

5.3 Finitely Generated Abelian Groups

Let A be an abelian group with a finite number of generators $\alpha_1, \alpha_2, \ldots, \alpha_m$. Then, every element α of A can be represented in the form

$$\alpha = g_1\alpha_1 + g_2\alpha_2 + \cdots + g_m\alpha_m$$

with integral coefficients $g_1, g_2, \ldots, g_m$. In general, this representation is not unique, since there may be relations of the form

$$c_1\alpha_1 + c_2\alpha_2 + \cdots + c_m\alpha_m = 0 \qquad (c_i \text{ integers})$$

among the generating elements.

Let M be the free abelian group of rank m. For an arbitrary basis $\xi_1, \ldots, \xi_m$ of M, we consider the mapping

$$g_1\xi_1 + \cdots + g_m\xi_m \to g_1\alpha_1 + \cdots + g_m\alpha_m$$

of M onto A. Clearly, this mapping is a homomorphism of M onto A. The kernel of this homomorphism consists of the subgroup L of all those elements

$$c_1\xi_1 + \cdots + c_m\xi_m$$

of M for which

$$c_1\alpha_1 + \cdots + c_m\alpha_m = 0.$$

By Theorem 2.3.1, we have

$$A \cong M/L.$$

From the results of section 5.2, we conclude that M/L, and hence A, are direct sums of cyclic groups, where the orders of the finite direct summands satisfy the divisibility conditions of Theorem 5.2.2. Thus, A has a direct decomposition

$$A = C_1 \oplus \cdots \oplus C_s \oplus U_1 \oplus \cdots \oplus U_r, \tag{5.10}$$

where $U_1, \ldots, U_r$ are infinite cyclic groups and $C_1, \ldots, C_s$ are finite cyclic groups whose orders $n_1, \ldots, n_s$ satisfy the conditions

$$n_{i+1} \equiv 0 \pmod{n_i} \quad \text{for} \quad i = 1, \ldots, s-1. \tag{5.11}$$

The decomposition (5.10) corresponds to the general case. If A is finite, then the U_j have to be omitted; whereas the C_i do not occur in case A is torsion-free.

We shall now prove that the numbers r and s in (5.10) and the orders, subject to the conditions (5.11), are uniquely determined by A. In other words: Let

$$A = C'_1 \oplus \cdots \oplus C'_{s'} \oplus U'_1 \oplus \cdots \oplus U'_{r'},$$

be another direct decomposition, where $U'_1, \ldots, U'_{r'}$, are infinite cyclic groups and $C'_1, \ldots, C'_{s'}$ finite cyclic groups such that the order of C'_i divides the order of C'_{i+1} $(i = 1, \ldots, s'-1)$; then, we have $r = r'$, $s = s'$, and $|C_i| = |C'_i|$ for $i = 1, \ldots, s$. First, we observe that

$$T = C_1 \oplus \cdots \oplus C_s \tag{5.12}$$

is the torsion subgroup of A. Since the torsion subgroup is uniquely determined by A, so is the factor group A/T. This factor group is a free abelian group, and by Theorem 5.2.1 its rank r is uniquely determined by A. The number r is called the *rank* of A. If A is a free abelian group, this notation agrees with that in section 5.2. The condition $r = 0$ is necessary and sufficient for A to be finite.

The numbers $n_1, \ldots, n_s$ are called the *torsion coefficients* of A. Next, we show that they are uniquely determined by A.

Decompose each direct summand C_i in (5.12) into a direct sum of primary cyclic groups. This yields a decomposition of T into a direct sum of cyclic groups whose orders are powers of primes. In section 5.1, we observed that a cyclic group of prime power order is directly indecomposable. So, we obtain a direct decomposition of T into directly indecomposable summands. By Theorem 4.4.1, the orders of the indecomposable direct summands are uniquely determined. This shows that the prime powers dividing the torsion coefficients do not depend on a particular direct decomposition of A. But then, the torsion coefficients themselves are uniquely determined by A, for the conditions (5.11) indicate how $n_1, \ldots, n_s$ are composed from the given prime powers.

In the case of a finite abelian group, however, it is not necessary to make use of the general uniqueness theorem for direct decompositions. We have to show that, for all decompositions of the finite abelian group T into a direct sum of

[1] In our new notation n_i corresponds to e_{h+i} in section 5.2.

cyclic groups of prime power order, the orders of the direct summands are the same but for their order. Without making use of Theorem 4.4.1, we proceed as follows.

First, T has a unique decomposition into a direct sum of primary abelian groups. The primary direct summands are the Sylow p-groups belonging to the distinct prime divisors of $|T|$. Hence, we can confine ourselves to primary abelian groups.

Now, let P be a finite abelian p-group and

$$P = \langle \pi_1 \rangle \oplus \langle \pi_2 \rangle \oplus \cdots \oplus \langle \pi_t \rangle \tag{5.13}$$

a decomposition of P into the direct sum of cyclic groups. Let π_i be of order p^{k_i} $(i = 1, \ldots, t)$. By a suitable numbering, we may assume that

$$0 < k_1 \leqq k_2 \leqq \cdots \leqq k_t.$$

We shall show that the numbers p^{k_i} are uniquely determined by the orders of certain subgroups of P that, for their part, are defined by P alone without reference to the direct decomposition (5.13).

For a fixed power p^k, we consider the complex $p^k P$ of all elements ϱ of P that can be expressed in the form $\varrho = p^k \xi$ for some $\xi \in P$. It is obvious that $p^k P$ is a subgroup of P and that the definition of $p^k P$ is independent of the direct decomposition (5.13). Referring to (5.13), however, $p^k P$ can be described as follows: Put $k_0 = 0$ and suppose that for a suitable index j, $1 \leqq j \leqq t$, we have

$$k_{j-1} \leqq k < k_j.$$

Then,

$$p^k P = \langle p^k \pi_j \rangle \oplus \langle p^k \pi_{j+1} \rangle \oplus \cdots \oplus \langle p^k \pi_t \rangle. \tag{5.14}$$

If $k \geqq k_t$, we obtain $p^k P = 0$.

Let l_k denote the number of those terms of the sequence $k_1, k_2, \ldots, k_t$ that are equal to k. If no k_i is equal to k, then we put $l_k = 0$. If the numbers $l_1, l_2, \ldots$ are known, then $k_1, k_2, \ldots, k_t$ are uniquely determined. It is now easy to find expressions for $l_1, l_2, \ldots$ that involve only the orders of the subgroups $p^k P$. We have

$$|P| = p^{l_1 + 2 l_2 + 3 l_3 + \cdots}$$

and, by (5.14), we obtain

$$|pP| = p^{l_2 + 2 l_3 + 3 l_4 + \cdots}$$

$$|p^2 P| = p^{l_3 + 2 l_4 + 3 l_5 + \cdots}$$

$$\cdots\cdots\cdots\cdots\cdots\cdots$$

$$|p^k P| = p^{l_{k+1} + 2 l_{k+2} + 3 l_{k+3} + \cdots}$$

$$\cdots\cdots\cdots\cdots\cdots\cdots$$

The orders of the consecutive factor groups in the series

$$P \supset pP \supset p^2P \supset \cdots$$

are therefore

$$|P/pP| = p^{l_1+l_2+l_3+\cdots}$$

$$|pP/p^2P| = p^{l_2+l_3+\cdots}$$

$$\cdots\cdots\cdots\cdots\cdots$$

This shows that $l_1, l_2, \ldots$ are uniquely determined by A and are independent of any particular direct decomposition. This completes the proof of the uniqueness proposition.

We summarize our results:

5.3.1 *Every finitely generated mixed abelian group A has a direct decomposition*

$$A = T \oplus F,$$

where T is the torsion subgroup of A and F is a free abelian group. The torsion subgroup is finite and uniquely determined. Moreover, the rank of the free abelian direct summand is the same in all such direct decompositions. It is called the rank of A.

Every finite abelian group T admits a decomposition

$$T = C_1 \oplus C_2 \oplus \cdots \oplus C_s$$

into a direct sum of cyclic groups C_i such that $|C_i|$ divides $|C_{i+1}|$ for $i = 1, \ldots s-1$. The orders $|C_i|$ are uniquely determined and are called the torsion coefficients of T.

Every finite abelian group can be decomposed into a direct sum of primary cyclic groups. In all such decompositions, the primary cyclic summands have the same orders.

By a *basis* of an arbitrary finitely generated abelian group A we mean a system of generators of the cyclic direct summands in any direct decomposition of A into directly indecomposable subgroups. In the case of a free abelian group, this definition agrees with the previous one. If an element of a basis is of finite order, then its order is a power of a prime number. The rank of A and the prime powers that occur as the finite orders of the elements in a basis of A are called the *invariants* of A. By Theorem 5.3.1, two finitely generated abelian groups are isomorphic if and only if their invariants coincide. Conversely, if the rank and an arbitrary finite set of prime powers are given, then there always exists an abelian group with these invariants.

Let $p, q, \ldots$ be the distinct prime divisors of the order of the finite abelian group T, and let $p^{a_1}, p^{a_2}, \ldots q^{b_1}, q^{b_2}, \ldots$ be the orders of the elements of a basis of T. Then, the sequence

$$(p^{a_1}, p^{a_2}, \ldots, q^{b_1}, q^{b_2}, \ldots)$$

is called the *type* of T. It is expedient to arrange the exponents such that $a_1 \leqq a_2 \leqq \ldots, b_1 \leqq b_2 \leqq \ldots$ (or in the converse order).

A finite abelian group of type $(p, p, \ldots, p)$ is called an *elementary abelian group*.

5.4 Finite Cyclic Groups

In this section, we derive some simple facts on the orders of elements in finite cyclic groups.

Let $\langle \alpha \rangle$ be a cyclic group of order m. We determine the order of the subgroup generated by $h\alpha$. If $h = d$ is a divisor of m, we saw in section 1.4 that $\langle d\alpha \rangle$ is of order $f = md^{-1}$. For an arbitrary integer $h \neq 0$, we shall obtain the following result: $\langle h\alpha \rangle = \langle d\alpha \rangle$, where $d = (h, m)$ is the greatest common divisor of h and m. Indeed, since h is divisible by d, we have $h\alpha \in \langle d\alpha \rangle$ and hence $\langle h\alpha \rangle \subseteq \langle d\alpha \rangle$. On the other hand, there are integers u, v such that $d = um + vh$. So, we have

$$d\alpha = um\,\alpha + vh\,\alpha = vh\,\alpha,$$

and this implies that $d\alpha \in \langle h\alpha \rangle$ and $\langle d\alpha \rangle \subseteq \langle h\alpha \rangle$. Hence, $\langle h\alpha \rangle = \langle d\alpha \rangle$. This gives:

5.4.1 *In a cyclic group $\langle \alpha \rangle$ of order m, the element $h\alpha$ generates a subgroup of order $m(h, m)^{-1}$.*

It is now easy to answer another question: How many elements of a given order f are contained in a cyclic group $\langle \alpha \rangle$ of order m. Of course, there are no elements of order f unless f divides m. So, we assume that f is a divisor of m, $m = df$ say. By Theorem 5.4.1, the element $h\alpha$ is of order f if and only if $m(h, m)^{-1} = f$ or $(h, m) = d$. This shows that the number of elements of order f in $\langle \alpha \rangle$ is equal to the number of integers h, $1 \leqq h \leqq m$, that satisfy the condition $(h, m) = d$. Writing these numbers h in form $h = dk$, we obtain $1 \leqq k \leqq f$, and $(h, m) = (dk, df) = d$ is equivalent to $(k, f) = 1$. The number of integers k such that $1 \leqq k \leqq f$ and $(k, f) = 1$ is denoted by $\varphi(f)$ (Euler's function). So, we have the following result:

5.4.2 *If the natural number f is a divisor of m, then a cyclic group of order m contains $\varphi(f)$ elements of order f.*

In particular, a cyclic group $\langle \alpha \rangle$ of order m contains $\varphi(m)$ generating elements, i. e. elements of order m. These are the elements $r\alpha$ where $(r, m) = 1$.

Every automorphism σ of a cyclic group $\langle\alpha\rangle$ of order m carries α into a generating element, $s\alpha$ say. Hence, $\alpha\sigma = s\alpha$, where $(s, m) = 1$. Conversely, every mapping $\alpha \to \alpha\tau = t\alpha$, where $(t, m) = 1$, defines an automorphism τ of $\langle\alpha\rangle$. Carrying out the automorphisms σ and τ in succession, we obtain

$$\alpha(\sigma\tau) = (\alpha\sigma)\,\tau = (s\,\alpha)\,\tau = s(\alpha\tau) = st\,\alpha\,.$$

This gives the following result:

5.4.3 *The automorphism group of a cyclic group of order m is isomorphic to the multiplicative group of the prime residue classes mod m.*

By Theorem 5.4.2, we can easily derive an important property of the Euler function. Let f range over all positive divisors of m, including 1 and m. Then, take, for every f, the $\varphi(f)$ elements of order f in a cyclic group $\langle\alpha\rangle$ of order m. Clearly, in this way every element of $\langle\alpha\rangle$ occurs precisely once. Therefore:

5.4.4 $\sum_{f/m} \varphi(f) = m$ *for every natural number m, where f ranges over all positive divisors of m.*

Finally, we consider the solutions of the equation

$$t\xi = 0 \tag{5.15}$$

in a cyclic group of order m. Here, t is a given positive integer. We observe that for $u = (t, m)$ the equations

$$u\xi = 0 \tag{5.16}$$

and (5.15) have the same solutions. Indeed, $u\xi = 0$ implies that $t\xi = 0$ since u divides t. On the other hand, there are integers r and s such that $u = rt + sm$. Hence, we have

$$u\,\xi = rt\,\xi + sm\,\xi = rt\,\xi$$

so that $t\xi = 0$ implies that $u\xi = 0$. Thus, it is sufficient to deal with (5.15) under the condition that t divides m. If t is a divisor of m, then precisely those elements are solutions of (5.15) whose orders are divisors of t, including 1 and t. Thus, by Theorem 5.4.2, the number of solutions of (5.15) is given by $\sum_{d/t} \varphi(d)$. By Theorem 5.4.4, this sum is equal to t. This gives the following result:

5.4.5 *If t is a positive divisor of the order m of a cyclic group, then the equation* $t\xi = 0$ *has t solutions* ξ *in that group.*

Exercises

1. Decompose the group of the rational numbers x, $0 \leqq x < 1$, under addition mod 1 into a direct sum of primary groups.

2. Let $\eta_1, \ldots, \eta_m$ be a basis of a free abelian group M and put

$$\zeta_i = \sum_{k=1}^{m} v_{ik}\, \eta_k \qquad (i = 1, \ldots, m)$$

where the v_{ik} are integers. Find a necessary and sufficient condition on the matrix $[v_{ik}]$ for $\zeta_1, \ldots, \zeta_m$ to be a basis of M.

3. Let an abelian group be defined by a system of relations

$$\sum_{k=1}^{m} a_{ik}\, \alpha_k = 0 \qquad (i = 1, \ldots, l)$$

on the generating elements $\alpha_1, \ldots, \alpha_m$.

(a) Show that it is sufficient to consider the case $l \leqq m$.

(b) Find a necessary and sufficient condition on the $l \times m$ matrix $A = [a_{ik}]$ for the group to be finite and express the order in terms of the matrix A.

(c) Prove that another system of relations with the $l \times m$ matrix B defines the same group if and only if $B = UAV$, where U and V are square matrices with integral elements and determinants 1 or -1.

(d) Let

$$A = \begin{bmatrix} 5 & 4 & 1 & 5 \\ 7 & 6 & 5 & 11 \\ 2 & 2 & 10 & 12 \\ 10 & 8 & -4 & 4 \end{bmatrix}.$$

Represent the group as the direct sum of cyclic groups.

4. Let m_1 and m_2 be natural numbers with $(m_1, m_2) = 1$. Prove that $\varphi(m_1 m_2) = \varphi(m_1)\, \varphi(m_2)$.

5. Prove that every finite multiplicative group whose elements belong to a field is cyclic. (Hint: Use the fact that a polynomial of degree n with coefficients in a field F cannot have more than n roots in F.)

6. Let R be the multiplicative group of all roots of unity in the field of the complex numbers. Prove that every finitely generated subgroup of R is cyclic.

CHAPTER 6

EXTENSIONS OF GROUPS

6.1 Schreier's Theorem

Let two groups A and Γ be given. A group G is said to be *an extension of A by Γ* if it contains a normal subgroup A_0 isomorphic to A such that the factor group G/A_0 is isomorphic to Γ. In general, the type of an extension is by no means uniquely determined by A and Γ alone. For example, there are two distinct types of groups of order 4 (both A and Γ cyclic of order 2) and two distinct types of order 6 (A and Γ of orders 3 and 2, respectively). There always exists at least one extension of A by Γ, namely the direct product $A \times \Gamma$. The aim of extension theory is to give a survey of all possible extensions of A by Γ. This task, however, is so complicated that one is far from a general solution. But, in particular cases, it is possible to obtain a complete classification of all extensions; moreover, there are several general theorems on extensions.

We may obviously simplify the notation by assuming that any extension G of A by Γ contains A itself as a normal subgroup.

We shall assume that, whenever we speak of an extension G of A by Γ, a fixed isomorphism of G/A onto Γ is given. Two extensions G_1 and G_2 of A by Γ are said to be *equivalent* if there is an isomorphism of G_1 onto G_2 that maps A elementwise onto itself and carries each element of G_1/A into that element of G_2/A which corresponds to the same element of Γ. It may happen that two extensions of A by Γ are isomorphic without being equivalent.

We shall now look for additional parameters that can be used to determine an extension of A by Γ. Let G be such an extension. Then, G contains A as a normal subgroup, and we are given an isomorphism of G/A onto Γ. The elements of G will be denoted by lower case Roman letters and the elements of Γ by lower case Greek letters. Then, we have

$$G = \bigcup_{\xi \in \Gamma} r_\xi A$$

where the elements r_ξ, $\xi \in \Gamma$, form a transversal to A in G such that the coset $r_\xi A$ corresponds to the element ξ of Γ under the given isomorphism of G/A onto Γ. According to the isomorphism

$$r_\xi A \to \xi$$

of G/A onto Γ, we have $r_\xi A r_\eta A = r_{\xi\eta} A$. Thus, $r_\xi r_\eta$ is contained in the coset $r_{\xi\eta} A$. Hence,

$$r_\xi r_\eta = r_{\xi\eta} c_{\xi,\eta}, \qquad c_{\xi,\eta} \in A. \tag{6.1}$$

Transformation of A by the elements r_ξ yields automorphisms of A. The image of an element a of A under this automorphism will be denoted by a^ξ:

$$r_\xi^{-1} a r_\xi = a^\xi \qquad (a \in A,\ \xi \in \Gamma). \tag{6.2}$$

The elements $c_{\xi,\eta}$ $(\xi, \eta \in \Gamma)$ are called the *factor system* of the extension G corresponding to the transversal $\{r_\xi\}$. The automorphisms $a \to a^\xi$ of A are said to form the *automorphism system* of G corresponding to $\{r_\xi\}$. Both systems together are called the *parameter system* $\{c_{\xi,\eta};\ a \to a^\xi\}$ of Gcorresponding to the transversal $\{r_\xi\}$. Two arbitrary elements $r_\xi a$ and $r_\eta b$ $(a, b \in A)$ are multiplied as follows:

$$r_\xi a r_\eta b = r_\xi r_\eta a^\eta b = r_{\xi\eta} c_{\xi,\eta} a^\eta b \qquad (c_{\xi,\eta} a^\eta b \in A). \tag{6.3}$$

This shows that G is completely determined by the parameter system.

By (6.1) and (6.2), we have

$$(a^\xi)^\eta = (a^{\xi\eta})\, c_{\xi,\eta}, \tag{6.4}$$

where we use the notation $x^y = y^{-1} x y$ for arbitrary $x, y \in A$. Applying the associative law to r_ξ, r_η, r_ζ, we obtain

$$r_\xi(r_\eta r_\zeta) = r_\xi r_{\eta\zeta} c_{\eta,\zeta} = r_{\xi\eta\zeta} c_{\xi,\eta\zeta} c_{\eta,\zeta},$$

$$(r_\xi r_\eta)\, r_\zeta = r_{\xi\eta} c_{\xi,\eta} r_\zeta = r_{\xi\eta\zeta} c_{\xi\eta,\zeta} c_{\xi,\eta}^\zeta,$$

and therefore

$$c_{\xi,\eta\zeta} c_{\eta,\zeta} = c_{\xi\eta,\zeta} c_{\xi,\eta}^\zeta \qquad (\xi, \eta, \zeta \in \Gamma). \tag{6.5}$$

Thus, every parameter system satisfies the relations (6.4) and (6.5). Equations (6.5) are called the *associativity relations*.

So far, we have started from a given extension G of A by Γ and have seen that, for a fixed transversal $\{r_\xi\}$, this extension leads to a parameter system satisfying (6.4) and (6.5). We shall now show that, conversely, each parameter system for which (6.4) and (6.5) hold belongs to an extension of A by Γ.

Suppose that we are given:

(a) a system of elements $c_{\xi,\eta}$ of A indexed by the ordered pairs ξ, η of elements of Γ;

(b) a system of automorphisms $a \to a^\xi$ corresponding to the elements ξ of Γ.

We assume that these elements satisfy the conditions (6.4) and (6.5). To every element ξ of Γ, we assign a symbol r_ξ. Then, let G denote the set of all symbols $r_\xi a$, where $a \in A$. Equation (6.3) suggests the following definition of the product of two such symbols:

$$r_\xi a r_\eta b = r_{\xi\eta} c_{\xi,\eta} a^\eta b \qquad (\xi, \eta \in \Gamma;\ a, b \in A).$$

We shall prove that G is a group with respect to this operation. The associative law follows from a simple calculation. Using (6.4), we obtain

$$\begin{aligned}(r_\xi a r_\eta b)\, r_\zeta c &= r_{\xi\eta}(c_{\xi,\eta} a^\eta b)\, r_\zeta c \\ &= r_{\xi\eta\zeta} c_{\xi\eta,\zeta} c_{\xi,\eta}^\zeta (a^\eta)^\zeta b^\zeta c. \\ &= r_{\xi\eta\zeta} c_{\xi\eta,\zeta} c_{\xi,\eta}^\zeta (a^{\eta\zeta})^{c_{\eta,\zeta}} b^\zeta c,\end{aligned}$$

$$\begin{aligned}r_\xi a(r_\eta b r_\zeta c) &= r_\xi a r_{\eta\zeta} c_{\eta,\zeta} b^\zeta c \\ &= r_{\xi\eta\zeta} c_{\xi,\eta\zeta} a^{\eta\zeta} c_{\eta,\zeta} b^\zeta c \\ &= r_{\xi\eta\zeta} c_{\xi,\eta\zeta} c_{\eta,\zeta} (a^{\eta\zeta})^{c_{\eta,\zeta}} b^\zeta c.\end{aligned}$$

By (6.5), the right-hand sides of the two equations coincide.

Before proceeding further, we derive some consequences of (6.4) and (6.5). Let ε denote the unit element of Γ. From (6.4), it follows for $\xi = \eta = \varepsilon$ that

$$(a^\varepsilon)^\varepsilon = (a^\varepsilon)^{c_{\varepsilon,\varepsilon}}.$$

If a ranges over A, then so does a^ε. Therefore,

$$a^\varepsilon = a^{c_{\varepsilon,\varepsilon}} \quad \text{for all} \quad a \in A. \tag{6.6}$$

Equations (6.5) and (6.6) for $\eta = \zeta = \varepsilon$ show that

$$c_{\xi,\varepsilon} c_{\varepsilon,\varepsilon} = c_{\xi,\varepsilon} c_{\xi,\varepsilon}^\varepsilon = c_{\xi,\varepsilon} c_{\xi,\varepsilon}^{c_{\varepsilon,\varepsilon}};$$

hence

$$c_{\varepsilon,\varepsilon} = c_{\xi,\varepsilon}^{c_{\varepsilon,\varepsilon}}.$$

Since transformation by $c_{\varepsilon,\varepsilon}^{-1}$ leaves $c_{\varepsilon,\varepsilon}$ fixed, we obtain from the last equation

$$c_{\varepsilon,\varepsilon} = c_{\xi,\varepsilon}. \tag{6.7}$$

Finally, using (6.5) for $\xi = \eta = \varepsilon$, we obtain

$$c_{\varepsilon,\zeta} c_{\varepsilon,\zeta} = c_{\varepsilon,\zeta} c_{\varepsilon,\varepsilon}^\zeta$$

so that

$$c_{\varepsilon,\zeta} = c_{\varepsilon,\varepsilon}^\zeta. \tag{6.8}$$

We now show that $r_\varepsilon c_{\varepsilon,\varepsilon}^{-1}$ is a right unit element of G. Indeed, by (6.6) and (6.7), we obtain

$$r_\xi a r_\varepsilon c_{\varepsilon,\varepsilon}^{-1} = r_\xi c_{\varepsilon,\varepsilon} a^\varepsilon c_{\varepsilon,\varepsilon}^{-1} = r_\xi a.$$

Moreover, $r_{\xi^{-1}}(a^{\xi^{-1}})^{-1}c^{-1}_{\xi,\xi^{-1}}c^{-1}_{\varepsilon,\varepsilon}$ is a right inverse of $r_\xi a$; for

$$r_\xi a r_{\xi^{-1}}(a^{\xi^{-1}})^{-1}c^{-1}_{\xi,\xi^{-1}}c^{-1}_{\varepsilon,\varepsilon} = r_\varepsilon c^{-1}_{\varepsilon,\varepsilon}.$$

Consequently, G is a group.

Next, we show that G contains a normal subgroup $\bar{A}$ isomorphic to A, and that $G/\bar{A}$ is isomorphic to Γ. The elements

$$\bar{a} = r_\varepsilon c^{-1}_{\varepsilon,\varepsilon}a$$

form a subgroup $\bar{A}$ of G, and the mapping $\bar{a} \to a$ is an isomorphism of $\bar{A}$ onto A. For, by (6.6), we have

$$\begin{aligned}\bar{a}\bar{b} &= r_\varepsilon c^{-1}_{\varepsilon,\varepsilon}a r_\varepsilon c^{-1}_{\varepsilon,\varepsilon}b = r_\varepsilon c_{\varepsilon,\varepsilon}(c^{-1}_{\varepsilon,\varepsilon})^\varepsilon a^\varepsilon c^{-1}_{\varepsilon,\varepsilon}b\\ &= r_\varepsilon c_{\varepsilon,\varepsilon}(c^{-1}_{\varepsilon,\varepsilon})^{c_{\varepsilon,\varepsilon}}a^{c_{\varepsilon,\varepsilon}}c^{-1}_{\varepsilon,\varepsilon}b\\ &= r_\varepsilon c^{-1}_{\varepsilon,\varepsilon}ab = \overline{ab}.\end{aligned}$$

Put

$$\bar{r}_\xi = r_\xi e,$$

where e is the unit element of A. Then, (6.7) yields

$$\bar{r}_\xi \bar{a} = r_\xi e r_\varepsilon c^{-1}_{\varepsilon,\varepsilon}a = r_\xi c_{\xi,\varepsilon}c^{-1}_{\varepsilon,\varepsilon}a = r_\xi a. \tag{6.9}$$

Moreover, we have

$$\bar{r}^{-1}_\eta \bar{r}_\xi = r_{\eta^{-1}}c^{-1}_{\eta,\eta^{-1}}c^{-1}_{\varepsilon,\varepsilon}r_\xi e = r_{\eta^{-1}\xi}c_{\eta^{-1},\xi}(c^{-1}_{\eta,\eta^{-1}})^\xi(c^{-1}_{\varepsilon,\varepsilon^{-1}})^\xi. \tag{6.10}$$

From (6.9) and (6.10), it follows that the elements $\bar{r}_\xi$ form a left transversal to $\bar{A}$ in G. Using (6.8) and (6.9) ,we obtain

$$\bar{a}\bar{r}_\xi = r_\varepsilon c^{-1}_{\varepsilon,\varepsilon}a r_\xi e = r_\xi c_{\varepsilon,\xi}(c^{-1}_{\varepsilon,\varepsilon})^\xi a^\xi e = r_\xi a^\xi = \bar{r}_\xi\overline{a^\xi};$$

hence

$$\bar{r}^{-1}_\xi \bar{a}\bar{r}_\xi = \overline{a^\xi}.$$

This shows that $\bar{A}$ is a normal subgroup of G. Moreover, transformation by $\bar{r}_\xi$ induces an automorphism of $\bar{A}$, which corresponds to the given automorphism $a \to a^\xi$ of A under the isomorphism $\bar{a} \to a$ of $\bar{A}$ onto A. Finally, (6.9) gives

$$\bar{r}_\xi \bar{r}_\eta = r_\xi e r_\eta e = r_{\xi\eta}c_{\xi,\eta} = \bar{r}_{\xi\eta}\bar{c}_{\xi,\eta}.$$

Thus, $G/\bar{A}$ is isomorphic to Γ, and the factor system $\{\bar{c}_{\xi,\eta}\}$ of the extension G of $\bar{A}$ by Γ corresponds to $\{c_{\xi,\eta}\}$ under the isomorphism $\bar{a} \to a$.

Summarizing our results, we obtain:

6.1.1 Schreier's Theorem. *Every extension of A by Γ can be described by a parameter system satisfying the conditions (6.4) and (6.5). Conversely, every parameter* *system for which (6.4) and (6.5) hold defines a unique equivalence class of extensions of A by Γ.*

We shall now discuss how the parameter system depends on the choice of the transversal. The notation is the same as before. Any other transversal $\{r'_\xi\}$ of G with respect to A has the form

$$r'_\xi = r_\xi b_\xi, \qquad b_\xi \in A. \tag{6.11}$$

Let $\{c'_{\xi,\eta};\ a \to a^{\xi'}\}$ be the parameter system corresponding to $\{r'_\xi\}$. Then, we have

$$r'_\xi r'_\eta = r'_{\xi\eta} c'_{\xi,\eta}, \quad a^{\xi'} = r'^{-1}_\xi a r'_\xi.$$

Using (6.11), we obtain

$$r'_\xi r'_\eta = r_{\xi\eta} c_{\xi,\eta} b^\eta_\xi b_\eta = r'_{\xi\eta} b^{-1}_{\xi\eta} c_{\xi,\eta} b^\eta_\xi b_\eta,$$

$$r'^{-1}_\xi a r'_\xi = b^{-1}_\xi r^{-1}_\xi a r_\xi b_\xi = (a^\xi)^{b_\xi};$$

therefore

$$c'_{\xi,\eta} = b^{-1}_{\xi\eta} c_{\xi,\eta} b^\eta_\xi b_\eta, \tag{6.12}$$

$$a^{\xi'} = (a^\xi)^{b_\xi}. \tag{6.13}$$

Two parameter systems $\{c_{\xi,\eta};\ a \to a^\xi\}$ and $\{c'_{\xi,\eta};\ a \to a^{\xi'}\}$ are said to be *associate* if there are elements b_ξ in A such that (6.12) and (6.13) hold. It follows from our results above that being associate is an equivalence relation and that there is a one-to-one correspondence between the classes of associate parameter systems and the classes of equivalent extensions of A by Γ.

If each $c_{\xi,\eta}$ commutes with all elements of A, then (6.4) shows that the mapping

$$\xi \to (a \to a^\xi) \tag{6.14}$$

is a homomorphism of Γ into the automorphism group $\mathsf{A}(A)$ of A.

Let $\mathsf{J}(A)$ denote the group of the inner automorphism of A. For an arbitrary automorphism $a \to a\sigma$ and any inner automorphism $a \to a\tau_b = b^{-1}ab$ of A, we obtain

$$a(\sigma^{-1}\tau_b\sigma) = (b^{-1}(a\sigma^{-1})\, b)\,\sigma = (b\sigma)^{-1} a(b\sigma) = a\,\tau_{b\sigma}.$$

This shows that $\mathsf{J}(A)$ is a normal subgroup of $\mathsf{A}(A)$. The elements of the factor group $\mathsf{A}(A)/\mathsf{J}(A)$ are called the automorphism classes of A. By (6.4), the mapping (6.14) is always a homomorphism of Γ into the group of the automorphism classes of A. By (6.13), this homomorphism does not depend on the choice of the transversal of G with respect to A. Suppose now that A is a group without centre. Then, every homomorphism of Γ into the group of the automorphism classes of A defines a unique class of equivalent extensions of A by Γ. Indeed, if we choose, for each $\xi \in \Gamma$, a fixed automorphism

$a \to a^\xi$ in the corresponding automorphism class, then the automorphisms $a \to (a^\xi)^\eta$ and $a \to a^{\xi\eta}$ differ from one another only by an inner automorphism of A. Since A has no centre, the element $c_{\xi,\eta}$ in (6.4) is uniquely determined. Using the equation $(a^{\xi\eta})^\zeta = (a^\xi)^{\eta\zeta}$, it is easy to show that the elements $c_{\xi,\eta}$ satisfy the conditions (6.5). Moreover, another choice of the automorphisms in their classes yields a factor system $\{c'_{\xi,\eta}\}$ that is related to $\{c_{\xi,\eta}\}$ by equations of the form (6.12). Generalizing this approach to arbitrary groups A, R. Baer [1] arrives at another characterization of the classes of equivalent extensions of A by Γ.

6.2 Semi-Direct Products

We keep the notation of the last section.

If $c_{\xi,\eta} = e$ for all $\xi, \eta \in \Gamma$, then (6.1) shows that the transversal $\{r_\xi\}$ is a subgroup R of G such that

$$G = RA, \quad R \cap A = e. \tag{6.15}$$

Of course, R is isomorphic to Γ. We call R a *complement* of A in G, and the group G itself is said to be a *semi-direct product* of the subgroup R and the normal subgroup A. We also say that the extension G of A by Γ *splits* or is *complemented*.

By (6.12) and (6.13), every parameter system associate to $\{e, a \to a^\xi\}$ has the form

$$\{b_{\xi\eta}^{-1} b_\xi^\eta b_\eta, \quad a \to a^{\xi'}\}.$$

A factor system $\{c'_{\xi,\eta}\}$ is said to *split* if there are elements b_ξ in A such that

$$c'_{\xi,\eta} = b_{\xi\eta}^{-1} b_\xi^\eta b_\eta \quad \text{for all} \quad \xi, \eta \in \Gamma. \tag{6.16}$$

Thus, a factor system splits if and only if it belongs to some parameter system associate to $\{e, a \to a^\xi\}$. If the factor system (6.2) corresponds to the transversal $\{r'_\xi\}$, then the elements $r_\xi = r'_\xi b_\xi^{-1}$ form a complement of A in G. A semi-direct product $G = RA$ is uniquely determined by the automorphisms of A obtained by transforming A by the elements of R. If all these automorphisms of A coincide with the identity automorphism, then we obviously have $G = R \times A$.

Note that every subgroup conjugate to R is also a complement of A in G.

For a given factor set $\{c'_{\xi,\eta}\}$ it is, in general, extremely difficult to decide whether there exist elements b_ξ in A such that eqs. (6.16) hold. Therefore, it is highly important to have more manageable conditions for the existence of a complement. In this and in the next section we shall derive two such conditions.

A factor system $\{c_{\xi,\eta}\}$ is called *abelian* if any two of its elements commute.

6.2.1 *Let Γ be a finite group, $|\Gamma| = n$, and $\{c_{\xi,\eta}\}$ an abelian factor system. Then, $\{c^n_{\xi,\eta}\}$ is a splitting factor system.*

Proof. Put

$$\prod_{\xi \in \Gamma} c_{\xi,\eta} = b_\eta$$

and form the product of the relations (6.5) for all $\xi \in \Gamma$ and fixed η, ζ. This gives

$$b_{\eta\zeta} c^n_{\eta,\zeta} = b_\zeta b^\zeta_\eta \qquad (\eta, \zeta \in \Gamma).$$

Then, (6.16) shows that $\{c^n_{\eta,\zeta}\}$ is a splitting factor system.

The first of the two conditions announced above is the:

6.2.2 Schur–Zassenhaus Theorem. *Let A and Γ be finite groups whose orders are relatively prime. Then, every extension of A by Γ splits.*

Proof. Let $|A| = m$, $|\Gamma| = n$, and G be an arbitrary extension of A by Γ. We have to show that G contains a subgroup of order n.

We proceed by induction on the order of A. The case $|A| = 1$ is trivial. We assume that every extension of a group A_1 by Γ splits if the order of A_1 is relatively prime to n and less than m.

Let p be a prime divisor of m and S a Sylow p-subgroup of G. Clearly, S is contained in A. The number of distinct Sylow p-subgroups of G is equal to the index $|G:N|$ of the normalizer $N = \mathsf{N}(S)$ of S in G. All Sylow p-subgroups of G belong to A, and since $A \cap N$ is the normalizer of S in A, we obtain $|A:(A \cap N)| = |G:N|$. It follows that $|N:(A \cap N)| = |G:A| = n$. Now, $(A \cap N)/S$ is a normal subgroup of N/S of index n. Hence, by our inductive assumption, N/S contains a subgroup H/S of order n. The centre Z of S is non-trivial, and being a characteristic subgroup of S, the centre Z is a normal subgroup of H. Thus, S/Z is a normal subgroup of H/Z on index n. Again, by the inductive assumption, H/Z contains a subgroup U/Z of order n. Now, U is an extension of Z. Let $\{d_{\varrho,\sigma}\}$, $\varrho, \sigma \in U/Z$, be a factor system of this extension. Since the order of Z is relatively prime to n, there is an integer n_1 such that $nn_1 \equiv 1 \pmod{|Z|}$. Therefore, $d_{\varrho,\sigma} = (d^n_{\varrho,\sigma})^{n_1}$. By Theorem 6.2.1, the factor system $\{d^n_{\varrho,\sigma}\}$ splits so that $\{d_{\varrho,\sigma}\}$ itself splits. Consequently, U contains a complement of Z of order n. Hence, G contains a subgroup of order n.

For the proof of the next theorem we need the following:

Lemma. Let A be an arbitrary group and Γ a group of order n. Suppose that there is a system of automorphisms $a \to a^\xi$ of A corresponding to the elements ξ of Γ. Suppose, moreover, that A contains a system of elements b_ξ, indexed by the elements ξ of Γ, such that any two elements b_ξ^ζ, b_η^τ commute with each other and

$$b_{\xi\eta} = b_\xi^\eta b_\eta \quad \text{for every pair} \quad \xi,\eta \in \Gamma. \tag{6.17}$$

Then, A contains an element d such that

$$b_\eta^n = d(d^\eta)^{-1} \quad \text{for every} \quad \eta \in \Gamma. \tag{6.18}$$

Proof. Keep η fixed and multiply the relations (6.17) for all $\xi \in \Gamma$. This gives

$$\prod_{\xi\in\Gamma} b_{\xi\eta} = \Big(\prod_{\xi\in\Gamma} b_\xi\Big)^\eta b_\eta^n$$

or

$$\prod_{\xi\in\Gamma} b_\xi = \Big(\prod_{\xi\in\Gamma} b_\xi\Big)^\eta b_\eta^n.$$

Hence, $d = \prod_{\xi\in\Gamma} b_\xi$ satisfies (6.18).

The following theorem is an important supplement to Theorem 6.2.2.

6.2.3 Zassenhaus' Theorem. *Let A and Γ be finite groups whose orders are relatively prime, and G an extension of A by Γ. Then, any two complements of A in G are conjugate provided that at least one of the two groups A, Γ is soluble.*

Remark. The solubility condition is, in fact, redundant. For, since $(|A|, |\Gamma|) = 1$, at least one of the two groups has odd order; hence, is soluble by Theorem 2.8.1.

Proof. By Theorem 6.2.2, G is complemented. Let $|A| = m$, $|\Gamma| = n$.

First, we assume that A is abelian. Let $R = \{r_\xi\}$ and $Q = \{r'_\xi\}$ be two complements of A in G. Since both R and Q are transversals to A in G, there are elements b_ξ in A such that $r'_\xi = r_\xi b_\xi$ for all $\xi \in \Gamma$. From (6.12), we obtain

$$b_{\xi\eta} = b_\xi^\eta b_\eta.$$

Then, it follows from the lemma that there exists an element d in A such that

$$b_\eta^n = d(d^\eta)^{-1}.$$

Since $(m, n) = 1$, the congruence $nn_1 \equiv 1 \pmod m$ has a solution n_1. Putting $d_1 = d^{n_1}$, we have

$$b_\eta = b_\eta^{nn_1} = d_1(d_1^\eta)^{-1}$$

so that (6.2) gives

$$r'_\xi = r_\xi d_1(d_1^\xi)^{-1} = d_1^{-1} r_\xi d_1.$$

This shows that $Q = d_1^{-1} R d_1$. Thus, the theorem is true if A is abelian.

Next, suppose that A is soluble, say k-step metabelian. Since our proposition is true for abelian groups A, i. e. for $k = 1$, we may assume that the theorem holds for extensions of groups that are $(k-1)$-step metabelian. Let R and Q be two complements to A in G. We now apply our result above for the abelian case to the extension G/A' of the factor group A/A'. Accordingly, there exists an element xA' in A/A' such that

$$(xA')^{-1}(RA')(xA') = QA';$$

hence,

$$x^{-1}RxA' = QA'. \tag{6.19}$$

Since A' is $(k-1)$-step metabelian, the theorem holds for the group QA' as an extension of A' by Q. By (6.19), $x^{-1}Rx$ is another complement of A' in QA' so that $y^{-1}x^{-1}Rxy = Q$ for some element y in A'. Consequently, R and Q are conjugate in G.

Finally, we suppose that Γ is soluble. Let R and Q be two complements of A in G, and assume that G/A has a chief series of length l. We choose a minimal normal subgroup U of R. Since R is soluble, it follows from Theorem 4.2.3 that U is an elementary abelian p-group. The group UA can also be written in the form VA, where V is a normal subgroup of Q isomorphic to U, namely $V = Q \cap UA$.

If $l = 1$, we have $U = R$ and $V = Q$. Thus, R and Q are Sylow p-subgroups of G, so that R and Q are conjugate in G. Suppose then that $l > 1$. We may assume that the theorem is true for extensions of A by a group that has a chief series of length less than l. It follows that there exists an element x in A such that $x^{-1}Ux = V$. Put $Q_1 = xQx^{-1}$. Then, both R and Q_1 are contained in the normalizer N of U in G. Thus, N is an extension of $N \cap A$ by Γ, and hence N/U is an extension of $(N \cap A)U/U$ in which R/U and Q_1/U are complements of $(N \cap A)U/U$. Since R/U has a chief series of length $l-1$, there is an element y in N such that $y^{-1}Ry = Q_1$. Consequently,

$$Q = (xy)^{-1}R(yx),$$

and the proof of Theorem 6.2.3 is complete.

6.3 Extensions of Abelian Groups

We keep the notation of the two previous sections and assume that A is abelian.

In section 6.1, we remarked that in this case the mapping

$$\xi \to (a \to a^{\xi}) \quad \text{for} \quad \xi \in \Gamma$$

is a homomorphism of Γ into the automorphism group of A. From (6.13), we conclude that the automorphism system does not depend on the choice of the transversal. Hence, if A is abelian, there is a one-to-one correspondence between the classes of equivalent extensions of A by Γ and the classes of associate factor systems.

Let $\{c_{\xi,\eta}\}$ and $\{d_{\xi,\eta}\}$ be two factor systems belonging to the same automorphism system. Then, the products

$$f_{\xi,\eta} = c_{\xi,\eta}\, d_{\xi,\eta} \tag{6.20}$$

are again a factor system corresponding to the same automorphism system; for, if both $\{c_{\xi,\eta}\}$ and $\{d_{\xi,\eta}\}$ satisfy the associativity relations (6.5), then so does the system $\{f_{\xi,\eta}\}$. Moreover, if $\{c'_{\xi,\eta}\}$ is associate to $\{c_{\xi,\eta}\}$ and $\{d'_{\xi,\eta}\}$ associate to $\{d_{\xi,\eta}\}$, then one readily verifies that $\{c'_{\xi,\eta}\, d'_{\xi,\eta}\}$ is associate to $\{c_{\xi,\eta}\, d_{\xi,\eta}\}$. This shows that, in fact, (6.20) defines a product of classes of associate factor systems. It is obvious that this product is associative. The unit element is the class of all splitting factor systems. The class that contains the factor system $\{c_{\xi,\eta}^{-1}\}$ is the inverse of the class of $\{c_{\xi,\eta}\}$. Thus, the classes of associate factor systems belonging to one and the same automorphism system form an abelian group, which is called the *second cohomology group* and is denoted by $H^2(\Gamma, A)$.

The name is due to the fact that, in the case of an abelian normal subgroup A, the theory of the factor systems can be regarded as part of the so-called homology theory, which involves topological concepts. We refer the reader to [9] and [45].)

If Γ and A are finite, then $H^2(\Gamma, A)$ is also finite, because there are altogether only finitely many factor systems. If A is of exponent t, then $c_{\xi,\eta}^t = e$, so that the t-th power of every factor system splits. This gives the following result:

6.3.1 *Let Γ and A be finite groups and A abelian. Then, the exponent of $H^2(\Gamma, A)$ divides the greatest common divisor of $|\Gamma|$ and the exponent of A.*

Let $p_1, p_2, \ldots, p_s$ be the distinct prime divisors of $|A|$. Then, A has a direct decomposition.

$$A = A^{(1)} \times A^{(2)} \times \cdots \times A^{(s)} \tag{6.21}$$

where $A^{(i)}$ denotes the Sylow p_i-subgroup of A $(i = 1, 2, \ldots, s)$. Accordingly, the elements of a factor system $\{c_{\xi,\eta}\}$ have unique representations

$$c_{\xi,\eta} = c_{\xi,\eta}^{(1)} \times c_{\xi,\eta}^{(2)} \times \cdots \times c_{\xi,\eta}^{(s)} \qquad (c_{\xi,\eta}^{(i)} \in A^{(i)})$$

as the product of their $A^{(i)}$-components. Each $A^{(i)}$ is a characteristic subgroup of A, so that every automorphism system of A yields automorphism systems of the Sylow subgroups. The relations (6.5) hold if and only if the corresponding relations are satisfied for each $A^{(i)}$-component. Moreover, the equation

$$c'_{\xi,\eta} = b_{\xi\eta}^{-1} b_{\xi}^{\eta} b_{\eta} c_{\xi,\eta}$$

is equivalent to the s equations

$$c'^{(i)}_{\xi,\eta} = b^{(i)-1}_{\xi\eta} b^{(i)\eta}_{\xi} b^{(i)}_{\eta} c^{(i)}_{\xi,\eta} \qquad (i = 1, 2, \ldots, s)$$

for the direct components. Thus, the factor systems $\{c'_{\xi,\eta}\}$ and $\{c_{\xi,\eta}\}$ are associate if and only if their direct components are associate. Since the $A^{(i)}$-components of a product of two factor systems are equal to the product of their $A^{(i)}$-components, we obtain the following result:

6.3.2 *If* $A = A^{(1)} \times A^{(2)} \times \cdots \times A^{(s)}$ *is the direct decomposition of the finite abelian group A into its Sylow subgroups, then*

$$H^2(\Gamma, A) = H^2(\Gamma, A^{(1)}) \times H^2(\Gamma, A^{(2)}) \times \cdots \times H^2(\Gamma, A^{(s)}).$$

It may happen that some of the direct factors $H^2(\Gamma, A^{(i)})$, or perhaps all of them, are equal to the unit element, i. e. equal to the class of the splitting factor systems. These direct factors are to be cancelled in the above direct decomposition. By Theorem 6.3.1, $H^2(\Gamma, A^{(i)})$ is always equal to the unit element if p_i does not divide $|\Gamma|$.

By Theorem 6.3.2, the question whether a factor system splits is reduced to the same question for extensions of abelian p-groups. Our next aim is to obtain a similar reduction for the group Γ.

Let G be an extension of the finite abelian group A by the finite group Γ, and let U denote a subgroup of G that contains A. Then, U is an extension of A by some subgroup Λ of Γ. We choose a transversal Σ to Λ in Γ, assuming that Λ itself is represented by the unit element ε of Γ. Then, every element ξ of Γ has a unique representation as a product

$$\xi = \underline{\xi}\bar{\xi}, \qquad \underline{\xi} \in \Lambda, \ \bar{\xi} \in \Sigma. \tag{6.22}$$

We choose a transversal $\{r_\alpha\}$ to A in U; here, α ranges over the elements of Λ. We assume that $r_\varepsilon = e$. Let $\{c_{\alpha,\beta}\}$ be the factor system of the extension U of A by Λ that corresponds to the transversal $\{r_\alpha\}$.

We now extend $\{r_\alpha\}$ to a special transversal to A in G. For every $\varrho \neq \varepsilon$ in Σ, we choose an arbitrary element r_ϱ in the coset of A that corresponds to ϱ. For an arbitrary element ξ of Γ, we put

$$r_\xi = r_{\underline{\xi}} r_{\bar{\xi}}, \tag{6.23}$$

where $\underline{\xi}$ and $\bar{\xi}$ are defined by (6.22). In case $\xi \in \Lambda$, we have $\underline{\xi} = \xi$, $\bar{\xi} = \varepsilon$, and for $\xi \in \Sigma$, we obtain $\underline{\xi} = \varepsilon$, $\bar{\xi} = \xi$. Since $r^\varepsilon = e$, (6.23) agrees with our previous choice of the representatives r_α for $\alpha \in \Lambda$ and r_ϱ for $\varrho \in \Sigma$. Let $\{c_{\xi,\eta}\}$ denote the factor system of the extension G of A by Γ that corresponds to the transversal $\{r_\xi\}$.

For $\alpha \in \Lambda$, $\xi \in \Gamma$, we have

$$c_{\alpha,\xi} = r_{\alpha\xi}^{-1} r_\alpha r_\xi = r_{\alpha\underline{\xi}\overline{\xi}}^{-1} r_\alpha r_\xi = (r_{\alpha\underline{\xi}} r_{\overline{\xi}})^{-1} r_\alpha r_{\underline{\xi}} r_{\overline{\xi}} = r_{\overline{\xi}}^{-1}(r_{\alpha\underline{\xi}}^{-1} r_\alpha r_{\underline{\xi}})\, r_{\overline{\xi}};$$

hence

$$c_{\alpha,\xi} = (c_{\alpha,\underline{\xi}})^{\overline{\xi}}. \tag{6.24}$$

Moreover,

$$r_{\underline{\xi}} r_{\overline{\xi}} r_\eta = r_{\underline{\xi}} r_{\overline{\xi}\eta}\, c_{\overline{\xi},\eta} = r_{\xi\eta}\, c_{\underline{\xi},\overline{\xi}\eta}\, c_{\overline{\xi},\eta}$$

and, on the other hand,

$$r_{\underline{\xi}} r_{\overline{\xi}} r_\eta = r_\xi r_\eta = r_{\xi\eta} c_{\xi,\eta};$$

therefore, .

$$c_{\xi,\eta} = c_{\overline{\xi},\eta}\, c_{\underline{\xi},\overline{\xi}\eta}$$

so that, by (6.24),

$$c_{\xi,\eta} = c_{\overline{\xi},\eta}\, c_{\underline{\xi},\underline{\overline{\xi}\eta}}^{\overline{\overline{\xi}\eta}}. \tag{6.25}$$

Using (6.25), and the associativity relations

$$c_{\eta,\zeta} = c_{\xi,\eta\zeta}^{-1} c_{\xi,\eta}^{\zeta} c_{\xi\eta,\zeta},$$

we obtain

$$c_{\eta,\zeta} = c_{\xi,\eta\zeta}^{-1} c_{\xi,\eta}^{\zeta} c_{\overline{\xi\eta},\zeta}\, c_{\underline{\xi\eta},\underline{\overline{\xi\eta}\zeta}}^{\overline{\overline{\xi\eta}\zeta}}.$$

For fixed η, ζ, we now multiply the last equation over all $\xi \in \Sigma$. Putting $|G:U| = n$, we obtain

$$c_{\eta,\zeta}^{n} = \prod_{\xi \in \Sigma} \left(c_{\xi,\eta\zeta}^{-1}\, c_{\xi,\eta}^{\zeta}\, c_{\overline{\xi\eta},\zeta}\, c_{\underline{\xi\eta},\underline{\overline{\xi\eta}\zeta}}^{\overline{\overline{\xi\eta}\zeta}}\right). \tag{6.26}$$

We now assume that $n = |G:U|$ is relatively prime to $|A|$. Then, the congruence $nm \equiv 1 \pmod{|A|}$ has a solution m, relatively prime to $|A|$. The mapping

$$a \to a^m \quad \text{for every} \quad a \in A$$

is an automorphism of A, which commutes with every automorphism of A. If η is fixed, and ξ ranges over all elements of Σ, then $\overline{\xi\eta}$ also ranges over all elements of Σ. Owing to these facts, the m-th power of (6.26) can be written as follows:

$$c_{\xi,\eta} = \Big(\prod_{\xi \in \Sigma} c_{\xi,\eta\zeta}\Big)^{-m} \Big(\Big(\prod_{\xi \in \Sigma} c_{\xi,\eta}\Big)^{m}\Big)^{\zeta} \Big(\prod_{\xi \in \Sigma} c_{\xi,\zeta}\Big)^{m} \Big(\prod_{\xi \in \Sigma} c_{\underline{\xi\eta},\underline{\overline{\xi\eta}\zeta}}^{\overline{\overline{\xi\eta}\zeta}}\Big)^{m}.$$

Putting

$$d_\zeta = \Big(\prod_{\xi \in \Sigma} c_{\xi,\varsigma}\Big)^{m},$$

we now obtain

$$c_{\eta,\zeta} = d_{\eta\zeta}^{-1} d_\eta^\zeta d_\zeta \Big(\prod_{\xi \in \Sigma} c_{\underline{\xi\eta},\underline{\xi\eta\zeta}}^{\overline{\overline{\xi\eta}\zeta}} \Big)^m . \tag{6.27}$$

This equation means that the factor system $\{c_{\eta,\zeta}\}$ is associate to

$$\Big\{ \Big(\prod_{\xi \in \Sigma} c_{\underline{\xi\eta},\underline{\xi\eta\zeta}}^{\overline{\overline{\xi\eta}\zeta}} \Big)^m \Big\},$$

Note that, in the last factor system, there occur only elements of the subsystem $\{c_{\alpha,\beta}\}$ for $\alpha, \beta \in A$ and their images under certain automorphisms of A. In particular, $c_{\alpha,\beta} = e$ for all $\alpha, \beta \in A$ implies that the whole factor system $\{c_{\eta,\zeta}\}$ $(\eta, \zeta \in \Gamma)$ splits.

If A is a finite abelian p-group and U a Sylow p-subgroup of G, the condition $(|G\colon U|, |A|) = 1$ is satisfied. This gives the following result: A finite extension G of an abelian p-group A splits if and only if a Sylow p-subgroup of G is a splitting extension of A.

We now consider an arbitrary finite abelian group A and its decomposition (6.21) into a direct product of its Sylow subgroups. Let G be an extension of A by the finite group Γ and denote by $\{c_{\xi,\eta}\}$ the corresponding factor system. If p_i is a prime divisor of $|A|$ that does not divide $|\Gamma|$, then the Sylow p_i-component $\{c_{\xi,\eta}^{(i)}\}$ splits by Theorem 6.3.1. If the prime divisor p_j of $|A|$ divides $|\Gamma|$, then $\{c_{\xi,\eta}^{(j)}\}$ splits if and only if a Sylow p_j-subgroup of G is a splitting extension of $A^{(j)}$. So, we have proved:

6.3.3 Gaschütz' Theorem. *An extension G of a finite abelian group A by a finite group Γ splits if and only if, for every common prime divisor p of $|A|$ and $|\Gamma|$, a Sylow p-subgroup G_p of G is a splitting extension of $G_p \cap A$.*

6.4 Extensions by Cyclic Groups

Let G denote an extension of an arbitrary group A by a cyclic group $\Gamma = \langle\gamma\rangle$.

If $\langle\gamma\rangle$ has infinite order, then G always splits. For, if r is any element of G not contained in A, then the powers r^k $(k = 0, \pm 1, \pm 2, \ldots)$ form a transversal to A in G. For a fixed choice of r, the group G is uniquely determined by the automorphism $a \to a^\gamma = r^{-1}ar$ of A. If r is replaced by another element rb, $b \in A$, then the corresponding automorphism of A is $a \to (a^\gamma)^b$. Hence, there is a one-to-one correspondence between the classes of equivalent extensions of A by $\langle\gamma\rangle$ and the automorphism classes of A as defined at the end of section 6.1.

Now, suppose that $\langle\gamma\rangle$ is a cyclic group of order n. For any element r of the coset of A that corresponds to γ, we have the decomposition

$$G = A \cup rA \cup r^2A \cup \cdots \cup r^{n-1}A$$

of G into cosets of A. The element $r^n = c$ belongs to A. Again, the whole automorphism system corresponding to this transversal is given by the automorphism $a \to a^\gamma = r^{-1}ar$ of A. We obtain

$$a^{\gamma^n} = r^{-n}\,ar^n = c^{-1}\,ac$$

and

$$c^\gamma = r^{-1}r^n r = c.$$

Thus, for a given representative r, an extension G of A by $\langle\gamma\rangle$ is defined by the automorphism $a \to a^\gamma = r^{-1}ar$ of A and an element c of A, such that

$$a^{\gamma^n} = c^{-1}ac, \quad c^\gamma = c. \tag{6.28}$$

Conversely, suppose that we are given an automorphism $a \to a^\gamma$ of A and an element c of A satisfying the conditions (6.28). We shall see that there exists an extension G of A by $\langle\gamma\rangle$ to which the given automorphism and the element c belong in the sense just described. For $0 \leqq i, j \leqq n-1$, we put

$$c_{\gamma^i,\gamma^j} = \begin{cases} e & \text{if} \quad i+j \leqq n-1, \\ c & \text{if} \quad i+j \geqq n. \end{cases}$$

Moreover, to any element γ^i of $\langle\gamma\rangle$ we assign an automorphism $a \to a^{\gamma^i}$ of A as follows:

$$a^{\gamma^0} = a^\varepsilon = a, \quad a^{\gamma^i} = (a^{\gamma^{i-1}})^\gamma \quad \text{for} \quad i = 1, 2, \ldots, n-1.$$

It is easy to verify that the factor system $\{c_{\gamma^i,\gamma^j}\}$ and the automorphism system $a \to a^{\gamma^i}$ satisfy the conditions (6.4) and (6.5). Consequently, there exists an extension G of A by $\langle\gamma\rangle$ corresponding to this parameter system. If the coset of A corresponding to γ is represented by r, then we have $a^\gamma = r^{-1}ar$, and a simple calculation shows that $r^n = c$.

If r is replaced by rb for some $b \in A$, then $a \to a^\gamma$ is replaced by $a \to (a^\gamma)^b$ and c by

$$(rb)^n = cb^{\gamma^{n-1}}b^{\gamma^{n-2}} \cdots b^\gamma b.$$

In section 6.1, we observed that two extensions of A by Γ may be isomorphic without being equivalent. In the case of an extension by a finite cyclic group Γ, it is easy to discuss a more general isomorphism class. Two extensions G_1 and G_2 of A by Γ are called A-isomorphic if there exists an isomorphism of G_1 onto G_2 that maps A elementwise onto itself. Of course, the factor group G_1/A and G_2/A are isomorphic, but in contrast to the case of equivalent extensions, this isomorphism is arbitrary.

Let G be an extension of A by the cyclic group $\langle\gamma\rangle$ of order n defined by $a \to a^\gamma$ and c. We obtain an A-isomorphic extension if we replace r by $\bar{r} = r^i b$, where $(i, n) = 1$ and b is an arbitrary element of A. Corresponding to this substitution, a^γ is replaced by

$$\bar{r}^{-1} a \bar{r} = (a^{\gamma^i})^b$$

and c by

$$\bar{c} = (r^i b)^n = c^i b^{\gamma^{(n-1)i}} b^{\gamma^{(n-2)i}} \cdots b^{\gamma^i} b.$$

We now give a complete survey of all extensions of a cyclic group $A = \langle a \rangle$ of order m by a cyclic group $\Gamma = \langle \gamma \rangle$ of order n. In this case, we have

$$a^\gamma = a^g, \quad c = a^k,$$

where g and k are certain non-negative integers. Since $c^{-1}ac = a$, the conditions (6.28) are equivalent to

$$g^n \equiv 1 \pmod{m}, \quad kg \equiv k \pmod{m}. \tag{6.29}$$

Thus, every extension of $\langle a \rangle$ by $\langle \gamma \rangle$ is generated by two elements a and r and has the defining relations

$$a^m = e, \quad r^n = a^k, \quad r^{-1}ar = a^g$$

where the non-negative integers k and g satisfy the conditions (6.29). Conversely, any two such integers k and g define an extension of $\langle a \rangle$ by $\langle \gamma \rangle$.

6.5 Partially Complemented Extensions

The conditions that an extension G of A by Γ contains a complement of A reduces the factor system to the trivial case $c_{\xi,\eta} = e$, but it is very restrictive. We shall show that a weaker condition also leads to some reduction of the factor system.

Suppose that G contains a proper subgroup S such that $G = SA$. Such a subgroup S is called a *partial complement* of A in G, and G is called a *partially complemented* extension. Put $C = S \cap A$. Clearly, S is a complement if and only if $C = e$. In any case, C is a normal subgroup of S. We can choose a transversal $\{s_\xi\}$ to A in G whose elements belong to S. However, if $C \neq e$, this choice is not unique. Using the transversal $\{s_\xi\}$, we obtain

$$G = \bigcup_{\xi \in \Gamma} s_\xi A$$

and

$$s_\xi s_\eta = s_{\xi\eta} c_{\xi,\eta} \qquad (c_{\xi,\eta} \in A), \tag{6.30}$$

$$s_\xi^{-1} a s_\xi = a^\xi \quad \text{for} \quad a \in A.$$

Equation (6.30) shows that the elements $c_{\xi,\eta}$ belong to the intersection $C = S \cap A$. Moreover,

$$\Gamma \cong G/A = SA/A \cong S/(S \cap A) = S/C.$$

Thus, S is an extension of C by Γ. The corresponding factor system is $\{c_{\xi,\eta}\}$, and the automorphism system is $c \to c^\xi = s_\xi^{-1} c s_\xi$ for $c \in C$.

Now, C is a proper subgroup of A, so that we have obtained a reasonable reduction, because the elements of the factor system belong to a smaller group.

In order to obtain G, provided that S has been constructed, one has to extend the automorphism system $c \to c^\xi$ of C to an automorphism system $a \to a^\xi$ of A in such a way that

$$a^\xi = s_\xi^{-1} a s_\xi \quad \text{for all} \quad a \in G$$

and

$$(a^\xi)^\eta = (a^{\xi\eta})^{c_{\xi,\eta}}.$$

Exercises

1. Prove that every extension of an arbitrary group by a free group splits.

2. Show that for every prime number p there are precisely two non-isomorphic types of groups of order $2p$.

3. Let $\{c_{\xi,\eta}\}$ be a factor system of an extension of the multiplicative group of all non-zero complex numbers by a finite group Γ of order n such that the automorphism system is the identity. Show that there exists a factor system associate to $\{c_{\xi,\eta}\}$ whose elements are n-th roots of unity.

CHAPTER 7

PERMUTATION GROUPS

7.1 Basic Concepts

A one-to-one mapping of a non-empty set Ω onto itself is called a permutation of Ω. We shall refer to the elements of Ω as *symbols*. If Ω is a finite set consisting of n symbols, then every permutation of Ω is called a permutation of *degree* n. The image of a symbol ξ under a permutation x will be denoted by ξx. We also use the notation

$$x = \begin{pmatrix} \xi \\ \xi x \end{pmatrix} \qquad (\xi \in \Omega). \tag{7.1}$$

We say that ξ is *fixed* under the permutation x if $\xi x = \xi$.

A permutation group G on Ω is a group whose elements are permutations of Ω such that the group operation is the product of permutations as defined in Example 1 in section 1.1, i. e. if x and y are two elements of G, then xy is the permutation obtained by carrying out x and y in succession, first x and then y. So, we have

$$\xi(xy) = (\xi x)\, y \quad \text{for all} \quad \xi \in \Omega.$$

The unit element is the identity permutation, and $\xi x = \eta$ implies that $\eta x^{-1} = \xi$.

The group of all $n!$ permutations of degree n is called the *symmetric group* of degree n and will be denoted by S_n (cf. Example 1 in section 1.1).

In the case of permutations of a finite set Ω, it is often more expedient to use another notation instead of (7.1). Let $\xi_1, \xi_2, \ldots, \xi_l$ be distinct symbols of Ω. By a *cycle*

$$(\xi_1, \xi_2, \ldots, \xi_l)$$

of length l we mean the permutation that carries ξ_1 into ξ_2, ξ_2 into $\xi_3, \ldots, \xi_{l-1}$ into ξ_l, and ξ_l into ξ_1, while all elements of Ω other than $\xi_1, \xi_2, \ldots, \xi_l$, if there are any, remain fixed. Of course,

$$(\xi_1, \xi_2, \ldots, \xi_l) = (\xi_2, \xi_3, \ldots, \xi_l, \xi_1) = \cdots = (\xi_l, \xi_1, \ldots, \xi_{l-1}). \tag{7.2}$$

It is easy to see that

$$(\xi_1, \xi_2, \ldots, \xi_{l-1}, \xi_l)^{-1} = (\xi_l, \xi_{l-1}, \ldots, \xi_2, \xi_1), \quad (\xi_1, \ldots, \xi_l)^l = e,$$

and l is precisely the order of a cycle of length l. Obviously, two cycles commute if they are disjoint, i. e. if there is no symbol that occurs in both. A cycle of length 2 is called a *transposition*.

It is evident that every permutation of a finite set can be written as a product of disjoint cycles; for instance,

$$\begin{pmatrix} 1 & 2 & 3 & 4 & 5 & 6 & 7 & 8 & 9 & 10 & 11 & 12 \\ 8 & 4 & 9 & 5 & 2 & 6 & 3 & 1 & 10 & 7 & 12 & 11 \end{pmatrix} = (2, 4, 5)\,(9, 10, 7, 3)\,(1, 8)\,(11, 12).$$

Clearly, this decomposition is unique except for the order in which the cycles are written. Occasionally, it is convenient to include also the symbols that remain fixed. For this end, we use cycles of length one to denote the fixed elements. Thus, the above permutation can also be written as follows

$$(6)\,(1, 8)\,(11, 12)\,(2, 4, 5)\,(9, 10, 7, 3).$$

Suppose that the permutation x of degree n is expressed as a product of s disjoint cycles including the cycles of length one if there are symbols that remain fixed, and let $l_1, l_2, \ldots, l_s$ be the lengths of the cycles. With a suitable numbering, we may assume that $l_1 \leqq l_2 \leqq \cdots \leqq l_s$. Evidently, $l_1 + l_2 + \cdots + l_s = n$. The sequence $(l_1, l_2, \ldots, l_s)$ is called the *type* of the permutation x. The above permutation is of type $(1, 2, 2, 3, 4)$. Since disjoint cycles commute and a cycle of length l has order l, we conclude that the order of a permutation of type $(l_1, l_2, \ldots, l_s)$ is equal to the least common multiple of $l_1, l_2, \ldots, l_s$. If the decomposition of a permutation of degree n into disjoint cycles contains k_l cycles of length l ($l = 1, 2, \ldots$), where $k_j = 0$ if no cycle of length j occurs, then we obviously have

$$k_1 + 2k_2 + 3k_3 + \cdots = n. \tag{7.3}$$

Thus, the number of distinct types among the permutations of degree n is equal to the number of solutions of (7.3) by non-negative integers k_l.

For two permutations,

$$x = \begin{pmatrix} \xi \\ \xi x \end{pmatrix}, \quad a = \begin{pmatrix} \xi \\ \xi a \end{pmatrix} \qquad (\xi \in \Omega),$$

we get

$$a^{-1}xa = \begin{pmatrix} \xi a \\ \xi \end{pmatrix} \begin{pmatrix} \xi \\ \xi x \end{pmatrix} \begin{pmatrix} \xi \\ \xi a \end{pmatrix} = \begin{pmatrix} \xi a \\ \xi \end{pmatrix} \begin{pmatrix} \xi \\ \xi x \end{pmatrix} \begin{pmatrix} \xi x \\ (\xi x)\,a \end{pmatrix} = \begin{pmatrix} \xi a \\ (\xi x)\,a \end{pmatrix}.$$

This gives the following rule: $a^{-1}xa$ is obtained by replacing each ξ by ξa in both rows of (7.1). If

$$x = (\xi_1, \ldots, \xi_k)(\eta_1, \ldots, \eta_l) \cdots (\zeta_1, \ldots, \zeta_m) \tag{7.4}$$

is the decomposition of x into disjoint cycles, then $a^{-1}xa$ has the decomposition

$$a^{-1}xa = (\xi_1 a, \ldots, \xi_k a)(\eta_1 a, \ldots, \eta_l a) \cdots (\zeta_1 a, \ldots, \zeta_m a).$$

The latter is obtained by replacing each symbol in (7.4) by its image under a. This shows, in particular, that x and $a^{-1}xa$ are of the same type. Conversely, if x and y are two permutations of the same type, then one can easily find a permutation a such that $y = a^{-1}xa$. Indeed, write

$$x = (\xi_1, \ldots, \xi_k)(\eta_1, \ldots, \eta_l) \cdots (\zeta_1, \ldots, \zeta_m),$$

$$y = (\xi_1', \ldots, \xi_k')(\eta_1', \ldots, \eta_l') \cdots (\zeta_1', \ldots, \zeta_m'),$$

then

$$a = \begin{pmatrix} \xi_1, \ldots, \xi_k, & \eta_1, \ldots, \eta_l, & \ldots, & \zeta_1, \ldots, \zeta_m \\ \xi_1', \ldots, \xi_k', & \eta_1', \ldots, \eta_l', & \ldots, & \zeta_1', \ldots, \zeta_m' \end{pmatrix} \tag{7.5}$$

has the required property. This gives the following result:

7.1.1 *Two permutations of the symmetric group S_n are conjugate if and only if they are of the same type. The number of conjugacy classes in S_n is equal to the number of solutions of the equation*

$$k_1 + 2k_2 + 3k_3 + \cdots = n$$

by non-negative integers k_i.

By taking $y = x$, the above method yields the normalizer of x in S_n. All permutations a such that $a^{-1}xa = x$ can be found as follows. Write a in the form (7.5). The upper row corresponds to the decomposition of x into disjoint cycles arranged in such a way that their lengths do not decrease. The lower row consists of the same cycles, their order being also subject to the condition that the lengths do not decrease; but, for any fixed length l, the cycles of length l may be arranged in an arbitrary order; moreover, each cycle of length l may be written in l different ways as indicated by (7.2). Suppose that x contains k_l cycles of length l. Combining the substitutions just described, we obtain $k_l!\,l^{k_l}$ distinct ways of writing the cycles of length l in the lower row of (7.5). So we have proved:

7.1.2 *Let x be a permutation of degree n whose decomposition into disjoint cycles contains k_l cycles of length l ($l = 1, 2, \ldots$). Then, the order of the normalizer of x in S_n is equal to*

$$k_1!\,1^{k_1} k_2!\,2^{k_2} k_3!\,3^{k_3} \cdots$$

Now, let Ω be an arbitrary non-empty set and G a group of permutations of Ω. As we saw in section 3.1, we obtain a partition of Ω into orbits of G. If there is only a single orbit, i. e. if any two elements of Ω are connected under G, then G is called *transitive*. G is said to be *intransitive* if there are at least two orbits.

Suppose that G is intransitive and let Γ be an orbit of G. Then, to every permutation x in G, there corresponds a permutation $x^{(\Gamma)}$ of the set Γ. The mapping $x \to x^{(\Gamma)}$ is a homomorphism of G onto a group $G^{(\Gamma)}$ of permutations of Γ, which is obviously transitive on Γ. The groups $G^{(\Gamma)}$ are called the *transitive constituents* of G.

Suppose that G is transitive. For a given symbol α of Ω, let G_α denote the *stabilizer subgroup* of α, i. e. the subgroup of all those permutations in G that leave α fixed. Of course, it may happen that $G_\alpha = e$. For every $\xi \in \Omega$, we choose a permutation a_ξ in G that carries α into ξ. Such permutations a_ξ exist, because G is transitive. It is easy to see that the system $\{a_\xi\}$ is a right transversal to G_α in G, i. e.

$$G = \bigcup_{\xi \in \Omega} G_\alpha a_\xi . \tag{7.6}$$

Indeed, the coset $G_\alpha a_\xi$ consists precisely of those permutations of G that carry α into ξ. For, if x belongs to G_α, then $\alpha(x a_\xi) = (\alpha x)\, a_\xi = \alpha a_\xi = \xi$. Conversely, if $\alpha y = \xi$, then $\alpha y a_\xi^{-1} = \alpha$; therefore, $y a_\xi^{-1} \in G_\alpha$ or $y \in G_\alpha a_\xi$.

It is easily verified that the stabilizer subgroup G_β of any other symbol β is conjugate to G_α, namely

$$G_\beta = a_\beta^{-1} G_\alpha a_\beta . \tag{7.7}$$

It follows from Theorem 3.1.2 that the order of a permutation group of finite degree is divisible by the least common multiple of the numbers of elements in its orbits. In particular, the order of a transitive permutation group is divisible by its degree.

A transitive permutation group G is called *regular* if $G_\alpha = e$ for some $\alpha \in \Omega$. If G is regular, it follows from (7.7) that $G_\beta = e$ for all $\beta \in \Omega$.

If ξ and η are any two symbols and G is transitive, then G contains at least one element x such that $\xi x = \eta$. If G is regular, x is the only element that carries ξ into η. For suppose that $\xi y = \eta$. Then, we have $\xi x = \xi y$; therefore, $\xi x y^{-1} = \xi$, so that $x y^{-1} \in G_\xi$. But this implies $x = y$, because $G_\xi = e$. So, we have:

7.1.3 *A permutation group is regular if and only if for any two symbols ξ and η it contains one and only one permutation that carries ξ into η.*

From (7.6), we obtain:

7.1.4 *A transitive permutation group of finite degree is regular if and only if its order is equal to its degree.*

Suppose that G is abelian. Then, (7.7) shows that $G_\alpha = G_\beta$ for any two symbols α, β. Consequently, G_α leaves all symbols fixed, i. e. $G_\alpha = e$. Hence:

7.1.5 *Every transitive abelian permutation group is regular.*

A permutation r of finite degree n is said to be *regular* if any power r^k $(k = 1, 2, \ldots)$ is either equal to the identity permutation or leaves no symbol fixed. If r is regular and $r \neq e$, then every symbol occurs in some cycle of the decomposition of r into disjoint cycles, and all cycles are of the same length. For, assume that among the disjoint cycles there is one of length l and another of length m, where $l < m$; then, r^l would be distinct from e, but would leave at least l symbols fixed. Thus, a regular permutation of degree n is of type $(l, l, \ldots, l)$, where the number of cycles is nl^{-1}. Conversely, every product of k disjoint cycles of length l is obviously regular provided that its degree is kl.

It is evident that every regular permutation group consists only of regular permutations. If all the permutations in a group G of degree n are regular, then G contains at most one permutation that carries α into any given symbol β. For, suppose that $a, b \in G$ are such that $\alpha a = \alpha b = \beta$; then, ab^{-1} is regular and leaves α fixed so that $ab^{-1} = e$. This shows that the order of G cannot exceed n. In case G is transitive, its order is divisible by n; hence $|G| = n$. Therefore, G is regular. If G is not transitive, then its transitive constituents are regular and isomorphic to G. Hence, in this case, $|G|$ is a proper divisor of n. So, we obtain:

7.1.6 *If all the permutations in a group G of degree n are regular, then $|G|$ divides n. If G is transitive, then it is regular.*

The next theorem deals with the centre of transitive groups.

7.1.7 *The centre of every transitive permutation group of degree n consists only of regular permutations. Hence, the order of the centre is a divisor of n.*

Proof. Let z be an element of the centre $\mathbf{Z}(G)$ of a transitive permutation group G of degree n. Suppose that z^k leaves α fixed, i. e. $z^k \in G_\alpha$. Since all stabilizer subgroups G_β are conjugate to G_α and $z^k \in \mathbf{Z}(G)$, it follows that $z^k \in G_\beta$ for all symbols β; hence, $z^k = e$. This proves the first part of our assertion. The second part then follows from Theorem 7.1.6.

Let c be any permutation that commutes with every element of a transitive group G of degree n. Then, c is contained in the centre of $\langle G, c\rangle$. Hence, it follows from Theorem 7.1.7 that c is regular. Combining this result with Theorem 7.1.6, we have:

7.1.8 *Let G be a transitive subgroup of a permutation group F of degree n. Then, the centralizer of G in F consists of regular permutations only and its order divides n.*

Let G be a transitive permutation group on Ω and G_α the stabilizer subgroup of the symbol α. Suppose that G contains a subgroup U such that $G_\alpha \subset U \subset G$. We emphasize that the conditions $G_\alpha \neq U$ and $U \neq G$ are crucial for our discussion. Let

$$U = \bigcup_{\varrho \in \Lambda} G_\alpha a_\varrho$$

be the decomposition of U into right cosets of G_α. Here, ϱ ranges over a proper subset Λ of Ω, which consists precisely of those symbols into which α is carried by the permutations in U; and a_ϱ is a permutation that maps α onto $\varrho \in \Lambda$. Since G_α is a proper subgroup of U, the set Λ contains more than one symbol. The group U permutes the symbols of Λ among themselves and is transitive on Λ.

For an arbitrary permutation z of G, let Λz denote the set of all symbols λz, where $\lambda \in \Lambda$. If x is not contained in U, then the sets Λ and Λx are disjoint. For, suppose the contrary that $\xi \in \Lambda \cap \Lambda x$. Then, we have $\xi = \varrho x$ for some symbol $\varrho \in \Lambda$; hence, $\xi = \alpha a_\xi = \alpha a_\varrho x$. Hence, $a_\varrho x a_\xi^{-1} \in G_\alpha$, so that x belongs to the complex $a_\varrho^{-1} G_\alpha a_\xi$, which is a subset of U. But this contradicts the assumption $x \notin U$. This result can be generalized as follows: If Ux and Uy are two distinct right cosets of U, then the sets Λx and Λy are disjoint. For, if the intersection $\Lambda x \cap \Lambda y$ were not empty, then $\Lambda \cap \Lambda y x^{-1}$ too would be non-empty so that by our previous result $yx^{-1} \in U$, which contradicts the assumption $Ux \neq Uy$. If

$$G = U \cup Ux \cup Uy \cup \cdots$$

is the decomposition of G into right cosets of U, then our last result shows that $\Lambda, \Lambda x, \Lambda y, \ldots$ are all the distinct sets of the form Λa for $a \in G$. All these sets are images of Λ under certain permutations, so that all of them contain the same number of symbols. Moreover, $\Lambda \cup \Lambda x \cup \Lambda y \cup \cdots = \Omega$.

These arguments show that the existence of a subgroup U with $G_\alpha \subset U \subset G$ implies the following particular property of the permutations of G: The set Ω is partitioned into at least two disjoint subsets $\Lambda, \Lambda x, \Lambda y, \ldots$ All these subsets contain the same number of symbols, and this number is greater than one. If a permutation a of G carries any symbol of an arbitrary subset Λx into a symbol of the same subset, then a maps the whole subset Λx onto itself. But, if a carries a symbol of Λx into a symbol of another subset Λy, then a maps the whole subset Λx onto Λy. We may say briefly, that the subsets $\Lambda, \Lambda x, \Lambda y, \ldots$ are not torn apart under the permutations of G. Such a permutation group G is said to be *imprimitive*, and the subsets $\Lambda, \Lambda x, \Lambda y, \ldots$ are called the *systems of imprimitivity*. If G is a transitive group for which no such partition of Ω is possible, G is said to be *primitive*. It follows that, in the case of a primitive group G, all the stabilizer subgroups G_α are maximal.

The converse of the last assertion is also true: Suppose that the transitive permutation group G is imprimitive and that G_α is an arbitrary stabilizer subgroup of G. Then, G contains a subgroup U such that $G_\alpha \subset U \subset G$. For, let Λ be the system of imprimitivity that contains α, and U the largest subgroup of G that maps Λ onto itself. Since Λ contains at least two symbols, G_α is a proper subgroup of U; and since there are at least two systems of imprimitivity, U is a proper subgroup of G. So, we have the following theorem:

7.1.9 *A transitive permutation group G is primitive if and only if all the stabilizer subgroups G_α are maximal.*

Each element of an imprimitive group G induces a permutation on the systems of imprimitivity. Thus, there is a homomorphism of G onto a transitive group of permutations of the systems of imprimitivity. In the above notation, the kernel of this homomorphism is the intersection of all subgroups conjugate to U.

7.1.10 *A normal subgroup $T \neq e$ of a primitive group G is transitive.*

Proof. Suppose that T is not transitive. Then, there exists an orbit Λ of T that contains more than one element and is a proper subset of Ω. The stabilizer subgroup G_α of an element α of Λ is a proper subgroup of $G_\alpha T$, and $G_\alpha T$ is a proper subgroup of G; but this contradicts the assumption that G is primitive.

The next theorem dates from the early stages of group theory.

7.1.11 Galois' Theorem. *The degree of a primitive soluble permutation group G is a prime power p^m. The group G contains a single minimal normal subgroup M. The order of M is p^m. If the degree of G is a prime number, then G is at most two-step metabelian.*

Proof. It follows from Theorem 4.2.2 that every minimal normal subgroup of G is elementary abelian so that its order is a power of a prime number. Let M be a minimal normal subgroup of G and $|M| = p^m$, where p is a prime number. By Theorem 7.1.10, M is transitive; hence, by Theorem 7.1.5, M is regular. Thus, the degree of G is p^m. Suppose that G contains another minimal normal subgroup M_1. Then, we have $M \cap M_1 = e$; hence, $MM_1 = M \times M_1$. Consequently- $M \times M_1$ is abelian and hence regular. This leads to the contradiction $|M \times M_1| = |M|$. By Theorem 7.1.8, the centralizer of M in G coincides with M. Thus, G/M is isomorphic to a group of automorphisms of M. If $|M| = p$, it follows from Theorem 5.4.3 that G/M is abelian. Hence, G is at most two-step metabelian. This completes the proof.

A permutation group G is said to be k-ply transitive if, for any two ordered sets $\{\alpha_1, \ldots, \alpha_k\}$, $\{\beta_1, \ldots, \beta_k\}$ containing k symbols each, there is at least one permutation in G which carries α_i into β_i, $i = 1, \ldots, k$. For G to be k-ply transitive, it is sufficient that a fixed ordered set $\{\alpha_1, \ldots, \alpha_k\}$ is mapped onto every ordered set $\{\beta_1, \ldots, \beta_k\}$. Thus, transitive groups in the sense of our previous definition are at least simply transitive.

If a group is at least doubly transitive, then it is primitive. For, if G were imprimitive, then we could choose two symbols α_1, α_2 of the same system of imprimitivity and a symbol β of another. Then, G would contain a permutation carrying α_1 into α_1 and α_2 into β, which contradicts the imprimitivity of G.

If G is k-ply transitive, then the stabilizer subgroup G_α of any symbol α is obviously $(k-1)$-ply transitive. Conversely, if G is transitive and G_α $(k-1)$-ply transitive, then G is k-ply transitive.

7.2 Representations of Groups by Permutations

Let G be an abstract group. A *permutational representation* of degree n of G is defined as a homomorphism of G into the symmetric group S_n or the image of G under such a homomorphism. If the homomorphism is an isomorphism, then the representation is said to be *faithful*. In this section, we shall describe all permutational representations of a given group G. Evidently, we may confine ourselves to representations by transitive permutation groups.

We start with a special case.

7.2.1 *Every group is isomorphic to a regular permutation group.*

Proof. Let G be a given group. We shall construct a regular permutation group isomorphic to G. The symbols on which this permutation group operates are the elements of G. To an arbitrary element a of G, we assign the permutation

$$\varrho_a = \begin{pmatrix} x \\ xa \end{pmatrix} \qquad (x \in G)$$

of the elements of G. Then, we have

$$\varrho_a \varrho_b = \begin{pmatrix} x \\ xa \end{pmatrix} \begin{pmatrix} x \\ xb \end{pmatrix} = \begin{pmatrix} x \\ xa \end{pmatrix} \begin{pmatrix} xa \\ (xa)\,b \end{pmatrix} = \begin{pmatrix} x \\ x(ab) \end{pmatrix} = \varrho_{ab}\,.$$

Thus, the mapping $a \to \varrho_a$ $(a \in G)$ is a homomorphism, and being one-to-one it is an isomorphism of G onto the group G^* of the permutations ϱ_a. Clearly, G^* is transitive. Moreover, G^* is regular, for $xa = x$ for any $x \in G$ implies that $a = e$.

This completes the proof.

The group G^* is called the *regular permutational representation* of G or, more precisely, the right regular representation.

We now determine the centralizer $^*G$ of G^* in the group of all permutations of the elements of G. A permutation

$$\sigma = \begin{pmatrix} x \\ x' \end{pmatrix} \qquad (x, x' \in G)$$

in $^*G$ satisfies the relation $\sigma\varrho_a = \varrho_a\sigma$ for all $\varrho_a \in G^*$, i. e.

$$\begin{pmatrix} x \\ x' \end{pmatrix}\begin{pmatrix} x \\ xa \end{pmatrix} = \begin{pmatrix} x \\ xa \end{pmatrix}\begin{pmatrix} x \\ x' \end{pmatrix}$$

or

$$\begin{pmatrix} x \\ x' \end{pmatrix}\begin{pmatrix} x' \\ x'a \end{pmatrix} = \begin{pmatrix} x \\ xa \end{pmatrix}\begin{pmatrix} xa \\ (xa)' \end{pmatrix}.$$

This means that $x'a = (xa)'$ for all $a, x \in G$. Taking $a = x^{-1}$, we obtain $(xx^{-1})' = e' = b$ for all $x \in G$. This gives $x'x^{-1} = b$, so that $x' = bx$ for all $x \in G$. Thus, the group $^*G$ consists of permutations of the form

$$\sigma = \begin{pmatrix} x \\ bx \end{pmatrix}.$$

It is easily verified that all permutations of this form commute with every ϱ_a. Thus, $^*G$ consists of all such permutations. Putting

$$\sigma_a = \begin{pmatrix} x \\ a^{-1}x \end{pmatrix} \qquad (x \in G)$$

we obtain $\sigma_a\sigma_b = \sigma_{ab}$. Consequently, $^*G$ is regular and isomorphic to G. The group $^*G$ is called the *left regular permutational representation* of G. A simple calculation shows that, conversely, G^* is the centralizer of $^*G$ in the symmetric group on the elements of G.

It is easy to construct other transitive permutational representations of G. Let H be a proper subgroup of G and

$$G = \bigcup_{r \in R} Hr$$

the decomposition of G into right cosets of H. To the element a of G, we assign the permutation

$$\gamma_a = \begin{pmatrix} Hr \\ Hra \end{pmatrix} \qquad (r \in R)$$

of the cosets of H. Then, we have

$$\gamma_a\gamma_b = \begin{pmatrix} Hr \\ Hra \end{pmatrix}\begin{pmatrix} Hr \\ Hrb \end{pmatrix} = \begin{pmatrix} Hr \\ Hra \end{pmatrix}\begin{pmatrix} Hra \\ Hrab \end{pmatrix} = \gamma_{ab},$$

so that the permutations $\gamma_a (a \in G)$ form a permutational representation Γ of G. Since the coset H is carried into all other cosets Ha, the permutation group Γ is transitive. We say that the representation Γ is *induced* by H. For $H = e$, we obtain G^*.

Let us determine the kernel of the homomorphism of G onto Γ. From $Hra = Hr$ for all $r \in R$, it follows that $r^{-1}Hra = r^{-1}Hr$, which means $a \in r^{-1}Hr$ for all $r \in R$. Conversely, if a is contained in the intersection of all subgroups conjugate to H, then γ_a is the identity permutation. This shows that

$$D = \bigcap_{r \in R} r^{-1}Hr$$

is the kernel of the homomorphism in question. The group D can be characterized as the maximal normal subgroup of G that is contained in H.

We now prove that every transitive permutational representation of G can be obtained in the way just described. Let Γ be a transitive permutational representation of G. Let Ω denote the set of symbols on which Γ operates. We choose an arbitrary symbol α in Ω and denote by Γ_α the stabilizer subgroup of α. Let H be the subgroup of G that corresponds to Γ_α under the homomorphism of G onto Γ. We shall see that Γ coincides with the permutational representation of G induced by H, provided that the symbols in Ω and the cosets of H are suitably matched.

For every $\xi \in \Omega$, we choose a permutation π_ξ in Γ that carries α into ξ. This is possible, because Γ is transitive. The set $\{\pi_\xi\}$ for all $\xi \in \Omega$ is a transversal to Γ_α in Γ so that

$$\Gamma = \bigcup_{\xi \in \Omega} \Gamma_\alpha \pi_\xi$$

is the decomposition of Γ into cosets of Γ_α. In G, we choose a system $\{r_\xi\}$ of pre-images of $\{\pi_\xi\}$ with respect to the homomorphism of G onto Γ. Then, $\{r_\xi\}$ is a transversal to H in G:

$$G = \bigcup_{\xi \in \Omega} Hr_\xi .$$

Now, it turns out that in the correspondence $\xi \leftrightarrow Hr_\xi$ the symbols in Ω are permuted under Γ in the same way as the cosets of H under the permutational representation induced by H. Let g be an arbitrary element of G and let γ denote its image under the homomorphism of G onto Γ. Suppose that γ carries ξ into η. In the permutational representation of G induced by H, the permutation that corresponds to g carries Hr_ξ into $Hr_\xi g$. The corresponding cosets in Γ are $\Gamma_\alpha \pi_\xi$ and $\Gamma_\alpha \pi_\xi \gamma$. The permutations in the coset $\Gamma_\alpha \pi_\xi \gamma$ carry α into η, so that $\Gamma_\alpha \pi_\xi \gamma = \Gamma_\alpha \pi_\eta$. Consequently, $Hr_\xi g = Hr_\eta$.

Summarizing our results we obtain:

7.2.2 *Every proper subgroup H of G induces a transitive permutational representation of G by assigning to any element a of G thepermutation*

$$\begin{pmatrix} Hr \\ Hra \end{pmatrix}$$

of the right cosets of H. All transitive permutational representations of G can be obtained in this way.

If the symbol α that we used in the last proof is replaced by β, then the new stabilizer subgroup Γ_β is conjugate to Γ_α. Consequently, the pre-image K of Γ_β in G is conjugate to H. Conversely, if the subgroup K of G is conjugate to H, then its image in Γ is conjugate to Γ_α so that it is the stabilizer subgroup of some $\beta \in \Omega$. This gives the following result: If H and K are conjugate in G, and only in this case, a one-to-one correspondence can be established between the cosets of H and K, respectively, and the symbols of Ω such that the permutations on these sets are the same. In other words: Precisely if H and K are conjugate subgroups of G, multiplication of the cosets of H and K, respectively, by elements of G induces the same permutation on both cosets.

7.3 Amalgams

Let A, B, H be three groups such that A and B contain subgroups H_A and H_B, respectively, both isomorphic to H. We assume that isomorphisms of H_A and H_B onto H are given and kept fixed in what follows. Such a system A, B, H is called an *amalgam*. We raise the question whether an amalgam can be embedded in a group, i. e. we ask whether there exists a group G with the following properties:

(a) G contains subgroups $\bar{A}$ and $\bar{B}$ such that $\bar{A} \cong A$, $\bar{B} \cong B$ and $G = \langle \bar{A}, \bar{B} \rangle$;

(b) there is an isomorphism α of A onto $\bar{A}$ and an isomorphism β of B onto $\bar{B}$ such that the images of H_A under α and of H_B under β coincide and are equal to the intersection $\bar{A} \cap \bar{B}$.

To put it in a less accurate but perhaps more suggestive way we ask whether A and B can be embedded in a group G such that their intersection is H. The condition $G = \langle \bar{A}, \bar{B} \rangle$ is not essential for our problem. For, if we have constructed a group G_1 that has the required property and is larger than $\langle \bar{A}, \bar{B} \rangle$, then the subgroup $G = \langle \bar{A}, \bar{B} \rangle$ of G_1 is also a solution of our problem.

We shall prove that our embedding problem always has at least one solution.

To simplify the notation, we assume that H is a common subgroup of A and B. We choose a left transversal S to H in A. Then, every $a \in A$ has a unique representation in the form

$$a = sh, \quad s \in S, \quad h \in H.$$

We write

$$s = a^{\sigma}, \quad h = a^{-\sigma+1}.$$

We choose a left transversal T to H in B, so that every $b \in B$ has a unique decomposition

$$b = th, \quad t \in T, \quad h \in H$$

whose factors are similarly denoted by

$$t = b^{\tau}, \quad h = b^{-\tau+1}.$$

Note that, in general, the functions σ, $-\sigma+1$, τ, $-\tau+1$ are not homomorphisms.

Let Ω denote the set of all triplets

$$(s, t, h), \quad s \in S, \quad t \in T, \quad h \in H.$$

We define mappings of Ω into itself that will soon turn out to be permutations. For any $a \in A$, we define a mapping $\varrho(a)$ of Ω into itself as follows:

$$(s, t, h)^{\varrho(a)} = (s', t', h'), \quad \text{where } s'h' = sha, \ t' = t.$$

Using the functions defined above we can write this in the form

$$(s, t, h)^{\varrho(a)} = ((sha)^{\sigma}, t, (sha)^{-\sigma+1}).$$

This mapping can be described as follows: For a fixed element t, there is a one-to-one correspondence between the triplets (s, t, h) and the elements sh of A. The mapping $\varrho(a)$ carries the triplet corresponding to sh into the triplet corresponding to sha. This shows that, on the triplets with a fixed t, $\varrho(a)$ induces the same permutation as a in the right regular permutational representation of A.

Similarly, for any $b \in B$, we define a mapping $\varrho(b)$ of Ω into itself:

$$(s, t, h)^{\varrho(b)} = (s, (thb)^{\tau}, (thb)^{-\tau+1}).$$

As before, on the triplets with a fixed element s, the $\varrho(b)$ induce permutations that correspond to the right regular permutational representation of B.

For an element h^* of H, we have

$$(shh^*)^{\sigma} = s, \quad (thh^*)^{\tau} = t;$$

hence,

$$(s, t, h)^{\varrho(h^*)} = (s, t, hh^*).$$

Thus, for the mappings $\varrho(h^*)$, it is irrelevant whether h^* is regarded as an element of A or of B.

From $(sha)^\sigma\,(sha)^{-\sigma+1} = sha$, it follows that, for $a, a' \in A$,

$$(s,\ t,\ h)^{\varrho(a)\,\varrho(a')} = ((sha)^\sigma,\ t,\ (sha)^{-\sigma+1})^{\varrho(a')} = ((shaa')^\sigma,\ t,\ (shaa')^{-\sigma+1}) = (s,\ t,\ h)^{\varrho(aa')}.$$

Since this holds for every triplet in Ω, we have

$$\varrho(a)\ \varrho(a') = \varrho(aa').$$

This shows that the mapping $a \to \varrho(a)$ is a homomorphism of A onto the group $\varrho(A)$ of all $\varrho(a)$, $a \in A$. Moreover, the mapping $a \to \varrho(a)$ is an isomorphism of A onto $\varrho(A)$. For, suppose that $\varrho(a)$ leaves (s, t, h) fixed; then,

$$(sha)^\sigma = s, \quad (sha)^{-\sigma+1} = h$$

so that $sha = sh$; hence $a = e$. In the same way, it follows that the set $\varrho(B)$ of all $\varrho(b)$, $b \in B$, is a group of permutations of Ω isomorphic to B.

A solution of our embedding problem is the subgroup G generated by $\varrho(A)$ and $\varrho(B)$ in the symmetric group on Ω. Clearly, (a) is satisfied, because G is generated by its subgroups $\varrho(A)$ and $\varrho(B)$, which are isomorphic to A and B, respectively. It remains to show that G also satisfies (b). As we saw above, for $h \in H$, the permutation $\varrho(h)$ is contained in $\varrho(A) \cap \varrho(B)$. Suppose that, on the other hand, $\varrho(a) \in \varrho(B)$ for some $a \in A$. Then, $\varrho(a)$ does not change the first element s in any triplet (s, t, h), i. e. $(sha)^\sigma = s$; it follows that $(sha)^{-\sigma+1} = ha$ belongs to h; hence, $a \in H$. So, we obtain $\varrho(A) \cap \varrho(B) = \varrho(H)$. Consequently, (b) is satisfied. This gives the following result:

7.3.1 *Every amalgam can be embedded in a group.*

Note that our construction makes use of the transversals S and T. A closer investigation shows that, in general, the structure of G essentially depends on the choice of these transversals (cf. [50]).

Let us discuss two special cases in which the structure of G does not depend on the choice of one or of both transversals.

We first assume that H lies in the centre of A. We shall prove that in this case the group G does not depend on the choice of the transversal T to H in B. If H belongs to the centre of A, then

$$(sha)^\sigma = (sah)^\sigma = (sa)^\sigma, \quad (sha)^{-\sigma+1} = (sa)^{-\sigma+1}\,h. \tag{7.8}$$

Now, let T and T' be two left transversals to H in B. If $b \in B$ has the decomposition

$$b = th = t'h'; \quad t \in T, \quad t' \in T'; \quad h, h' \in H,$$

we write

$$t' = b^{\tau'}, \quad h' = b^{-\tau'+1},$$

and denote the set of all triplets (s, t', h), $t' \in T'$ by Ω'. Let the permutations $\varrho(a)$, $\varrho(b)$ be defined as above; in the symmetric group on Ω, they generate the subgroup G. Similarly, we define permutations $\varrho'(a)$, $\varrho'(b)$ of Ω' and denote by G' the subgroup generated by the $\varrho'(a)$ and $\varrho'(b)$ in the symmetric group on Ω'.

We now define a mapping φ of Ω into Ω':

$$(s, t, h)^{\varphi} = (s, (th)^{\tau'}, \quad (th)^{-\tau'+1}).$$

This mapping can also be described as follows:

$$(s, t, h)^{\varphi} = (s', t', h')$$

if and only if $s = s'$, $th = t'h'$. This shows that φ is a one-to-one mapping of Ω onto Ω'. The inverse mapping φ^{-1} is defined by

$$(s, t', h')^{\varphi^{-1}} = (s, (t'h')^{\tau}, (t'h')^{-\tau+1}).$$

We compare the permutations $\varphi^{-1}\varrho(a)\,\varphi$ and $\varrho'(a)$ of Ω'.

As a consequence of

$$t(sa)^{-\sigma+1}h = th(sa)^{-\sigma+1} = t'h'(sa)^{-\sigma+1} = t'(sa)^{-\sigma+1}h'$$

and of (7.8), we obtain

$$\begin{aligned}(s,' t', h')^{\varphi^{-1}\varrho(a)\varphi} &= (s, t, h)^{\varrho(a)\varphi} = ((sha)^{\sigma}, t, (sha)^{-\sigma+1})^{\varphi} \\ &= ((sa)^{\sigma}, t, (sa)^{-\sigma+1}h)^{\varphi} = ((sa)^{\sigma}, t', (sa)^{-\sigma+1}h').\end{aligned}$$

On the other hand, $s' = s$ and (7.8) show that

$$(s', t', h')^{\varrho'(a)} = ((s'h'a)^{\sigma}, t', (s'h'a)^{-\sigma+1}) = ((s'a)^{\sigma}, t', (s'a)^{-\sigma+1}h') = (s', t', h')^{\varphi^{-1}\varrho(a)\varphi}.$$

Since this holds for each triplet (s', t', h'), we have

$$\varphi^{-1}\varrho(a)\,\varphi = \varrho'(a). \tag{7.9}$$

Comparing $\varphi^{-1}\varrho(b)\,\varphi$ and $\varrho'(b)$, we obtain

$$\begin{aligned}(s', t', h')^{\varphi^{-1}\varrho(b)\varphi} &= (s, t, h)^{\varrho(b)\varphi} = (s, (thb)^{\tau}, (thb)^{-\tau+1})^{\varphi} \\ &= (s, (thb)^{\tau'}, (thb)^{-\tau'+1}) = (s', (t'h'b)^{\tau'}, (t'h'b)^{-\tau'+1}) = (s', t', h')^{\varrho'(b)};\end{aligned}$$

hence,

$$\varphi^{-1}\varrho(b)\,\varphi = \varrho'(b). \tag{7.10}$$

It follows from (7.9) and (7.10) that G and G' are isomorphic.

If H is contained in the centre of A as well as of B, it follows that G is independent of the choice of both transversals S and T, i. e. G is uniquely determined by A, B, H. In this case, every $\varrho(a)$ commutes with every $\varrho(b)$. For, in addition to (7.8), we now have

$$(thb)^\tau = (tbh)^\tau = (tb)^\tau, \ (thb)^{-\tau+1} = (tb)^{-\tau+1} h.$$

This gives

$$\begin{aligned}(s, t, h)^{\varrho(a)\varrho(b)} &= ((sha)^\sigma, t, (sha)^{-\sigma+1})^{\varrho(b)} = ((sa)^\sigma, t, (sa)^{-\sigma+1} h)^{\varrho(b)} \\ &= ((sa)^\sigma, (t(sa)^{-\sigma+1} hb)^\tau, (t(sa)^{-\sigma+1}hb)^{-\tau+1}) \\ &= ((sa)^\sigma, (tb)^\tau, (tb)^{-\tau+1} (sa)^{-\sigma+1} h) \\ &= ((sa)^\sigma, (tb)^\tau, (sa)^{-\sigma+1} (tb)^{-\tau+1} h) = (s, t, h)^{\varrho(b)\varrho(a)},\end{aligned}$$

so that $\varrho(a)\ \varrho(b) = \varrho(b)\ \varrho(a)$. In this case, the group G is called the direct product of A and B with the amalgamated subgroup H.

Let E be a property of groups such that every subgroup of G inherits the property E from G. Suppose that A and B have the property E. Then, one can ask whether an amalgam A, B, H can be embedded in a group G with the property E. If E is the property of being finite or abelian, then our arguments above answer the question in the affirmative. A similar result can be proved if E is the property of being soluble. For the proof and further investigations on the embedding of amalgams, we refer the reader to [50].

7.4 The Symmetric and Alternating Groups of Finite Degree

In this section, it will be convenient to take for the set Ω whose permutations are studied the set of the natural numbers $1, 2, \ldots, n$.

Every cycle can be written as a product of transpositions, viz.

$$(\alpha_1, \alpha_2, \ldots, \alpha_r) = (\alpha_1, \alpha_2)(\alpha_1, \alpha_3) \cdots (\alpha_1, \alpha_r). \tag{7.11}$$

Now, every permutation is a product of cycles; hence it can be represented as a product of transpositions. This representation is not unique but, as we shall see later, in all representations of a given permutation as a product of transpositions their number is always even or always odd.

All transpositions generate the whole symmetric group S_n, but only a small part of them is really needed. For example, the $n - 1$ transpositions

$$(1, 2), (1, 3), \ldots, (1, n) \tag{7.12}$$

are sufficient to generate S_n. For, we have $(1, \alpha)(1, \beta)(1, \alpha) = (\alpha, \beta)$. The transpositions

$$(1, 2), (2, 3), \ldots, (n-1, n) \tag{7.13}$$

form another system of generators of S. For, we have

$$(1, 2)(2, 3)(1, 2) = (1, 3),$$
$$(1, 3)(3, 4)(1, 3) = (1, 4),$$
$$\cdots\cdots\cdots\cdots\cdots$$

so that all the transpositions (7.12) can be expressed by the system (7.13)

The group S_n can even be generated by two suitably chosen elements, for instance, by

$$a = (1, 2), \qquad b = (1, 2, \ldots, n).$$

It is readily verified that

$$b^{-1}ab = (2,3),\ b^{-2}ab^2 = (3,4),\ \ldots,\ b^{-(n-2)}ab^{n-2} = (n-1, n).$$

Thus, the system (7.13) of generators can be expressed by a and b.

By applying the permutation

$$a = \begin{pmatrix} 1 & 2 & \cdots & n \\ \alpha_1 & \alpha_2 & \cdots & \alpha_n \end{pmatrix}$$

to the product

$$\Delta = \prod_{\substack{i,k=1 \\ i>k}}^{n} (i - k),$$

we obtain

$$\Delta^a = \prod_{\substack{i,k=1 \\ i>k}}^{n} (\alpha_i - \alpha_k).$$

It is clear that $\Delta^a = \Delta$ or $\Delta^a = -\Delta$, because except for the sign the factors in both products are all the differences between distinct members in the sequence $1, \ldots, n$. We put

$$\Delta^a = \chi(a)\,\Delta$$

where $\chi(a) = 1$ or -1. There exist permutations a for which $\chi(a) = -1$, for example $a = (n-1, n)$. The number $\chi(a)$ is called the *character* of the permutation a. If b is another permutation, we have

$$\Delta^{ab} = (\Delta^a)^b = \chi(a)\Delta^b = \chi(a)\,\chi(b)\Delta$$

and on the other hand

$$\Delta^{ab} = \chi(ab)\,\Delta.$$

Therefore,

$$\chi(a)\,\chi(b) = \chi(ab). \tag{7.14}$$

Thus, the mapping $a \to \chi(a)$ is a homomorphism of S_n onto the cyclic group of order 2 consisting of 1 and -1. The kernel A_n of this homomorphism is called the *alternating group* of degree n. Clearly, A_n is a normal subgroup of index 2 in S_n.

As we remarked above, the character of $(n-1, n)$ is equal to -1. Since in S_n all transpositions are conjugate to each other, it follows from

$$\chi(b^{-1}ab) = \chi(b^{-1})\,\chi(a)\,\chi(b) = \chi(a)$$

that the character of every transposition is equal to -1. This means that all the transpositions do not belong to A_n, but to the other coset of A_n in S_n. Moreover, in every representation of a permutation a as a product of transpositions, the number of transpositions is always either even or odd according as $\chi(a) = 1$ or $\chi(a) = -1$. Therefore, the permutations a for which $\chi(a) = 1$, i. e. the permutations in A_n, are called *even* permutations. The permutations in the other coset of A_n are said to be *odd* permutations.

From (7.11), we conclude that a cycle of length r is even or odd according as r is odd or even.

Let

$$a = \begin{pmatrix} 1 & 2 & \cdots n \\ \alpha_1 & \alpha_2 & \cdots \alpha_n \end{pmatrix}$$

be a given permutation. Two numbers α_i, α_k in the sequence $\alpha_1, \alpha_2, \ldots, \alpha_n$ are said to form an *inversion* if $i < k$ and $\alpha_i > \alpha_k$. If $J(a)$ denotes the total number of inversions in the sequence $\alpha_1, \alpha_2, \ldots, \alpha_n$, then we have

$$\chi(a) = (-1)^{J(a)}.$$

For, by carrying out $J(a)$ transpositions of adjacent numbers one can remove all inversions, so that we arrive at the sequence $1, 2, \ldots n$. The example in section 7.1 contains 25 inversions.

The alternating group A_n is $(n-2)$-ply transitive, for the permutations

$$\begin{pmatrix} 1 & 2 & \cdots n-2 & n-1 & n \\ \alpha_1 & \alpha_2 & \cdots \alpha_{n-2} & \alpha_{n-1} & \alpha_n \end{pmatrix}, \qquad \begin{pmatrix} 1 & 2 & \cdots n-2 & n-1 & n \\ \alpha_1 & \alpha_2 & \cdots \alpha_{n-2} & \alpha_n & \alpha_{n-1} \end{pmatrix}$$

differ from each other by a single transposition, so that one of them is contained in A_n.

As we saw above, every permutation of degree n can be represented as a product of the transpositions $(1, 2), (1, 3), \ldots, (1, n)$. Thus, every even permutation can be written as a product of an even number of those transpositions. We have $(1, \alpha)(1, \beta) = (1, \alpha, \beta)$. This shows that every even permutation can be represented as a product of cycles of length 3, and the cycles of the form $(1, \alpha, \beta)$ are sufficient for this purpose. Moreover, we obtain

$$(1, 2, \beta)(1, 2, \alpha)(1, 2, \beta)^{-1} = (1, \alpha, \beta).$$

This shows that the $n - 2$ cycles

$$(1, 2, 3), (1, 2, 4), \ldots, (1, 2, n)$$

generate the alternating group.

Since the factor group S_n/A_n is abelian, A_n contains the commutator subgroup of S_n. We prove:

7.4.1 *The commutator subgroup of the symmetric group S_n coincides with the alternating group A_n.*

By our previous remark, it remains to show that the commutator subgroup of S_n is not a proper subgroup of A_n. We have

$$(1, \alpha)^{-1} (1, \beta)^{-1} (1, \alpha) (1, \beta) = (1, \beta, \alpha).$$

Thus, every cycle of length 3 is a commutator. Since the cycles of length 3 generate the group A_n, our assertion follows.

With the exception of the case $n = 4$, our last proposition follows from a more general result.

7.4.2 *For $n = 3$ and $n \geqq 5$, the alternating group A_n is the only non-trivial normal subgroup of S_n. The symmetric group S_4 contains two non-trivial normal subgroups; one of them is A_n, the other is isomorphic to the four-group.*

Proof. Let $T \neq e$ be a normal subgroup of S_n. We consider two cases.

(a) T contains a permutation a such that in the decomposition of a into a product of mutually disjoint cycles there is at least one cycle of length $\geqq 3$. Then, we have

$$a = (\alpha, \beta, \gamma, \ldots) \ldots$$

By interchanging α and β, we obtain the permutation

$$b = (\beta, \alpha, \gamma, \ldots) \ldots$$

that is also contained in T, since it is conjugate to a. Consequently, $a^{-1}b = (\alpha, \beta, \gamma)$ belongs to T. As T is a normal subgroup of S_n, it follows that T contains all cycles of length 3; hence, $A_n \subseteq T$.

(b) It remains to consider the case that every permutation $c \neq e$ of T is a product of disjoint transpositions. Then, we have

$$c = (\alpha_1, \beta_1)(\alpha_2, \beta_2) \cdots (\alpha_k, \beta_k).$$

If $2k < n$, then T would also contain some permutation

$$d = (\gamma, \beta_1)(\alpha_2, \beta_2) \cdots (\alpha_k, \beta_k)$$

where γ is distinct from $\alpha_1, \beta_1, \ldots, \alpha_k, \beta_k$. Consequently, the product $cd = (\alpha_1, \gamma, \beta_1)$ would belong to T. But this is the case dealt with in (a). Hence, we have $2k = n$ which implies $k \geqq 2$. For $k > 2$, we come back to (a), because in this case T also contains

$$c^* = (\alpha_1, \beta_2)(\alpha_2, \beta_3) \cdots (\alpha_{k-1}, \beta_k)(\alpha_k, \beta_1),$$

so that

$$c^* c = (\alpha_1, \ldots, \alpha_k)(\beta_1, \ldots, \beta_k)^{-1}$$

belongs to T. Finally, we have to consider $k = 2$, i. e. $n = 4$. In this case, the permutations

$$e, (1, 2)(3, 4), \ (1, 3)(2,4), \ (1, 4)(2, 3)$$

form another non-trivial normal subgroup of S_4.

This completes the proof. We use the result for the proof of the following theorem.

7.4.3 *The alternating groups A_n are simple except for $n = 4$.*

Proof. $A_2 = e$, A_3 is of order 3. By Theorem 7.4.2, A_4 is not simple.

We may now assume that $n \geqq 5$. Suppose that A_n contains a non-trivial normal subgroup H. We choose an arbitrary odd permutation u to obtain the decomposition

$$S_n = A_n \cup A_n u$$

of S_n into cosets of A_n. By Theorem 7.4.2, H is not normal in S_n so that $u^{-1}Hu = H_1$ is distinct from H and a normal subgroup of A_n. Moreover, H and H_1 form a complete class of conjugate subgroups of S_n. Hence, it follows from Theorem 7.4.2 that $H \cap H_1 = e$. Since both H and H_1 are normal subgroups of A_n, we have $HH_1 = H \times H_1$. This direct product is a normal subgroup of S_n which is contained in A_n. Theorem 7.4.2 then gives

$$A_n = H \times H_1. \tag{7.15}$$

Now A_n is primitive, because it is $(n - 2)$-ply transitive. Hence, it follows from Theorem 7.1.10 that both H and H_1 are transitive. Thus, 7.1.8 shows that $|H| = |H_1| \leqq n$. But this contradicts (7.15), because for $n \geqq 5$

$$|A_n| = \frac{1}{2} n! > n^2 \geqq |H \times H_1|.$$

Finally, we use the theory of permutation groups to prove a result on arbitrary finite groups.

7.4.4 *Let G be a group of order $2u$, where u is an odd number. Then, G contains a normal subgroup of order u.*

Proof. By Theorem 3.2.2, G contains an element a of order 2. The permutation ϱ_a that corresponds to a in the regular permutational representation G^* of G is a product of u mutually disjoint transpositions. Consequently, ϱ_a is an odd permutation. The even permutations in G^* form a normal subgroup T^* of G^*, which does not coincide with G^* since ϱ_a is odd. So, we have $|G^*:T^*| = 2$. Hence, the subgroup T of G that corresponds to T^* is a normal subgroup of order u.

7.5 The Wreath Product and the Sylow Subgroups of the Symmetric Groups

In this section, we investigate the Sylow subgroups of the symmetric groups of finite degree. The description of their structure provides an opportunity of dealing with another method of constructing new groups from given ones. In recent times, this method has become a powerful tool in various branches of group theory.

Let A be a group of permutations of a set Ω_1 and B a group of permutations of a set Ω_2. We assume that both Ω_1 and Ω_2 are not empty and disjoint. As before, the image of a symbol ξ_1 of Ω_1 under the permutation a of A is denoted by $\xi_1 a$; similarly, $\xi_2 b$ is the image of $\xi_2 \in \Omega_2$ under $b \in B$. By $\Omega_1 \times \Omega_2$ we mean the set of all pairs (ξ_1, ξ_2), where $\xi_1 \in \Omega_1$, $\xi_2 \in \Omega_2$. Let W denote the set of all permutations w of $\Omega_1 \times \Omega_2$ that have the following form:

$$(\xi_1, \xi_2)\, w = (\xi_1 a(\xi_2), \xi_2 b), \qquad a(\xi_2) \in A, \quad b \in B. \tag{7.16}$$

The image of the second symbol in each pair is determined by the permutation b and is independent of the first symbol. The image of the first symbol in each pair, however, does not only depend on that symbol, but also on the second symbol. The first symbols in all pairs with one and the same second symbol ξ_2 are mapped according to the permutation $a(\xi_2)$ of A. In general, $a(\xi_2) \neq a(\xi_2')$ if $\xi_2' \neq \xi_2$. Thus, w is determined by an element b of B and a function $a(\xi_2)$ with domain Ω_2 and range A.

If $|A|$, $|B|$ and the number $|\Omega_2|$ of symbols in Ω_2 are finite, then W consists of

$$|B|\,|A|^{|\Omega_2|} \tag{7.17}$$

distinct permutations.

If w' is another permutation in W,

$$(\xi_1, \xi_2)\, w' = (\xi_1 a'(\xi_2), \xi_2 b'),$$

then the product ww' also belongs to W; for, we have

$$(\xi_1, \xi_2)\, ww' = (\xi_1 a(\xi_2), \xi_2 b)\, w' = (\xi_1 a(\xi_2)\, a'(\xi_2 b), \xi_2 bb'), \tag{7.18}$$

which is obviously a permutation of the same kind. The product ww' is the identity permutation if we take

$$b' = b^{-1},\ a'(\xi_2) = a(\xi_2 b^{-1})^{-1}.$$

This shows that W is a group. It is called the *complete* or *unrestricted wreath product* of A and B. We write $W = A\mathrm{Wr}B$.

Apart from trivial cases, the wreath product is not commutative. We shall now prove that the associative law holds. Let C be a group of permutations of a non-empty set Ω_3. Then, the wreath product $(A\mathrm{Wr}B)\mathrm{Wr}C$ consists of all permutations of $(\Omega_1 \times \Omega_2) \times \Omega_3$ of the following form:

$$((\xi_1, \xi_2), \xi_3) \rightarrow ((\xi_1, \xi_2)\, w(\xi_3), \xi_3 c) = ((\xi_1 a(\xi_2, \xi_3), \xi_2 b(\xi_3)), \xi_3 c)\,.$$

Here, $b(\xi_3)$ is an arbitrary function defined on Ω_3 whose values are elements of B; and the function $a(\xi_2, \xi_3)$ is a mapping of $\Omega_2 \times \Omega_3$ into A. On the other hand, $A\mathrm{Wr}(B\mathrm{Wr}C)$ is the set of all permutations

$$(\xi_1, (\xi_2, \xi_3)) \rightarrow (\xi_1 a(\xi_2, \xi_3), (\xi_2 b(\xi_3), \xi_3 c))$$

of the set $\Omega_1 \times (\Omega_2 \times \Omega_3)$. By identifying the sets $(\Omega_1 \times \Omega_2) \times \Omega_3$ and $\Omega_1 \times (\Omega_2 \times \Omega_3)$, we find that $A\mathrm{Wr}(B\mathrm{Wr}C) = (A\mathrm{Wr}B)\mathrm{Wr}C$.

If we assign to the permutation w in (7.16) the element b, then (7.18) shows that we obtain a homomorphism of $A\mathrm{Wr}B$ onto B. The kernel K of this homomorphism consists of all permutations k of $A\mathrm{Wr}B$ for which

$$(\xi_1, \xi_2)\, k = (\xi_1 a(\xi_2), \xi_2)\,.$$

The structure of K can be described as follows: There is a one-to-one correspondence between the elements k in K and the systems

$$\{a(\xi_2)\} \qquad (\xi_2 \in \Omega_2) \tag{7.19}$$

of elements of A. For each $\xi_2 \in \Omega_2$, we take a group A_{ξ_2} isomorphic to A and form the cartesian product

$$D = \prod_{\xi_2 \in \Omega_2} A_{\xi_2}\,.$$

Then, we can establish a one-to-one correspondence between the systems (7.19) and the elements of D in a natural way, by regarding $a(\xi_2)$ as the A_{ξ_2}-component of an element of D. Consequently, there is a one-to-one correspondence between the two groups K and D. Let k' be another element of K and denote the components of the corresponding element in D by $a'(\xi_2)$. Then, using (7.18) for $b = b' = e$, we find that the element of D corresponding to kk' has the components $a(\xi_2)a'(\xi_2)$. Thus, K and D are isomorphic.

Let $\bar{B}$ denote the subgroup of all those elements of $A\mathrm{Wr}B$ that leave the first symbol of every pair (ξ_1, ξ_2) fixed. These elements are characterized by the property that $a(\xi_2)$ is the identity permutation for all $\xi_2 \in \Omega_2$. By (7.18), $\bar{B}$ is isomorphic to B. It is evident that $\bar{B}$ is a transversal to K in $A\mathrm{Wr}B$, i. e.

$$A\mathrm{Wr}B = K\bar{B}\,.$$

Thus, $A\mathrm{Wr}B$, regarded as an abstract group, is a splitting extension of K (or D) by B. Let $\bar{b}_1 \in \bar{B}$, $k \in K$ and denote by b_1 the element of B that corresponds to $\bar{b}_1$ under the above isomorphism; then, we obtain

$$(\xi_1, \xi_2)\,\bar{b}_1 k \bar{b}_1^{-1} = (\xi_1, \xi_2 b_1)\,k\bar{b}_1^{-1} = (\xi_1 a(\xi_2 b_1),\ \xi_2 b_1)\bar{b}_1^{-1} = (\xi_1 a(\xi_2 b_1),\ \xi_2)\,. \quad (7.20)$$

So far, we have defined the wreath product of permutation groups. To define $A\mathrm{Wr}B$ for abstract groups A and B, we replace A and B by their right regular permutational representations. Then, the elements of A take the place of the symbols in Ω_1, and instead of the permutations

$$\begin{pmatrix} \xi_1 \\ \xi_1 a^* \end{pmatrix} \qquad (\xi_1 \in \Omega_1;\ a^* \in A)$$

of Ω_1, we have to consider the permutations

$$\begin{pmatrix} a \\ aa^* \end{pmatrix} \qquad (a, a^* \in A)$$

of the elements of A. The same applies to Ω_2 and B. In view of what we observed above, $A\mathrm{Wr}B$ can now be described as follows: To each $b \in B$ we assign a group A_b isomorphic to A and we form the cartesian product

$$D = \prod_{b \in B} A_b$$

of these groups. Then, $A\mathrm{Wr}B$ is isomorphic to the group of all formal products db, $d \in D$, $b \in B$, which are multiplied according to the rule

$$d_1 b_1 d_2 b_2 = d_1 d_2^{b_1} b_1 b_2, \qquad (d, d_1 \in D;\ b, b_1 \in B)\,, \quad (7.21)$$

where $d \to d^{b_1}$ is the automorphism of D defined by

$$(a_b)^{b_1} = a_{bb_1} \qquad (b, b_1 \in B)\,. \quad (7.22)$$

We observe that (7.22) is the adaptation of (7.20) to the present notation. Since $d_2^{b_1}$ corresponds to $b_1 d_2 b_1^{-1}$, the composition of the automorphisms has to be defined by

$$d^{b_1 b_2} = (d^{b_2})^{b_1}. \quad (7.23)$$

That the elements db form a group with respect to the operation (7.21) and the rules (7.22) and (7.23) can also be verified by a simple computation or can be deduced from the general theory of group extensions. The group D was defined as the cartesian or unrestricted direct product of the groups A_b; therefore, we speak of the unrestricted wreath product $A\mathrm{Wr}B$. If D is replaced by the restricted direct product, then the construction can be carried out in the same way to yield the so-called restricted wreath product $A\mathrm{wr}B$.

To find the Sylow p-subgroups of the symmetric group S_n, we first determine their order, i. e. the highest power of p that divides $n!$. For any real number ϱ, let $[\varrho]$ be the unique integer such that $[\varrho] \leqq \varrho < [\varrho] + 1$. For $k = 1, 2, \ldots,$ there are $[n/p^k]$ integers in the sequence $1, 2, \ldots, n$ that are divisible by p^k, namely the numbers xp^k, $1 \leqq x \leqq [n/p^k]$. The highest power of p dividing $n!$ can now be found as follows: First each multiple of p in the sequence $1, 2, \ldots, n$ yields one factor p; after that each multiple of p^2 contains one additional factor p, then each multiple of p^3 has one factor p not yet counted, etc. Consequently, $n!$ contains exactly

$$m(n) = \left[\frac{n}{p}\right] + \left[\frac{n}{p^2}\right] + \left[\frac{n}{p^3}\right] + \cdots \tag{7.24}$$

factors p. In particular, we have

$$m(p^r) = p^{r-1} + p^{r-2} + \cdots + p + 1 \qquad (r > 0). \tag{7.25}$$

Evidently,

$$m(p^{r+1}) = pm(p^r) + 1. \tag{7.26}$$

Expressing n in the scale of p, we obtain

$$n = x_0 + x_1 p + x_2 p^2 + \cdots + x_t p^t \qquad (0 \leqq x_i < p).$$

It follows from (7.24) that

$$m(n) = x_t(p^{t-1} + \cdots + p + 1) + x_{t-1}(p^{t-2} + \cdots + p + 1) + \cdots + x_2(p + 1) + x_1.$$

Hence, (7.25) gives

$$m(n) = x_t m(p^t) + x_{t-1} m(p^{t-1}) + \cdots + x_2 m(p^2) + x_1. \tag{7.27}$$

This representation of $m(n)$ shows by a simple enumeration that the Sylow p-subgroups of S_n can easily be described in terms of the Sylow p-subgroups P_r of the symmetric group S_{p^r}. To see this, we partition the integers $1, 2, \ldots, n$ into

x_t subsets, each containing p^t integers,

x_{t-1} subsets, each containing p^{t-1} integers,

. . . .

x_1 subsets, each containing p integers,

and

x_0 subsets, each consisting of a single integer.

We assume that these subsets are disjoint. For each subset of p^r integers ($r = 1, \ldots, t$), we take a Sylow p-subgroup P_r of the symmetric group S_{p^r} on these integers. So, we obtain x_r copies of P_r ($r = 1, \ldots, t$), which will be denoted by $P_r^{(i)}$ ($i = 1, \ldots, x_r$). Since all these Sylow p-subgroups operate on disjoint sets, they commute elementwise with each other. It follows that the subgroup generated by the $P_r^{(i)}$ ($r = 1, \ldots, t$; $i = 1, \ldots, x_r$) in S_n is the direct product

$$\prod_{r=1}^{t} \prod_{i=1}^{x_r} P_r^{(i)}.$$

By (7.27), this direct product is of order $p^{m(n)}$, so that it is a Sylow p-subgroup of S_n.

It remains to consider the Sylow p-subgroups P_r of S_{p^r}. A Sylow p-subgroup of S_p, operating on $1, \ldots, p$, is the subgroup of order p generated by the cycle $(1, 2, \ldots, p)$.

In the group S_{p^2} on $1, 2, \ldots, p^2$, we consider the p cycles

$$\begin{aligned} a_1 &= (1, 2, \ldots, p), \\ a_2 &= (p+1, p+2, \ldots, 2p), \\ &\ldots\ldots\ldots\ldots\ldots\ldots \\ a_p &= (p^2 - p + 1, p^2 - p + 2, \ldots, p^2) \end{aligned}$$

of order p. The subgroup of S_{p^2} generated by $a_1, a_2, \ldots, a_p$ is the direct product

$$\langle a_1 \rangle \times \langle a_2 \rangle \times \cdots \times \langle a_p \rangle.$$

Then, we put

$$b^{-1} = (1, p+1, \ldots, p^2 - p + 1)(2, p+2, \ldots, p^2 - p + 2) \cdots (p, 2p, \ldots, p^2).$$

Clearly, b^{-1} is of order p. The notation b^{-1} will turn out to be in accordance with our previous notation for wreath products. One easily verifies that $ba_ib^{-1} = a_{i+1}$, where we set $a_i = a_j$ if $i \equiv j \pmod{p}$. The last remark suggests indexing the a_i by the powers of b. Accordingly, we write

$$a_i = a_{b^i},$$

so that we obtain

$$a_{b^i}^b = b a_{b^i} b^{-1} = a_{b^{i+1}}.$$

The last equation corresponds to (7.22) for $B = \langle b \rangle$. This shows that the subgroup $P = \langle a_1, b \rangle$ of S_{p^2} is isomorphic to the wreath product of two cyclic groups of order p. The order of P is $pp^r = p^{p+1}$. Thus, P is a Sylow p-subgroup of S_{p^2}.

We shall now prove that a Sylow p-subgroup P_r of S_{p^r} is isomorphic to the wreath product of r cyclic groups of order p. We just saw that this is true for $r = 2$. Using induction on r, it remains to show that $P_{r+1} = P_r \operatorname{wr} Z$ where Z is a cyclic group of order p.

We partition the numbers $1, 2, \ldots, p^{r+1}$ into p subsets each of which contains p^r numbers:

$$
\begin{aligned}
&1, 2, \ldots, p^r;\\
&p^r + 1, p^r + 2, \ldots, 2p^r;\\
&\cdots\cdots\cdots\cdots\cdots\cdots\\
&p^{r+1} - p^r + 1, p^{r+1} - p^r + 2, \ldots, p^{r+1}.
\end{aligned}
$$

In each symmetric group on such a subset, we choose a Sylow p-subgroup. So, we obtain p subgroups of $S_{p^{r+1}}$ isomorphic to P_r; they are denoted by $P_r^{(1)}, \ldots, P_r^{(p)}$. Since these groups operate on disjoint sets, they commute elementwise with one another. Hence, the group they generate in $S_{p^{r+1}}$ is the direct product

$$P_r^{(1)} \times P_r^{(2)} \times \cdots \times P_r^{(p)}.$$

Now, $S_{p^{r+1}}$ contains the permutation

$$c^{-1} = \prod_{j=1}^{p^r} (j, p^r + j, 2p^r + j \ldots, (p-1)p^r + j)$$

which is a product of p^r disjoint cycles of length p; hence it is of order p. As above, we use the powers of c to index the groups $P_r^{(i)}$ by putting

$$P_r^{(i)} = P_r^{(c^i)}.$$

Then, one easily checks that

$$cP_r^{(c^i)}c^{-1} = P_r^{(c^{i+1})}.$$

Comparing this equation with (7.22), we see that the subgroup of $S_{p^{r+1}}$ generated by $P_r^{(1)}$ and c is $P_r^{(1)}$ wr $\langle c\rangle$. By (7.17) and (7.26), we obtain

$$|P_r \operatorname{wr} \langle c\rangle| = |\langle c\rangle||P_r|^{|\langle c\rangle|} = pp^{m(p^r)p} = p^{m(p^{r+1})}.$$

This shows that $P_r^{(1)}$ wr $\langle c\rangle$ is a Sylow p-subgroup of $S_{p^{r+1}}$, so that

$$P_{r+1} = P_r \operatorname{wr} \langle c\rangle \qquad (c^p = e)$$

which is the required result.

Summarizing our results, we obtain:

7.5.1 *Any Sylow p-subgroup P_r of the symmetric group S_{p^r} of degree p^r is isomorphic to the wreath product of r groups of order p. Let*

$$n = x_0 + x_1 p + x_2 p^2 + \cdots + x_t p^t \qquad (0 \leqq x_i < p)$$

be the representation of n in the scale of p; then, the Sylow p-subgroups of the symmetric group S_n of degree n are isomorphic to the direct product of x_1 copies of P_1, x_2 copies of P_2, ..., x_t copies of P_t.

7.6 Finite Rotation Groups

A *rotation group* is defined as a group of rotations of a three-dimensional Euclidean space around axes through a fixed point 0. We shall determine all finite rotation groups.

Among the rotations we include the identity e which leaves all points fixed. If we wish to emphasize that a rotation is not the identity, then we speak of a proper rotation. In what follows we use a few simple facts from geometry which may readily be visualized; for proofs, we refer to textbooks on geometry.

Let Σ be a sphere with centre 0. Every rotation around 0 induces a one-to-one mapping of Σ onto itself. For a point α of Σ and a rotation d, the image of α under d is denoted by αd. Every proper rotation leaves two points of Σ fixed, namely the points where its axis pierces Σ. These two points are called the *poles* of the rotation. The identity is the only rotation that leaves more than two points of Σ fixed. Let α_1, α_2 and β_1, β_2 be two pairs of antipodal points of Σ, e. g. the poles of two rotations; if a rotation d carries α_1 into β_1, then d also carries α_2 into β_2.

Let G be a finite rotation group of order $n > 1$. Suppose that α is a pole of G; then, there is a rotation $d \neq e$ in G such that $\alpha d = \alpha$. It is easy to see that the image αx of α under an arbitrary rotation x of G is also a pole of G. Indeed, we have $\alpha x(x^{-1}dx) = \alpha\, dx = \alpha x$, which shows that αx is a pole of $x^{-1}dx$. Therefore, G induces a group of permutations of its poles. Let $\Gamma_1, \ldots, \Gamma_k$ be the orbits of this permutation group. The number of poles in Γ_i is denoted by m_i. Let $\alpha \in \Gamma_i$. The rotations that leave α fixed together with e form a subgroup Z_i of G. We choose rotations $t_2, \ldots, t_{m_i}$ of G that carry α into the remaining $m_i - 1$ poles of Γ_i. Then, $e, t_2, \ldots, t_{m_i}$ form a right transversal to Z_i in G, so that we have the decomposition

$$G = Z_i \cup Z_i t_2 \cup \cdots \cup Z_i t_{m_i} \tag{7.28}$$

of G into cosets of Z_i. The coset $Z_i t_j$ consists precisely of all rotations in G that carry α into αt_j. From (7.28), we obtain $n = m_i n_i$, where n_i denotes the order of Z_i.

Clearly, Z_i is the stabilizer subgroup of α in G. Hence, the stabilizer subgroup of αt_j is $t_j^{-1} Z_i t_j$. Thus, each pole in Γ_i is a pole of $n_i - 1$ proper rotations of G.

We now count the number of pairs (x, α) consisting of a proper rotation x of G and a pole α of x. This can be done in two different ways. On the one hand, G contains $n - 1$ proper rotations each of which has two poles. Hence, the number in question is $2(n - 1)$. On the other hand, every pole of Γ_i remains fixed under $n_i - 1$ proper rotations. Since Γ_i contains m_i poles, we have:

$$2(n-1) = \sum_{i=1}^{k} m_i(n_i - 1).$$

Dividing this equation by $n = m_i n_i$, we obtain

$$2\left(1 - \frac{1}{n}\right) = \sum_{i=1}^{k}\left(1 - \frac{1}{n_i}\right)$$

or

$$k - 2 + \frac{2}{n} = \sum_{i=1}^{k} \frac{1}{n_i}. \tag{7.29}$$

It is evident that the orders n_i of the subgroups Z_i satisfy the inequalities $2 \leqq n_i \leqq n$ $(i = 1, \ldots, k)$. By these inequalities, it follows from (7.29) that the only possible values of k are 2 and 3.

I. $k = 2$. In this case, (7.29) takes the form

$$\frac{2}{n} = \frac{1}{n_1} + \frac{1}{n_2}.$$

Since $n_i \leqq n$, we find that $n_1 = n_2 = n$. All proper rotations of G have the same poles so that G consists of rotations around one and the same axis. Let φ be the smallest positive rotation angle that occurs among the rotations in G, and let a be a rotation through φ. For an arbitrary rotation b in G whose angle is ψ, we determine an integer l such that $l\varphi \leqq \psi < (l+1)\varphi$. Then, ba^{-l} is a rotation through the angle $\psi - l\varphi$, which is non-negative and less than φ. Owing to the choice of φ, it follows that $\psi - l\varphi = 0$. This gives $b = a^l$. Consequently, G is a cyclic group of order n generated by the rotation a through the angle $2\pi/n$.

II. $k = 3$. In this case, (7.29) has the form

$$1 + \frac{2}{n} = \frac{1}{n_1} + \frac{1}{n_2} + \frac{1}{n_3}. \tag{7.30}$$

For a suitable numbering, we may assume that

$$(2 \leqq)\ n_1 \leqq n_2 \leqq n_3 (\leqq n).$$

We conclude that $n_1 = 2$ and $n_2 \leqq 3$, because otherwise the right-hand side of (7.30) would be at most equal to 1. Then, it is easy to see that there are only four systems of natural numbers n_1, n_2, n_3 satisfying (7.30):

$$\text{(a)} \quad n_1 = 2, \quad n_2 = 2; \quad n_3 = \frac{n}{2};$$

$$\text{(b)} \quad n_1 = 2, \quad n_2 = 3, \quad n_3 = 3, \quad n = 12;$$

$$\text{(c)} \quad n_1 = 2, \quad n_2 = 3, \quad n_3 = 4, \quad n = 24;$$

$$\text{(d)} \quad n_1 = 2, \quad n_2 = 3, \quad n_3 = 5, \quad n = 60.$$

We shall now discuss these four cases separately.

IIa. Here, $m_3 = 2$ so that Γ_3 consists of two poles α_1, α_2. Every rotation leaves both poles fixed or interchanges them. Those rotations that leave both α_1 and α_2 fixed form a subgroup Z of G. The decomposition of G into cosets of Z has the form $G = Z \cup Zu$, where u is an arbitrary rotation interchanging α_1 and α_2. Since all rotations in Z have the same axis, it follows in the same way as above that Z is cyclic, but in the present case of order $n/2$. The square of every rotation zu in the coset Zu is the identity. For, on the one hand, $(zu)^2$ belongs to Z so that both α_1 and α_2 remain fixed. On the other hand, zu interchanges α_1 and α_2 and is therefore a rotation around an axis that does not pass through α_1 and α_2; hence, the poles β_1, β_2 of zu are distinct from α_1, α_2. It follows that α_1, α_2, β_1, β_2 remain fixed under $(zu)^2$; this shows that $(zu)^2 = e$. In particular, $u^2 = e$. From $(zu)^2 = e$, we see that $u^{-1}zu = z^{-1}$. These groups are called *dihedral groups*.

The name refers to the following geometrical characterization of such groups. Put $n/2 = m$ and let Δ_m denote a regular m-gon with centre 0 whose plane is perpendicular to the axis through α_1 and α_2. We regard Δ_m as a degenerate polyhedron having two faces. Then, G can be characterized as the group of all rotations about 0 that carries Δ_m into itself. Z consists of the m rotations around the axis through α_1 and α_2 whose angles are multiples of $2\pi/m$. The coset Zu consists of m rotations through half-turns around axes in the plane of Δ_m. If m is even, then these axes join opposite vertices and the midpoints of opposite sides of Δ_m. If m is odd, then the axes are the lines joining each vertex to the midpoint of the opposite side.

As an abstract group, a dihedral group of order $n = 2m$ is generated by two elements a, u, subject to the relations $a^m = e$, $u^2 = e$, $u^{-1}au = a^{-1}$.

In the remaining three cases, the orders of the groups are uniquely determined. We shall see that the groups themselves are unique. In all these cases, each of the orbits Γ_1, Γ_2, Γ_3 contains more than two poles. It follows that the identity is the only rotation that leaves each pole in some orbit fixed. Consequently, we obtain faithful permutational representations of G by the permutations induced on the poles in any orbit. This allows us to use the notation G also for these permutation groups.

IIb. We regard G as a group of permutations of the 4 poles in Γ_2. Let S_4 denote the symmetric group on these 4 poles. Then, $n = 12$ and $|S_4:G| = 2$ so that G is a normal subgroup of S_4. Hence, by Theorem 7.4.2, G is isomorphic to the alternating group A_4.

Let T be a regular tetrahedron inscribed in Σ such that the vertices coincide with poles in Γ_2. Then, G consists of all rotations about 0 that carry T into itself. Therefore, G is also called the *tetrahedral group*. The poles in Γ_3 are the antipodes of those in Γ_2. The 6 poles in Γ_1 are the projections from 0 onto Σ of the midpoints of the edges of T.

IIc. We consider the permutations of the 8 poles in Γ_2 under G. These poles belong to rotations of order 3. Since the poles in Γ_1 and Γ_3 belong to rotations of order 2 and 4, respectively, we conclude that both poles of every rotation of order 3 lie in Γ_2. Thus, the 8 poles in Γ_2 form 4 pairs of antipodal poles, and these pairs are systems of imprimitivity under the permutations induced by G. Thus, we obtain a representation of G by permutations of the 4 pairs of antipodal poles.

Next, we prove that this representation is faithful. Suppose that $u \neq e$ is a rotation that leaves the 4 pairs fixed. Then, both poles in each pair remain fixed or are interchanged by u. Consequently, u^2 leaves all the 8 poles fixed so that $u^2 = e$. Since u is of order 2, its poles are not in Γ_2 (but in Γ_1). Therefore, u interchanges the poles in each of our 4 pairs, for otherwise u would leave more than two points fixed. Denoting the 4 pairs of antipodal poles in Γ_2 by α_1, α_2; β_1, β_2; γ_1, γ_2; δ_1, δ_2, we obtain

$$u = (\alpha_1, \alpha_2)\,(\beta_1, \beta_2)\,(\gamma_1, \gamma_2)\,(\delta_1, \delta_2)\,.$$

Since the 4 pairs are systems of imprimitivity under G, we have $d^{-1}ud = u$ for every $d \in G$. In particular, u commutes with any rotation a of order 3. But, then, ua would be of order 6, whereas the orders of the proper rotations in G are 2, 3, 4.

It follows that G is isomorphic to a permutation group of degree 4. But, since $|G| = 24$, G is isomorphic to the symmetric group S_4.

To obtain a geometrical characterization of G we consider a regular octahedron Ω circumscribed about Σ such that its faces touch Σ at the 8 poles in Γ_2. Then, G consists of all rotations that carry Ω into itself. Therefore, G is said to be the *octahedral group*. It is easy to see that G induces the symmetric group on the 4 lines joining the centres of opposite faces of Ω (corresponding to the 4 pairs of antipodal poles in Γ_2). The rotation through $\pi/2$ around an axis through two opposite vertices of Ω yields a cyclic permutation of the 4 lines. A rotation through a half-turn around an axis joining the midpoints of two opposite edges of Ω interchanges exactly two of the 4 lines. A cycle of length 4 and a transposition generate S_4. The poles in Γ_1 and Γ_3 are, respectively, the projections from 0 onto Σ of the midpoints of the edges and of the vertices of Ω.

The centres of the faces of a regular octahedron form the vertices of a cube. Thus, a cube has the same symmetry group as an octahedron.

IId. We consider the permutations induced by G on the 30 poles in Γ_1. Each of these poles belongs to a unique rotation of order 2. Since the poles in Γ_2 and Γ_3 belong to rotations of order 3 and 5, respectively, we see that both poles of every rotation of order 2 are in Γ_1. So, we obtain a partition of the 30 poles in Γ_1 into 15 pairs of antipodal poles, which are systems of imprimitivity under G. It is easy to see that the identity is the only rotation that leaves all the 15 pairs fixed. Hence, G has a faithful representation by permutations of the 15 pairs.

Let $\alpha = (\alpha_1, \alpha_2)$ be a pair of antipodal poles in Γ_1. As we remarked above, G contains a single proper rotation a whose poles are α_1 and α_2. We have $a^2 = e$. Since Γ_1 is an orbit, G contains a rotation b that carries α_1 into α_2. Then, b evidently interchanges α_1 and α_2 so that $b^2 = e$. The rotation $b^{-1}ab$ leaves both α_1 and α_2 fixed, hence $b^{-1}ab = a$. Let β_1, β_2 be the poles of b. They belong to Γ_1, as b is of order 2. We have

$$\beta_1 a = \beta_1 ba = \beta_1 ab.$$

Consequently, $\beta_1 a$ is a pole of b. Since $\beta_1 a \neq \beta_1$, we find $\beta_1 a = \beta_2$, which implies that $\beta_2 a = \beta_1$. In other words: b interchanges the poles of a and a interchanges the poles of b. We put $ab = c$. Clearly, c is of order 2 and interchanges α_1 and α_2 as well as β_1 and β_2. The poles γ_1, γ_2 of c belong to Γ_1, since $c^2 = e$. Moreover,

$$\gamma_1 a = \gamma_1 ca = \gamma_1 ac;$$

hence, $\gamma_1 a$ is a pole of c. Thus, a interchanges γ_1 and γ_2. In the same way, we infer that b interchanges γ_1 and γ_2. The rotations a, b, c and e form a subgroup $F_{\alpha\beta\gamma}$ isomorphic to the four-group.

Every rotation of G that leaves at least one of the pairs

$$\alpha = (\alpha_1, \alpha_2), \quad \beta = (\beta_1, \beta_2), \quad \gamma = (\gamma_1, \gamma_2)$$

fixed belongs to $F_{\alpha\beta\gamma}$. For, if a rotation $\neq e$ leaves both poles in a pair fixed, then it coincides with a, b, or c. Suppose that d interchanges β_1 and β_2, say. Then, da leaves both β_1 and β_2 fixed so that $da = c$ or $da = b$, i. e. $d = a$ or $d = c$.

Since G induces a transitive group on the 15 pairs, there is a rotation s that carries $\alpha = (\alpha_1, \alpha_2)$ into $\beta = (\beta_1, \beta_2)$. It follows that $s^{-1}as = b$. Moreover, s does not belong to $F_{\alpha\beta\gamma}$. The rotations $s^{-1}as$, $s^{-1}bs$, $s^{-1}cs$ leave the pair $\beta = (\beta_1, \beta_2)$ fixed; hence, they coincide with a, b, c in some order. Therefore,

$$s^{-1} F_{\alpha\beta\gamma}\, s = F_{\alpha\beta\gamma}.$$

Further, s permutes α, β, γ and leaves none of them fixed because $s \notin F_{\alpha\beta\gamma}$. Thus, s induces a cyclic permutation on α, β, γ. Hence, s^3 leaves α, β, γ fixed, so that $s^3 \in F_{\alpha\beta\gamma}$. Since G contains no rotations of order 6, we have $s^3 = e$. Let

$$H_{\alpha\beta\gamma} = \langle F_{\alpha\beta\gamma}, s\rangle.$$

Since $|H_{\alpha\beta\gamma} : F_{\alpha\beta\gamma}| = 3$, the 15 pairs of poles in Γ_1 are partitioned into 5 systems of imprimitivity, each consisting of 3 pairs.

We obtain a representation of G as a group of permutations of these 5 triplets. The kernel of this representation is the intersection of all subgroups conjugate to $H_{\alpha\beta\gamma}$. If α', β', γ' is a triplet distinct from α, β, γ, then $F_{\alpha\beta\gamma} \cap F_{\alpha'\beta'\gamma'} = e$. Thus, the kernel in question is e or $\langle s\rangle$. We shall prove that $\langle s\rangle$ is not a normal subgroup of G so that the kernel is equal to e. Suppose that $\langle s\rangle$ is a normal subgroup of G. Then, s, considered as a permutation of the 15 pairs, is a product of 5 cycles of length 3 and each cycle consists of one of the above triplets:

$$s = (\alpha, \beta, \gamma)\,(\alpha', \beta', \gamma')\ \ldots .$$

Now, a leaves α, β, γ fixed. Therefore, $a^{-1}sa$ contains the cycle (α, β, γ), which shows that $a^{-1}sa = s$. This contradicts our previous result $s^{-1}as = b$.

Consequently, G is isomorphic to a permutation group of degree 5. This permutation group is of order 60 so that it is a normal subgroup of the symmetric group S_5. Now, it follows from Theorem 7.4.2 that G is isomorphic to the alternating group A_5.

The symmetry group of every regular polyhedron is a finite rotation group. The remaining two polyhedra are the icosahedron and the dodecahedron. The centres of the faces of a regular icosahedron form the vertices of a regular dodecahedron. Therefore, they have the same symmetry group. Consequently, G is the symmetry group of a regular icosahedron, the *icosahedral group*.

Exercises

1. Prove that every regular permutation of degree n is a power of some cycle of length n.

2. Consider permutations of the set $\Omega = \{1, 2, 3, 4, 5, 6\}$. Determine all the permutations x of Ω such that

(a) $x^2 = (1, 2, 3)(4, 5, 6)$,
(b) $x^2 = (1, 2, 3, 4)(5, 6)$.

Show that there is no permutation x of Ω such that $x^2 = (1, 2)(3, 4)(5, 6)$.

3. Let x be a permutation of a finite set Ω and k a given natural number, $k > 1$. Find a necessary and sufficient condition on the type of x such that there exists one and only one permutation y of Ω for which $y^k = x$.

4. Determine the normalizers of $(1, 2, 3, 4, 5, 6)$, $(1, 2)(3, 4)(5, 6)$, and $(1, 2, 3)(4, 5)$ in S_6.

5. If a transitive permutation group of finite degree contains a single transposition, then it is a symmetric group, or it is imprimitive. Prove this proposition and give examples for both cases.

6. Let a be a permutation of degree n which is the product of r mutually disjoint cycles (including cycles of length 1). Prove that $\chi(a) = (-1)^{n-r}$.

7. Let a be a permutation of $\{1, 2, \ldots, n\}$. Prove that a and a^{-1} have the same number of inversions.

8. A finite permutation group can happen to be imprimitive in more than one way. Let Λ and M be two systems of imprimitivity. Show that $\Lambda \cap M$ is also a system of imprimitivity provided that $\Lambda \cap M$ contains more than a single symbol.

9. Show that the symmetric group S_4 has two essentially distinct permutational representations of degree 6; i. e., one representation cannot be obtained from the other by changing the notation of the symbols.

10. Let G be a transitive, imprimitive permutation group of finite degree containing a regular abelian subgroup H of the same degree. Prove that G contains an intransitive normal subgroup D such that $H \cap D \neq e$.

11. Let G^* and $^*G$ denote the two regular representations of a finite group G. Prove that G^* and $^*G$ are conjugate in the group of all permutations of the elements of G. Determine the intersection $G^* \cap {}^*G$.

12. Six of the diagonals of the faces of a cube form the edges of a regular tetrahedron. What does this mean for the corresponding symmetry groups?

CHAPTER 8

MONOMIAL GROUPS AND THE TRANSFER

8.1 Monomial Representations of Groups

A permutation

$$s = \begin{pmatrix} x_i \\ x_{is} \end{pmatrix} \qquad (i = 1, \ldots, n)$$

of n indeterminates $x_1, \ldots, x_n$ can be regarded as a linear substitution. Putting

$$x = \begin{bmatrix} x_1 \\ \vdots \\ x_n \end{bmatrix}, \quad x^s = \begin{bmatrix} x_{1s} \\ \vdots \\ x_{ns} \end{bmatrix}, \quad P_s = [\delta_{is,k}] \qquad (i, k = 1, \ldots, n),$$

where i indicates the rows and k the columns of the $n \times n$ matrix P_s and

$$\delta_{j,k} = \begin{cases} 1 & \text{for } j = k \\ 0 & \text{for } j \neq k, \end{cases}$$

we obtain

$$x^s = P_s x.$$

The matrix P_s is called the *permutation matrix* corresponding to the permutation s. For two permutations s and t, we have $P_{st} = P_s P_t$. As is well known, the determinant $|P_s|$ is equal to the character $\chi(s)$ in the sense of section 7.4.

We shall now generalize the permutations of $x_1, \ldots, x_n$ as follows: Let A be a multiplicative group and suppose that for every $\alpha \in A$ the products αx_i are defined. Moreover, we assume that $(\alpha\beta) x_i = \alpha(\beta x_i)$ for arbitrary $\alpha, \beta \in A$ and $\varepsilon x_i = x_i$ for the unit element ε of A. A familiar situation arises when A consists of numbers. In what follows, however, A may be an arbitrary multiplicative group; in particular, we do not assume that A is abelian.

By a *monomial substitution* $\bar{s}$ of $x_1, \ldots, x_n$ with coefficients in A, we mean a substitution of the following form

$$\bar{s} = \begin{pmatrix} x_i \\ \alpha_i x_{is} \end{pmatrix} \qquad (\alpha_i \in A; \quad i = 1, \ldots, n), \tag{8.1}$$

where $x_{1s}, \ldots, x_{ns}$ is a permutation of $x_1, \ldots, x_n$. The product of $\bar{s}$ and a monomial substitution

$$\bar{t} = \begin{pmatrix} x_i \\ \beta_i x_{it} \end{pmatrix} \qquad (\beta_i \in \Lambda; \quad i = 1, \ldots, n)$$

is the monomial substitution

$$\bar{s}\bar{t} = \begin{pmatrix} x_i \\ \alpha_i x_{is} \end{pmatrix} \begin{pmatrix} x_i \\ \beta_i x_{it} \end{pmatrix} = \begin{pmatrix} x_i \\ \alpha_i x_{is} \end{pmatrix} \begin{pmatrix} \alpha_i x_{is} \\ \alpha_i \beta_{is} x_{ist} \end{pmatrix} = \begin{pmatrix} x_i \\ \alpha_i \beta_{is} x_{ist} \end{pmatrix}. \tag{8.2}$$

Evidently, all monomial substitutions of $x_1, \ldots, x_n$ with coefficients in Λ form a group M. If Λ is finite, then $|M| = n!|\Lambda|^n$.

The analogous generalization of the permutation matrices yields the *monomial matrices*

$$M_{\bar{s}} = [\alpha_i \delta_{is,k}] \qquad (i, k = 1, \ldots, n). \tag{8.3}$$

One readily verifies that $\bar{s} \to M_{\bar{s}}$ is an isomorphism of M onto the multiplicative group of the monomial matrices.

To any monomial substitution $\bar{s}$, we assign the permutation

$$\bar{s}^* = \begin{pmatrix} x_i \\ x_{is} \end{pmatrix} \qquad (i = 1, \ldots, n).$$

Clearly,

$$\bar{s}^* \bar{t}^* = (\bar{s}\bar{t})^*$$

so that $\bar{s} \to \bar{s}^*$ is a homomorphism of M onto the symmetric group S_n. The kernel $\bar{D}$ of this homomorphism consists of all monomial substitutions

$$\bar{d} = \begin{pmatrix} x_i \\ \alpha_i x_i \end{pmatrix} \qquad (\alpha_i \in \Lambda; \quad i = 1, \ldots, n).$$

The corresponding monomial matrices $M_{\bar{d}}$ are diagonal.

It follows that, for a subgroup $\bar{U}$ of M, the mapping

$$\bar{u} \to \bar{u}^*, \qquad \bar{u} \in \bar{U}$$

is a homomorphism of $\bar{U}$ onto a permutation group $\bar{U}^*$ whose kernel is $\bar{U} \cap \bar{D}$. The monomial group $\bar{U}$ is said to be transitive if $\bar{U}^*$ is a transitive permutation group.

By a *monomial representation* of a group G, we mean a representation of G by monomial substitutions.

Let H be a subgroup of G of finite index $|G:H| = n$, and let

$$G = Hr_1 \cup \cdots \cup Hr_n \tag{8.4}$$

denote the decomposition of G into right cosets of H. For an arbitrary element s of G, we obtain

$$r_i s = h_i r_{is} \qquad (i = 1, \ldots, n), \tag{8.5}$$

where $h_i \in H$. Clearly, $r_{1s}, \ldots, r_{ns}$ is a permutation of $r_1, \ldots, r_n$. If φ is a homomorphism of H into a multiplicative group A that carries h_i into $\alpha_i \in A$, then the mapping

$$s \to \bar{s} = \begin{pmatrix} x_i \\ \alpha_i x_{is} \end{pmatrix} \qquad (i = 1, \ldots, n)$$

gives rise to a monomial representation $\bar{G}$ of G with coefficients in A. For, let $t \in G$,

$$r_i t = k_i r_{it} \qquad (k_i \in H;\ i = 1, \ldots, n),$$

and denote by β_i the image of k_i under φ so that

$$\bar{t} = \begin{pmatrix} x_i \\ \beta_i x_{it} \end{pmatrix} \qquad (i = 1, \ldots, n).$$

Then,

$$r_i st = h_i r_{is} t = h_i k_{is} r_{ist}$$

and by (8.2)

$$\bar{s}\bar{t} = \begin{pmatrix} x_i \\ \alpha_i \beta_{is} x_{ist} \end{pmatrix} = \overline{st}.$$

The permutation group $\bar{G}^*$ is the permutational representation induced by H in the sense of Theorem 7.2.2. We know that $\bar{G}^*$ is transitive.

If we replace the indeterminates $x_1, \ldots, x_n$ by

$$y_i = \gamma_i x_i \qquad (\gamma_i \in A; \quad i = 1, \ldots, n),$$

then any monomial group on $x_1, \ldots, x_n$ induces an isomorphic monomial group on $y_1, \ldots, y_n$ and vice versa. Such monomial groups are said to be equivalent. A transitive group of monomial substitutions is called monic if, for every index i, $1 \leqq i \leqq n$, it contains a substitution that carries x_1 into x_i (not only into $\gamma_i x_i$ for some $\gamma_i \in A$). Let G° be a transitive group of monomial substitutions on $x_1, \ldots, x_n$; then, there is an equivalent group that is monic. Indeed, since G° is transitive, it contains substitutions that carry x_1 into $\varrho_i x_i\,(i = 2, \ldots, n)$. Then, the corresponding group on $y_1 = x_1$, $y_i = \varrho_i x_i$ $(i = 2, \ldots, n)$ is obviously monic.

A monomial representation $\bar{G}$ of G induced by H is monic if we choose the representative of H in (8.4) to be e. We shall now prove that, conversely, every transitive monic monomial representation $\bar{G}$ of G is induced by a suitable subgroup H of G.

Let H denote the subgroup of all elements h of G such that the monomial substitutions $\bar{h} \in \bar{G}$ carry x_1 into αx_1, $\alpha \in \Lambda$. Since $\bar{G}$ is monic, we can choose substitutions $\bar{r}_i$ that carry x_1 into x_i $(i = 2, \ldots, n)$. Let r_i denote a pre-image of $\bar{r}_i$ in G. Then,

$$G = Hr_1 \cup Hr_2 \cup \cdots \cup Hr_n \qquad (r_1 = e)$$

is the decomposition of G into right cosets of H. For any $h \in H$, the substitution $\bar{h}$ carries x_1 into some αx_1. Hence, the mapping $\varphi: h \to \alpha$ is a homomorphism of H into Λ. Now, let s be an arbitrary element of G and

$$\bar{s} = \begin{pmatrix} x_i \\ \alpha_i x_{is} \end{pmatrix} \qquad (i = 1, \ldots, n) \tag{8.6}$$

its image in $\bar{G}$. For $i = 1, \ldots, n$ we have

$$r_i s = h_i r_{is}, \qquad h_i \in H. \tag{8.7}$$

It remains to show that φ maps each h_i in (8.7) onto the corresponding α_i in (8.6). By (8.7), we obtain

$$\bar{r}_i \bar{s} = \bar{h}_i \bar{r}_{is};$$

hence, $\bar{h}_i = \bar{r}_i \bar{s} \bar{r}_{is}^{-1}$. This shows that $\bar{h}_i$ carries x_1 into $\alpha_i x_i$ so that φ in fact maps h_i onto α_i. This completes the proof.

Let us replace the coset representatives r_i in (8.4) by other representatives $q_i = c_i r_i$, $c_i \in H$. Then, (8.5) becomes

$$q_i s = c_i r_i s = c_i h_i r_{is} = c_i h_i c_{is}^{-1} q_{is}.$$

Denoting the image of c_i under φ by γ_i and putting $y_i = \gamma_i x_i$, we have

$$\hat{s} = \begin{pmatrix} y_i \\ \gamma_i \alpha_i \gamma_{is}^{-1} y_{is} \end{pmatrix}.$$

Thus, another transversal in (8.4) leads to a monomial representation $\hat{G}$ equivalent to $\bar{G}$. It follows that the monomial matrices $M_{\bar{s}}$ and $M_{\hat{s}}$ satisfy the relation

$$M_{\hat{s}} = C M_{\bar{s}} C^{-1}, \tag{8.8}$$

where C denotes the diagonal matrix with the diagonal elements $\gamma_1, \ldots, \gamma_n$.

8.2 The Transfer

We use the notation of the previous section.

The monomial matrices $M_{\bar{s}}$ defined by (8.3) satisfy the equation

$$M_{\bar{s}} M_{\bar{t}} = M_{\overline{st}}.$$

If Λ consists of numbers, the multiplication theorem for determinants gives

$$|M_{\bar{s}}||M_{\bar{t}}| = |M_{\bar{s}\bar{t}}|.$$

Hence, in this case the mapping $\bar{s} \to |M_{\bar{s}}|$ is a homomorphism of the monomial group M into the multiplicative group of all non-zero complex numbers. Writing $||M_{\bar{s}}||$ for the modulus of $|M_{\bar{s}}|$, we have

$$||M_{\bar{s}}||\, ||M_{\bar{t}}|| = ||M_{\bar{s}\bar{t}}||,$$

so that $\bar{s} \to ||M_{\bar{s}}||$ is a homomorphism of M onto some multiplicative group of positive numbers.

Without any restrictions an Λ, the determinants need not be defined. But, if Λ is abelian, we can still use the familiar formula

$$||M_{\bar{s}}|| = \prod_{i=1}^{n} \alpha_i = d_{\bar{s}}$$

to obtain

$$d_{\bar{s}} d_{\bar{t}} = d_{\bar{s}\bar{t}}.$$

This relation may be false if Λ is not abelian, but even in that case we have at least

$$d_{\bar{s}} d_{\bar{t}} \equiv d_{\bar{s}\bar{t}} \qquad (\mathrm{mod}\ \Lambda')$$

where Λ' denotes the commutator subgroup of Λ. Thus, the mapping

$$\bar{s} \to \Lambda' d_{\bar{s}}$$

is a homomorphism of the monomial group M into Λ/Λ'.

We shall discuss this homomorphism for the monomial representation of a group G induced by a subgroup H of it as defined in section 8.1. For our purpose, we may confine ourselves to the case that Λ coincides with H.

Let

$$G = Hr_1 \cup \cdots \cup Hr_n$$

be the decomposition of G into right cosets of H. Using a somewhat more detailed notation instead of that in (8.5), we have

$$r_i s = h_{i,s} r_{is} \qquad (h_{i,s} \in H;\ i = 1, \ldots, n).$$

These equations mean that multiplication by s on the right induces a monomial substitution of the transversal $\{r_i\}$ with coefficients in H. In this case, the expression corresponding to $\Lambda' d_{\bar{s}}$ is

$$v(s) = H' \prod_{i=1}^{n} h_{i,s},$$

where H' denotes the commutator subgroup of H. The mapping

$$s \to v(s)$$

is a homomorphism of G into H/H'; it is called the *transfer* of G into H. When we wish to indicate the groups involved, we write $v(s) = v_{G\to H}(s)$.

As we saw in section 8.1, another choice of the transversal yields an equivalent monomial representation. By (8.8), the expressions $||M_{\bar{s}}||$ remain unchanged. In our case, this means that $v(s)$ does not depend on the choice of the transversal.

The transfer is a highly efficient tool in the theory of finite groups. Therefore, we shall derive the main properties of the transfer once more using an approach that is independent of the theory of monomial representations.

Let H be a subgroup of G of finite index $n = |G:H|$. We choose a right transversal $\{r_i\}$ to H in G to obtain the decomposition

$$G = Hr_1 \cup \cdots \cup Hr_n$$

into right cosets of H. For any element s of G, we then have

$$r_i s = h_{i,s} r_{is} \qquad (i = 1, \ldots, n),$$

where $h_{i,s} \in H$ and $r_{1s}, \ldots, r_{ns}$ is a permutation of $r_1, \ldots, r_n$. The mapping

$$s \to v(s) = H' \prod_{i=1}^{n} h_{i,s}$$

where H' denotes the commutator subgroup of H is called the *transfer* of G into H.

8.2.1 *The transfer $s \to v(s)$ is a homomorphism of G into H/H'. The coset $v(s)$ is independent of the right transversal to H in G.*

Proof. Our assertion follows from a straightforward calculation. For $s, t \in G$, we have

$$v(s) = H' \prod_{i=1}^{n} h_{i,s}, \qquad v(t) = H' \prod_{i=1}^{n} h_{i,t};$$

hence,

$$v(s)\, v(t) = H' \prod_{i=1}^{n} (h_{i,s} h_{i,t}).$$

On the other hand,

$$r_i st = h_{i,s} r_{is} t = h_{i,s} h_{is,t} r_{ist}.$$

Since $1s, \ldots, ns$ is a permutation of $1, \ldots, n$, this gives

$$v(st) = H' \prod_{i=1}^{n} (h_{i,s} h_{is,t}) = H' \prod_{i=1}^{n} (h_{i,s} h_{i,t}) = v(s)\, v(t).$$

For another transversal $\{q_i\} = \{c_i r_i\}$, where $c_i \in H$, we have

$$q_i s = c_i r_i s = c_i h_{i,s} = c_i h_{i,s} c_{is}^{-1} q_{is}.$$

Computing the transfer, we find

$$H' \prod_{i=1}^{s} (c_i h_{i,s} c_{is}^{-1}) = H' \prod_{i=1}^{n} h_{i,s} = v(s).$$

Thus, $v(s)$ is in fact independent of the choice of the transversal. This completes the proof.

The elements s of G such that $v(s) = H'$ form a normal subgroup G_1 of G, which is called the *kernel* of the transfer of G into H. The transfer is an isomorphism of G/G_1 into H/H'. Thus, G/G_1 is abelian so that G_1 contains the commutator subgroup of G.

Let U and H be subgroups of G such that $U \subset H \subset G$. We assume that the indices $|G:H| = n$ and $|H:U| = m$ are finite. Then, the transfer is transitive in the following sense:

8.2.2 *For every* $s \in G$,

$$v_{G \to U}(s) = v_{H \to U}(v_{G \to H}(s)).$$

Proof. Let

$$G = Hr_1 \cup \cdots \cup Hr_n, \quad H = Ua_1 \cup \cdots \cup Ua_m$$

be the coset decompositions. Thus,

$$G = \bigcup_{i=1}^{m} \bigcup_{j=1}^{n} Ua_i r_j$$

is the decompositions of G into right cosets of U. For $s \in G$, we have

$$r_j s = h_{j,s} r_{js} \qquad (h_{j,s} \in H;\ j = 1, \ldots, n),$$

$$a_i h_{j,s} = u_{i,j,s} a_l \qquad (u_{i,j,s} \in U;\ i = 1, \ldots, m).$$

This shows that

$$a_i r_j s = u_{i,j,s} a_l r_{js}.$$

So, we obtain

$$v_{G \to U}(s) = U' \prod_{i=1}^{m} \prod_{j=1}^{n} u_{i,j,s}.$$

Moreover,

$$v_{G \to H}(s) = H' \prod_{j=1}^{n} h_{j,s}.$$

As we observed above, $v_{H \to U}(h') = U'$ if $h' \in H'$. Accordingly, we obtain

$$v_{H \to U}(v_{G \to H}(s)) = v_{H \to U}\left(H' \prod_{j=1}^{n} h_{j,s}\right) = v_{H \to U}\left(\prod_{j=1}^{n} h_{j,s}\right)$$

$$= \prod_{j=1}^{n} v_{H \to U}(h_{j,s}) = U' \prod_{j=1}^{n} \prod_{i=1}^{m} u_{i,j,s} = v_{G \to U}(s).$$

This completes the proof.

Since the transfer is independent of the transversal it is often convenient to compute $v(s)$ by means of a particular transversal depending on s.

We consider the permutational representation of G induced by its subgroup H of index n. The permutation corresponding to $s \in G$ is

$$\begin{pmatrix} Hr_i \\ Hr_i s \end{pmatrix} \qquad (i = 1, \ldots, n). \tag{8.9}$$

Suppose that this permutation is a product of t mutually disjoint cycles of lengths $l_1, \ldots, l_t$. If there are fixed elements, then the corresponding cycles of length 1 have to be included so that

$$\sum_{j=1}^{t} l_j = n = |G:H|. \tag{8.10}$$

The numbers l_j divide the order of s, because the order of the permutation (8.9) is the same as that of s or is a proper divisor of it.

Let us assume that

$$(Hr_1, Hr_2, \ldots, Hr_l)$$

is one of the cycles of the permutation (8.9). Then, we have

$$Hr_1 s = Hr_2,\ H_2 s = Hr_3,\ \ldots,\ Hr_{l-1} s = Hr_l,\ Hr_l s = Hr_1.$$

We now replace $r_1, r_2, r_3, \ldots, r_l$ by new representatives

$$r_1' = r_1,\ r_2' = r_1 s,\ r_3' = r_1 s^2,\ \ldots,\ r_l' = r_1 s^{l-1};$$

this gives

$$r_k' s = r_{k+1}' \qquad \text{for} \quad k = 1, \ldots, l-1.$$

$$r_l' s = r_1 s^l = r_1 s^l r_1^{-1} r_1.$$

Thus, the contribution of our l cosets to $v(s)$ is $H' r_1 s^l r_1^{-1}$. Here, s^l is the least positive power of s that is contained in $r_1^{-1} H r_1$.

By applying the same procedure to each cycle of the permutation (8.9), we obtain the following rule for computing the transfer.

Let $q_1, \ldots, q_t$ be representatives of an arbitrary coset in each cycle of the permutation (8.9). Then,

$$v(s) = H' \prod_{j=1}^{t} q_j s^{l_j} q_j^{-1}. \tag{8.11}$$

8.3 An Application of the Transfer

There are several theorems of the following type: If a finite group G contains a subgroup H satisfying certain conditions, then G contains a normal subgroup D such that

$$G = HD, \qquad H \cap D = e$$

or, in other words, G is a semi-direct product of H and D. Such a normal subgroup D is called a *normal complement* of H. One of the best known theorems of this kind is:

8.3.1 Burnside's Theorem. *If a Sylow p-subgroup P of the finite group G lies in the centre of its normalizer, then G contains a normal complement of P.*

Proof. From the conditions on P it follows, in particular, that P is abelian so that $P' = e$. We consider the transfer of G into P. By (8.11), we obtain for $x \in P$

$$v(x) = \prod_{j=1}^{t} q_j x^{l_j} q_j^{-1},$$

where, by (8.10),

$$\prod_{j=1}^{t} l_j = |G:P| = n\,.$$

Each factor $q_j x^{l_j} q_j^{-1}$ of $v(x)$ belongs to P and is conjugate in G to x^{l_j}. By Theorem 3.3.5, x^{l_j} and $q_j x^{l_j} q_j^{-1}$ are conjugate in the normalizer $\mathsf{N}(P)$ of P in G. Since P belongs to the centre of $\mathsf{N}(P)$, this shows that $q_j x^{l_j} q_j^{-1} = x^{l_j}$ $(j = 1, \ldots, t)$. So, we have $v(x) = x^n$ for every $x \in P$. As n is relatively prime to the order of P, we conclude that the transfer of G into P maps P onto itself. Accordingly the elements of P form a transversal to the kernel D of the transfer in G. $G = PD$, $P \cap D = e$, and the proof is complete.

We use Theorem 8.3.1 to prove:

8.3.2 *The order of a finite non-abelian simple group is divisible by 12 or by the cube of its smallest prime factor.*

Proof. Let p denote the smallest prime divisor of the order of the finite non-abelian simple group G. Suppose that the order of a Sylow p-subgroup P of G is p or p^2. Then, P is abelian. By Theorem 8.3.1, P does not belong to the centre of its normalizer $\mathsf{N}(P)$. If P is of order p, then the automorphism group of P is of order $p - 1$ (cf. Theorem 5.4.3); and if P is of order p^2, then the order of its automorphism group is $p(p-1)$ or $(p^2-1)(p^2-p) = p(p-1)^2(p+1)$, according as P is cyclic or elementary abelian (cf. Theorem 5.4.3 and section 12.1). For an odd prime number p, the orders of the automorphism groups are not divisible by any prime number greater than p. Thus, transformation of P by elements of $\mathsf{N}(P)$ cannot induce an automorphism of P other than the identity. For $p = 2$, the only order of an automorphism group that is divisible by a prime number >2 is $p(p-1)^2(p+1) = 6$. In this case, P is isomorphic to the four-group, and $|G|$ is divisible by 12.

CHAPTER 9

NILPOTENT AND SUPERSOLUBLE GROUPS

9.1 Higher Commutators

Let F be a free group of countably infinite rank. The letters x, y, z, sometimes also indexed, stand for free generators of F. The operation

$$[x, y] = x^{-1}y^{-1}xy$$

can be iterated in various ways to yield the so-called *higher commutators*, for instance,

$$[x, [y, z]], \quad [[x, y], z], \quad [[x_1, x_2], [x_3, x_4]].$$

Higher commutators of free generators of F are called *commutator polynomials*.

The weight of a commutator polynomial is defined as follows. Commutator polynomials of weight 1 are the free generators. If

$$f_1(x_1, \ldots, x_{w_1}), \; f_2(x_{w_1+1}, \ldots, x_{w_1+w_2})$$

are commutator polynomials of weights w_1 and w_2, respectively, then

$$[f_1(x_1, \ldots, x_{w_1}), f_2(x_{w_1+1}, \ldots, x_{w_1+w_2})] = f(x_1, \ldots, x_{w_1+w_2})$$

is a commutator polynomial of weight $w = w_1 + w_2$. For example,

$$w = 2: \; [x_1, x_2],$$

$$w = 3: \; [[x_1, x_2], x_3] \quad \text{and} \quad [x_1, [x_2, x_3]],$$

$$w = 4: \; [[x_1, x_2], [x_3, x_4]].$$

The so-called *left-normed commutators* $[x_1, \ldots, x_w]$ are defined inductively as follows:

$$[x_1, x_2],$$

$$[[x_1, x_2], x_3] = [x_1, x_2, x_3],$$

$$\cdots\cdots\cdots\cdots\cdots\cdots$$

$$[[x_1, \ldots, x_{w-1}], x_w] = [x_1, \ldots, x_w].$$

By substituting elements of a group G for the free generators in a commutator polynomial of weight w, we obtain commutators of weight w of G. Note that distinct free generators may be replaced by one and the same element of G.

We list some commutator identities, which can be verified by straightforward calculation:

$$[x, y] = [y, x]^{-1}. \tag{9.1}$$

Writing $x^y = y^{-1}xy$, we have

$$x^y = x[x, y], \tag{9.2}$$

$$[xy, z] = [x, z]^y [y, z] = [x, z][x, z, y][y, z], \tag{9.3}$$

$$[x, yz] = [x, z] [x, y]^z = [x, z][x, y][x, y, z]. \tag{9.4}$$

Since these identities hold in a free group, they remain valid if x, y, z are replaced by elements of an arbitrary group G. For any endomorphism γ of G, we have

$$[a, b]^\gamma = [a^\gamma, b^\gamma] \qquad (a, b \in G).$$

Hence, for an arbitrary commutator polynomial $f(x_1, \ldots, x_w)$,

$$f(a_1, \ldots, a_w)^\gamma = f(a_1^\gamma, \ldots, a_w^\gamma) \qquad (a_i \in G). \tag{9.5}$$

For two non-empty complexes A and B of the group G, let $[A, B]$ denote the subgroup of G generated by all commutators $a^{-1}b^{-1}ab$, where $a \in A$, $b \in B$. In most applications, A and B are subgroups of G. If A and B are admissible with respect to an operator domain Ω, then so is $[A, B]$.

Let $U_1, U_2, \ldots$ be arbitrary subgroups of G, not necessarily distinct from one another. For a commutator polynomial $g(x)$ of weight 1, i. e. $g(x) = x$, we define

$$g(U_i) = U_i \qquad (i = 1, 2, \ldots).$$

Now, let f_1, f_2 and $f = [f_1, f_2]$ be the commutator polynomials considered above. Suppose that the subgroups

$$f_1(U_1, \ldots, U_{w_1}),\ f_2(U_{w_1+1}, \ldots, U_{w_1+w_2})$$

of G have already been defined. Then, we put

$$f(U_1, \ldots, U_{w_1+w_2}) = [f_1(U_1, \ldots, U_{w_1}),\ f_2(U_{w_1+1}, \ldots, U_{w_1+w_2})].$$

In this way, we obtain for every system of subgroups and every commutator polynomial a well-defined subgroup of G.

Clearly, $f(U_1, \ldots, U_w)$ contains all commutators $f(u_1, \ldots, u_w)$, $u_i \in U_i$, but it is not necessarily generated by these commutators. However, we can prove:

9.1.1 *If $A_1, \ldots, A_w$ are normal subgroups of G, then $f(A_1, \ldots, A_w)$ is generated by the commutators $f(a_1, \ldots, a_w)$, $a_i \in A_i$.*

Proof. The assertion is trivial for commutator polynomials of weight 1. Suppose that the theorem is true for all commutator polynomials of weight less than the weight w of f. Now, we have $f = [f_1, f_2]$, where f_1 and f_2 are of weight less than w. By the inductive hypothesis, the subgroups $f_1(A_1, \ldots, A_{w_1})$ and $f_2(A_{w_1+1}, \ldots, A_{w_1+w_2})$ are generated by the corresponding commutators. In other words, if K_1 denotes the complex of all commutators

$$f_1(a_1, \ldots, a_{w_1}), \qquad a_i \in A_i$$

and K_2 is defined similarly for f_2, then we have

$$\langle K_1 \rangle = f_1(A_1, \ldots, A_{w_1}) = N_1,$$

$$\langle K_2 \rangle = f_2(A_{w_1+1}, \ldots, A_{w_1+w_2}) = N_2.$$

By (9.5), N_1 and N_2 are normal subgroups of G. Let N denote the normal subgroup of G generated by all the commutators $f(a_1, \ldots, a_w)$, $a_i \in A$. Then, we have

$$N = [K_1, K_2]; \tag{9.6}$$

moreover, $f(A_1, \ldots, A_w) = [N_1, N_2]$. Clearly,

$$N = [K_1, K_2] \subseteq [N_1, N_2].$$

On the other hand, (9.6) shows that $[k_1, k_2] \in N$ for any two elements k_1 of K_1 and k_2 of K_2. Since both K_1 and K_2 are normal complexes, it follows from (9.3) and (9.4) that $[n_1, n_2] \in N$ for all $n_1 \in N_1$, $n_2 \in N_2$. This means that $[N_1, N_2] \subseteq N$. Hence, $f(A_1, \ldots, A_w) = N$, and the theorem is proved.

The following substitution rule is an immediate consequence of Theorem 9.1.1.:

9.1.2 *Let $f(y_1, \ldots, y_k)$ be a commutator polynomial of weight k and let*

$$\varphi_i(x_{i,1}, \ldots, x_{i,w_i}) \qquad (i = 1, \ldots, k)$$

be k commutator polynomials of weights $w_1, \ldots, w_k$, respectively. Put

$$g(x_{1,1}, \ldots, x_{k,w_k}) = f(\varphi_1(x_{1,1}, \ldots, x_{1,w_k}), \ldots, \varphi_k(x_{k,1}, \ldots, x_{k,w_k}))$$

so that g is a commutator polynomial of weight $w_1 + \cdots + w_k$. If

$$A_{i,j} \qquad (i = 1, \ldots, k;\ j = 1, \ldots, w_i)$$

are normal subgroups of a group G, then

$$g(A_{1,1}, \ldots, A_{k,w_k}) = f(B_1, \ldots, B_k),$$

where

$$B_i = \varphi_i(A_{i,1}, \ldots, A_{i,w_i}) \qquad (i = 1, \ldots, k).$$

As an example, let us consider the commutator polynomial $[[x_1, x_2], [x_3, x_4]]$. For $A_1 = A_2 = A_3 = A_4 = G$, we obtain the normal subgroup $N = [[G, G], [G, G]]$ of G. By Theorem 9.1.1, N is generated by the commutators $[[a_1, a_2], [a_3, a_4]]$, where the a_i range over all elements of G. But, $[G, G] = G'$ is the commutator subgroup, so that by Theorem 9.1.2

$$[[G,G], [G,G]] = [G', G'] = G''.$$

It follows from (9.1) that

$$[A, B] = [B, A]$$

for arbitrary complexes A, B. If A_1, A_2, A_3 are normal subgroups of G, then Theorem 9.1.2 shows that

$$[A_1, A_2, A_3] = [[A_1, A_2], A_3] = [[A_2, A_1] A_3] = [A_2, A_1, A_3].$$

Later, we shall need the following theorem:

9.1.3 *Let, A, B, C be normal subgroups of G. Then, each of the normal subgroups $[A, B, C]$, $[B, C, A]$, $[C, A, B]$ is contained in the product of the other two.*

Proof. We write

$$[A, B, C] = P, \quad [B, C, A] = Q, \quad [C, A, B] = R$$

and $D = PQ$. By symmetry, it is sufficient to prove that $R \subseteq D$. Now, R is generated by the elements

$$[c, a, b] = [[c, a], b] = [b, [c, a]]^{-1} \qquad (a \in A, \ b \in B, \ c \in C).$$

Hence, we have to show that every element of the form

$$[b, [c, a]] = [b, c^{-1}a^{-1}ca]$$

belongs to D. By (9.4), we have

$$[b, ca] = [b, a][b, c][b, c, a]$$

and

$$[b, ca] = [b, aa^{-1}ca] = [b, a^{-1}ca][b, a][b, a, a^{-1}ca].$$

Now, $[b, c, a] \in Q$, and as Q is a normal subgroup, we obtain $[b, a, a^{-1}ca] = [b, a, c_1]$, where $c_1 \in C$, so that

$$[b, a, a^{-1}ca] \in [B, A, C] = [A, B, C] = P.$$

Thus, $[b, c, a]$ and $[b, a, a^{-1}ca]$ lie in D; hence,

$$[b, ca] \equiv [b, a][b, c] \equiv [b, a^{-1}ca][b, a] \qquad (\operatorname{mod} D)$$

or

$$[b, a^{-1}ca] \equiv [b, a][b, c][b, a]^{-1} \qquad (\operatorname{mod} D).$$

By (9.1) and (9.2), we obtain

$$[b, a][b, c][b, a]^{-1} = [b, c]\big[[b, c],\ [a, b]\big].$$

As A is a normal subgroup, we have

$$[a, b] = a^{-1}b^{-1}ab = a_1 \in A$$

so that

$$\big[[b, c],\ [a, b]\big] = \big[[b, c], a_1\big] = [b, c, a_1]$$

belongs to Q and hence to D. This shows that

$$[b, a^{-1}ca] \equiv [b, c] \qquad (\operatorname{mod} D). \tag{9.7}$$

As B is a normal subgroup and (9.7) holds for every $b \in B$, we obtain

$$[b, c^{-1}] = b^{-1}cbc^{-1} \in B$$

and

$$\big[[b, c^{-1}], a^{-1}ca\big] = [b, c^{-1}, a^{-1}ca] \equiv [b, c^{-1}, c] \qquad (\operatorname{mod} D). \tag{9.8}$$

Equation (9.4) implies that

$$[b, c^{-1}a^{-1}ca] = [b, a^{-1}ca][b, c^{-1}][b, c^{-1}, a^{-1}ca],$$

so that (9.7) and (9.8) yield

$$[b, c^{-1}a^{-1}ca] \equiv [b, c][b, c^{-1}][b, c^{-1}, c] \qquad (\operatorname{mod} D).$$

Using (9.4), we obtain

$$[b, c][b, c^{-1}][b, c^{-1}, c] = [b, c^{-1}c] = [b, e] = e$$

which means that

$$[b, c^{-1}a^{-1}ca] \equiv e \qquad (\operatorname{mod} D).$$

This completes the proof.

9.2 Central Series

In an arbitrary group G, we put

$$Z^0(G) = e,$$

$$Z^1(G) = \mathsf{Z}(G) = \text{centre of } G,$$

and more generally, we define $Z^i(G)$ inductively by

$$Z^i(G)/Z^{i-1}(G) = \text{centre of } G/Z^{i-1}(G) \qquad (i = 1, 2, \ldots).$$

The sequence

$$e = Z^0(G) \subseteq Z^1(G) \subseteq Z^2(G) \subseteq \cdots$$

is called the *upper central chain* of G. When there is no danger of confusion, we write Z^i instead of $Z^i(G)$. Since the centre is a characteristic subgroup, each $Z^i(G)$ is a characteristic subgroup of G.

If

$$Z^k(G) = Z^{k+1}(G) = \cdots$$

for some k, then $Z^k(G)$ is called the *hypercentre* of G.

The group G is said to be *nilpotent* if the upper central chain leads to G itself in a finite number of steps. More precisely, G is *nilpotent of class* c if $Z^c(G) = G$ but $Z^{c-1}(G) \neq G$. In this case, the characteristic series

$$e = Z^0(G) \subset Z^1(G) \subset Z^2(G) \subset \cdots \subset Z^{c-1}(G) \subset Z^c(G) = G \tag{9.9}$$

is called the *upper central series* of G.

As the factor groups of the upper central series are abelian, every nilpotent group is soluble. Every subgroup U/Z^{i-1} of the centre Z^i/Z^{i-1} of G/Z^{i-1} is a normal subgroup of G/Z^{i-1}. Thus, every subgroup U such that $Z^{i-1} \subseteq U \subseteq Z^i$ is a normal subgroup of G. Hence, by refining the upper central series of a finite nilpotent group to a composition series, we obtain a chief series. It follows that the chief indices of a finite nilpotent group are prime numbers.

By a *central series* of G, we mean a finite normal series

$$e = G_0 \subseteq G_1 \subseteq G_2 \subseteq \cdots \subseteq G_{m-1} \subseteq G_m = G \tag{9.10}$$

such that

$$G_i/G_{i-1} \text{ lies in the centre of } G/G_{i-1} \qquad (i = 1, \ldots, m). \tag{9.11}$$

Clearly, (9.11) is equivalent to the condition

$$[G_i, G] \subseteq G_{i-1} \qquad (i = 1, \ldots, m).$$

We shall show that the terms of the upper central series are maximal in the following sense: If G_i is the i-th term of an arbitrary central series (9.10) of G, then

$$G_i \subseteq Z^i(G) \qquad (i = 0, 1, \ldots, m). \tag{9.12}$$

This is certainly true for $i = 0$. We proceed by induction and assume that

$$G_{i-1} \subseteq Z^{i-1}(G) = Z^{i-1}.$$

Then, we have

$$G/Z^{i-1} \cong (G/G_{i-1})/(Z^{i-1}/G_{i-1}),$$

so that there is a homomorphism of G/G_{i-1} onto G/Z^{i-1}. This homomorphism maps G_i/G_{i-1} into the centre Z^i/Z^{i-1} of G/Z^{i-1}, because G_i/G_{i-1} belongs to the centre of G/G_{i-1}. The image of G_i/G_{i-1} is G_iZ^{i-1}/Z^{i-1}. Therefore, $G_iZ^{i-1} \subseteq Z^i$, hence, $G_i \subseteq Z^i$. Thus, (9.12) holds for $i = 0, 1, \ldots, m$, and we have the following result:

9.2.1 *A group G is nilpotent if and only if it has a central series of finite length. If G is nilpotent of class c, then the length of every central series is at least equal to c.*

Let U be an arbitrary subgroup of the nilpotent group G. If $Z^{i-1} \subseteq U$ for some i, then Z^i belongs to the normalizer of U in G. Indeed, every element of U/Z^{i-1} commutes with every element of the centre Z^i/Z^{i-1} of G/Z^{i-1}. Consequently, Z^i/Z^{i-1} belongs to the normalizer of U/Z^{i-1} in G/Z^{i-1}; hence, Z^i is contained in the normalizer of U in G. Since at least $Z^0 = e$ lies in every subgroup, we obtain the following theorem:

9.2.2 *Every proper subgroup U of a nilpotent group G is properly contained in its normalizer. If G is nilpotent of class c, then the sequence of the iterated normalizers of U ascends to G in at most c steps.*

In section 3.4, we observed that every finite p-group is nilpotent. It follows from Theorem 4.1.2 that also the direct product of a finite number of finite p-groups is nilpotent. On the other hand, Theorem 3.3.2 and Theorem 9.2.2 show that every Sylow subgroup of a finite nilpotent group is a normal subgroup. Hence, there is only one Sylow p-subgroup for every prime divisor p of the order. Any two Sylow subgroups corresponding to distinct primes commute elementwise, because they are normal subgroups intersecting trivially. This yields a characteristic property of finite nilpotent groups:

9.2.3 *A finite group is nilpotent if and only if it is the direct product of its Sylow subgroups.*

In an arbitrary group G, we define the *lower central chain*

$$G = Z_1(G) \supseteq Z_2(G) \supseteq Z_3(G) \supseteq \cdots$$

as follows:

$$Z_i(G) = [Z_{i-1}(G), G] \qquad (i = 2, 3, \ldots). \tag{9.13}$$

Evidently, the $Z_i(G)$ are fully invariant subgroups of G. When the meaning is obvious from the context, we write Z_i instead of $Z_i(G)$. It follows from (9.13) that

$$Z_{i-1}(G)/Z_i(G) \quad \text{belongs to the centre of} \quad G/Z_i(G).$$

If the lower central series terminates in the unit element after a finite number of steps, it is therefore a central series, the *lower central series*; hence, G is nilpotent.

Conversely, suppose that G is nilpotent and that (9.10) an arbitrary central series. Then, we have

$$Z_i(G) \subseteq G_{m-i+1} \qquad (i = 1, \ldots, m-1). \tag{9.14}$$

Clearly, this is true for $i = 1$. Assume that

$$Z_{i-1} \subseteq G_{m-(i-1)+1}. \tag{9.15}$$

As $G_{m-(i-1)+1}/G_{m-i+1}$ lies in the centre of G/G_{m-i+1}, we obtain

$$[G_{m-(i-1)+1}, G] \subseteq G_{m-i+1};$$

hence, (9.15) implies that

$$Z_i = [Z_{i-1}, G] \subseteq G_{m-i+1}.$$

Thus, (9.14) is proved. It follows that $Z_{m+1}(G) = e$ at the latest.

Let G be nilpotent of class c. If we replace the arbitrary central series (9.10) by the upper central series (9.9), then (9.14) shows that

$$Z_i(G) \subseteq Z^{c-i+1}(G) \qquad (i = 1, \ldots, c+1). \tag{9.16}$$

Therefore, $Z_{c+1}(G) = e$. Moreover, no term of the lower central series prior to Z_{c+1} is equal to e, because G has no central series of length less than c. Hence, the lower central series of G has the form

$$G = Z_1(G) \supset Z_2(G) \supset \cdots \supset Z_c(G) \supset Z_{c+1}(G) = e. \tag{9.17}$$

We observe that (9.14) states a minimal property of the terms of the lower central series similar to the maximal property of the terms of the upper central series expressed by (9.12). This accounts for the names. Summarizing our main results, we obtain:

9.2.4 *If G is nilpotent of class c, then the upper and the lower central series of G satisfy the relation* (9.16).

For a subgroup U of G, it follows from (9.13) by induction that

$$Z_i(U) \subseteq Z_i(G).$$

Moreover, if N is a normal subgroup of G, then

$$Z_i(G/N) = Z_i(G)N/N.$$

So we have:

9.2.5 *Every subgroup and every factor group of a nilpotent group G is nilpotent and its class does not exceed the class of c.*

We shall now derive further relations between the terms of the upper and lower central series. Let

$$H = H_1 \supseteq H_2 \supseteq H_3 \supseteq \cdots \tag{9.18}$$

be a sequence of subgroups of an arbitrary group G such that

$$[H_i, G] \subseteq H_{i+1} \qquad (i = 1, 2, \ldots). \tag{9.19}$$

It is easy to see that the H_i are normal subgroups of G; for, let $x_i \in H_i$, $y \in G$, then (9.18) and (9.19) imply that

$$x_i^{-1} y^{-1} x_i y \in [H_i, G] \subseteq H_{i+1} \subseteq H_i;$$

hence, $y^{-1} x_i y \in H_i$. Moreover, if

$$Z_i(G) \subseteq H_r$$

for some indices i, r, then

$$Z_{i+1}(G) = [Z_i(G), G] \subseteq [H_r, G] \subseteq H_{r+1}$$

and by induction

$$Z_{i+j}(G) \subseteq H_{r+j} \qquad (j = 1, 2, \ldots). \tag{9.20}$$

If G is nilpotent of class c, we may take

$$H_i = Z^{c-i+1}(G) \qquad (i = 1, \ldots, c+1) \tag{9.21}$$

to obtain a sequence (9.18) with the property (9.19) and $H_1 = G$. Since $G = Z_1(G) = H_1$, it follows from (9.20) that

$$Z_i(G) \subseteq H_i = Z^{c-i+1}(G).$$

This is again (9.16). But, now, we have the sharper result that $Z_i(G)$ cannot be contained in $Z^{c-1}(G)$. For, otherwise, (9.20) and (9.21) would imply that $Z_c(G) \subseteq Z^0(G) = e$, which contradicts the assumption that G is nilpotent of class c.

For an arbitrary group G, we obtain

$$[H_k, Z_i(G)] \subseteq H_{i+k} \qquad (i, k = 1, 2, \ldots). \tag{9.22}$$

We prove this relation by induction on i. By (9.19), we have

$$[H_k, Z_1(G)] = [H_k, G] \subseteq H_{1+k}$$

so that (9.22) holds for $i = 1$ and all k. Assume that

$$[H_k, Z_{i-1}] \subseteq H_{i-1+k} \tag{9.23}$$

for all k. Then, it follows from Theorem 9.1.3 that

$$\begin{aligned}[H_k, Z_i] = [Z_i, H_k] = \big[[Z_{i-1}, G], H_k\big] &= [G, Z_{i-1}, H_k] \\ \subseteq [Z_{i-1}, H_k, G][H_k, G, Z_{i-1}].\end{aligned}$$

Now, the inductive hypothesis (9.23) implies that

$$[Z_{i-1}, H_k, G] = [[Z_{i-1}, H_k], G] \subseteq [H_{i-1+k}, G] \subseteq H_{i+k},$$
$$[H_k, G, Z_{i-1}] = [[H_k, G], Z_{i-1}] \subseteq [H_{k+1}, Z_{i-1}] \subseteq H_{i+k}.$$

This completes the proof of (9.22).

Taking $H_k = Z_k(G)$, (9.22) shows that

$$[Z_i(G), Z_k(G)] \subseteq Z_{i+k}(G). \tag{9.24}$$

If G is nilpotent of class c, then we may take $H_k = Z^{c-k+1}(G)$ and then (9.22) implies that

$$[Z^{c-k+1}, Z_i] \subseteq Z^{c-i-k+1}$$

on the understanding that $Z^{-1} = Z^{-2} = \cdots = e$. Writing $c - k + 1 = j$, we obtain

$$[Z^j, Z_i] \subseteq Z^{j-i}.$$

In particular, $Z_i(G)$ and $Z^i(G)$ commute elementwise.

We now use (9.24) to derive a result about higher commutators of an arbitrary group.

9.2.6 *In an arbitrary group G, every commutator of weight w is contained in $Z_w(G)$.*

Proof. This is certainly true for commutators of weight 1, i. e. for the elements of G. We assume that our theorem holds for all commutators of weight less than w. If a is a commutator of weight w, then $a = [a_1, a_2]$ where a_1, a_2 are commutators of weights w_1, w_2, respectively, and $w_1 < w$, $w_2 < w$, $w_1 + w_2 = w$. By the inductive hypothesis, we have $a_1 \in Z_{w_1}$, $a_2 \in Z_{w_2}$. Hence, by (9.24)

$$a = [a_1, a_2] \in [Z_{w_1}, Z_{w_2}] \subseteq Z_{w_1+w_2} = Z_w,$$

which proves our assertion.

Equation (9.13) shows that $Z_k(G)$ is generated by commutators of weight k. Hence, $Z_i(Z_k(G))$ is generated by commutators of weight ik. Thus, Theorem 9.2.6 gives

$$Z_i(Z_k(G)) \subseteq Z_{ik}(G).$$

The n-th derived group $G^{(n)}$ of G is generated by commutators of weight 2^n. So, we conclude from Theorem 9.2.6 that

$$G^{(n)} \subseteq Z_{2^n}(G).$$

Now, suppose that G is nilpotent of class c and $G^{(n-1)} \neq e$, but $G^{(n)} = e$. Then, we obtain $2^{n-1} < c + 1$ or

$$n \leqq 1 + \log_2 c.$$

Hence;

9.2.7 *If G is nilpotent of class c, then G is n-step metabelian where $n \leqq 1 + \log_2 c$.*

As a counterpart to Theorem 9.2.6, we state:

9.2.8 *The left-normed commutators $[a_1, \ldots, a_w]$ of weight w generate $Z_w(G)$.*

Proof. Our assertion holds trivially for $w = 1$. Further, $Z_2(G) = [G, G]$, which means that $Z_2(G)$ is generated by the (left-normed) commutators of weight 2. We assume that Z_{w-1} is generated by the left-normed commutators of weight $w - 1$.

The definition $Z_w = [Z_{w-1}, G]$ shows that Z_w is generated by the commutators $[k, g]$, where $k \in Z_{w-1}$, $g \in G$. By the inductive hypothesis, we have

$$k = c_1^{\varepsilon_1} c_2^{\varepsilon_2} \cdots c_r^{\varepsilon_r} \quad (\varepsilon_i = \pm 1),$$

where the c_i are left-normed commutators of weight $w - 1$. By (9.3),

$$[k, g] = [c_1^{\varepsilon_1} c_2^{\varepsilon_2} \cdots c_r^{\varepsilon_r}, g]$$

is a product of commutators of the form $[c_i, g]^h$ or $[c_i^{-1}, g]^h$, where $h \in G$. As c_i is a left-normed commutator of weight $w - 1$, we have

$$[c_i, g]^h = [c_i^h, g^h] = [[g_1, \ldots, g_{w-1}], g^h] = [g_1, \ldots, g_{w-1}, g^h],$$

so that $[c_i, g]^h$ is a left-normed commutator of weight w. Again, (9.3) gives

$$e = [c_i c_i^{-1}, g] = [c_i, g]^{c_i^{-1}} [c_i^{-1}, g],$$

which shows that

$$[c_i^{-1}, g]^h = [[g_1', \ldots, g_{w-1}'], g_\omega']^{-1} = [g_1', \ldots, g_w']^{-1}.$$

Consequently, $[k, g]$ is a product of left-normed commutators of weight w and inverses of such commutators. This completes the proof.

The next theorem states an important property of nilpotent normal subgroups of an arbitrary group.

9.2.9 *Let A and B be nilpotent normal subgroups of an arbitrary group G. Then AB is also nilpotent. Moreover, if A and B are nilpotent of class a and b, respectively, then the class of AB does not exceed $a + b$.*

Proof. By Theorem 9.2.8, $Z_k(AB)$ is generated by the left-normed commutators $[z_1, \ldots, z_k]$, $z_i \in AB$. We write $z_i = x_i y_i$, where $x_i \in A$, $y_i \in B$.

First, we show that $[z_1, \ldots, z_k]$ can be expressed as a product of left-normed commutators $[u_1, \ldots, u_k]$ where all the u_i belong to A or to B. This is obvious for $k = 1$. Assume that there exists a representation

$$[z_1, \ldots, z_{k-1}] = c_1 \cdots c_t,$$

where the c_i are left-normed commutators of weight $k-1$ of elements of A or B. By (9.4), we obtain

$$\begin{aligned}[z_1, \ldots, z_{k-1}, z_k] &= [[z_1, \ldots, z_{k-1}], x_k y_k] \\ &= [c_1 \cdots c_t, x_k y_k] = [c_1 \cdots c_t, y_k][c_1 \cdots c_t, x_k]^{y_k} \\ &= [c_1 \cdots c_t, y_k][c_1' \cdots c_t', x_k'],\end{aligned} \tag{9.25}$$

where $c_i' = c_i^{y_k}$, $x_k' = x_k^{y_k}$. Since A and B are normal subgroups, the elements of each c_i' belong to A or to B; moreover, $x_k' \in A$. By applying (9.3) to (9.25), we obtain a representation of $[z_1, \ldots, z_k]$ as a product of left-normed commutators of weight k whose elements belong to A or B.

We have

$$[u_1, \ldots, u_k] = [u_1, \ldots, u_{k-1}]^{-1} u_k^{-1} [u_1 \ldots, u_{k-1}] u_k.$$

If $[u_1, \ldots, u_{k-1}] \in Z_i(A)$ and $u_k \in B$, then $[u_1, \ldots, u_k] \in Z_i(A)$ as $Z_i(A)$ is a normal subgroup of AB. But, if $[u_1, \ldots, u_{k-1}] \in Z_i(A)$ and $u_k \in A$, then $[u_1, \ldots, u_k] \in Z_{i+1}(A)$. Similarly, $[u_1, \ldots, u_{k-1}] \in Z_i(B)$ implies that $[u_1, \ldots, u_k]$ belongs to $Z_i(B)$ or $Z_{i+1}(B)$ according as $u_k \in A$ or $u_k \in B$.

The first step of the proof shows that the left-normed commutators $[u_1, \ldots, u_{a+b+1}]$, $u_i \in A$ or $u_i \in B$, generate $Z_{a+b+1}(AB)$. Now, $[u_1, \ldots, u_{a+b+1}]$ contains at least $a+1$ elements of A or $b+1$ elements of B. From the argument preceding this paragraph, it follows that $[u_1, \ldots, u_{a+b+1}]$ belongs to $Z_{a+1}(A)$ or $Z_{b+1}(B)$. But, since $Z_{a+1}(A) = Z_{b+1}(B) = e$, we obtain $[u_1, \ldots, u_{a+b+1}] = e$ so that $Z_{a+b+1}(AB) = e$. This completes the proof.

It follows from Theorem 9.2.9 that the product of all nilpotent normal subgroups of a finite group G is itself a nilpotent normal subgroup of G. This product is called the *Fitting subgroup* of G. Clearly the Fitting subgroup is a characteristic subgroup of G.

9.3 The Frattini Subgroup of a Finite Group

A proper subgroup of the group G is said to be *maximal* if it is not properly contained in another proper subgroup of G. In this section, we consider only finite groups so that maximal subgroups always exist.

The *Frattini subgroup* $\Phi(G)$ of a finite group G is the intersection of all maximal subgroups of G. From this definition, it is evident that $\Phi(G)$ is a characteristic subgroup of G. With the exception of the trivial case $G = e$, the elements of the Frattini subgroup can also be characterized as follows:

9.3.1 *A complex F of G belongs to $\Phi(G)$ if and only if it has the following property: If K is any complex of G such that $\langle K, F\rangle = G$, then $\langle K\rangle = G$.*

In other words, the Frattini subgroup consists precisely of those elements that are redundant in every system of generators of G.

Proof. Suppose that $F \subseteq \Phi(G)$ and that K is an arbitrary complex of G. Then, $G = \langle K, F \rangle$ implies that $G = \langle K \rangle$. For, otherwise, we could find a maximal subgroup M containing K; hence, $G = \langle M, F \rangle$, which is impossible because $F \subseteq \Phi(G) \subseteq M$. Conversely, assume that F is a complex such that from $G = \langle K, F \rangle$ for any complex K it follows that $G = \langle K \rangle$. Taking $K = M$, a maximal subgroup of G, we have $\langle M \rangle = M \neq G$, and hence $\langle M, F \rangle \neq G$ so that $\langle M, F \rangle = M$, which means that $F \subseteq M$. Thus, F is contained in every maximal subgroup of G, hence $F \subseteq \Phi(G)$.

Let N be a proper normal subgroup of G. A proper subgroup G_1 of G is called a partial complement to N in G, if $G = NG_1$. As a corollary of Theorem 9.3.1, we obtain:

9.3.2 *A normal subgroup N of G is contained in $\Phi(G)$ if and only if there is no partial complement to N in G.*

Proof. In Theorem 9.3.1, take $F = N$ and let K be an arbitrary subgroup of G. Then, we have $\langle K, N \rangle = KN$ and $\langle K \rangle = K$. It follows that $N \subseteq \Phi(G)$ and $KN = G$ together imply that $K = G$. Conversely, if $KN = G$ and $K \neq G$, then N does not belong to $\Phi(G)$.

The next two theorems relate the Frattini subgroup of G to that of homomorphic images of G.

9.3.3 *If the normal subgroup T of G is contained in $\Phi(G)$, then*

$$\Phi(G/T) = \Phi(G)/T.$$

Proof. The natural homomorphism of G onto G/T establishes a one-to-one correspondence between the subgroups of G containing T and the subgroups of G/T. In particular, the maximal subgroups of G that contain T correspond to the maximal subgroups of G/T. This proves our theorem.

If we drop the assumption that the kernel of the homomorphism belongs to $\Phi(G)$, then we have only a weaker assertion:

9.3.4 *If σ is any homomorphism of G, then*

$$\Phi(G)\,\sigma \subseteq \Phi(G\sigma).$$

Proof. Let N denote the kernel of σ. Then, we have $\Phi(G\sigma) \cong D/N$, where D is the intersection of all maximal subgroups of G that contain N. Hence, $\Phi(G) \subseteq D$ and

$$\Phi(G)\,\sigma \cong \Phi(G)\,N/N \subseteq D/N.$$

The next theorem contains a remarkable result on the structure of all Frattini subgroups.

9.3.5 *The Frattini subgroup of every finite group is nilpotent.*

Proof. Let P be an arbitrary Sylow subgroup of $\Phi(G)$. We have to prove that P is normal in $\Phi(G)$. As $\Phi(G)$ is a normal subgroup of G, every subgroup conjugate to P in G is contained in $\Phi(G)$ so that it is conjugate to P in $\Phi(G)$. Thus, for every element g of G there exists an element x of $\Phi(G)$ such that $g^{-1}Pg = x^{-1}Px$. Then, gx^{-1} belongs to the normalizer $\mathsf{N}(P)$ of P in G, and, hence, $g \in \mathsf{N}(P)x$. Since g is an arbitrary element of G, we conclude that $G = \mathsf{N}(P)\,\Phi(G)$. By Theorem 9.3.2, this implies that $\mathsf{N}(P) = G$. Thus, P is not only a normal subgroup of $\Phi(G)$, but even of G.

9.3.6 *Let G' denote the commutator subgroup and Z the centre of G. Then, $G' \cap Z \subseteq \Phi(G)$.*

Proof. Let us assume that M is a maximal subgroup of G that does not contain $G' \cap Z$. Then, we have $(G' \cap Z)M = G$. It follows that M is a normal subgroup of G such that G/M is abelian; hence, $G' \subseteq M$. Therefore, $G' \cap Z \subseteq M$, which contradicts our assumption. Thus, $G' \cap Z$ is contained in every maximal subgroup of G; hence, $G' \cap Z \subseteq \Phi(G)$.

9.3.7 *Let N be a normal subgroup and U an arbitrary subgroup of G. Then, $N \subseteq \Phi(U)$ implies that $N \subseteq \Phi(G)$.*

Proof. Assume that N is not contained in $\Phi(G)$. Then, it follows from Theorem 9.3.2 that $G = NG_1$, where G_1 is a proper subgroup of G. Since N is not contained in G_1, we conclude that N is contained in U but not in $G_1 \cap U$. Consequently, $G_1 \cap U$ is a proper subgroup of U. Now, $G = NG_1$ implies that $U = N(G_1 \cap U)$ so that $G_1 \cap U$ is a partial complement to N in U. By Theorem 9.3.2, this contradicts the condition $N \subseteq \Phi(U)$.

9.3.8 *If $G = A \times B$ is a direct product, then*

$$\Phi(G) = \Phi(A) \times \Phi(B).$$

Proof. By Theorem 9.3.7,

$$\Phi(A) \times \Phi(B) \subseteq \Phi(G).$$

To check whether $\Phi(G)$ is larger than $\Phi(A) \times \Phi(B)$, we put

$$G/(\Phi(A) \times \Phi(B)) = G^*$$

and determine $\Phi(G^*)$ by Theorem 9.3.3. This shows that

$$\Phi(G^*) = \Phi(G)/(\Phi(A) \times \Phi(B)).$$

On the other hand, we have $G^* = A^* \times B^*$, where $A^* \cong A/\Phi(A)$, $B^* \cong B/\Phi(B)$. By Theorem 9.3.3, we have $\Phi(A^*) = \Phi(B^*) = e$. Hence, the intersection of all maximal subgroups of G^* containing A^* is equal to A^*, and a similar result holds for B^*. Thus, $A^* \cap B^* = e$ implies that $\Phi(G^*) = e$ and, hence, $\Phi(G) = \Phi(A) \times \Phi(B)$.

Later in this section, we need a generalization of Theorem 9.3.5:

9.3.9 *Let T and D be normal subgroups of G such that $D \subseteq T \cap \Phi(G)$ and T/D is nilpotent. Then, T is nilpotent.*

Proof. We may assume that $|T| > 1$. Let p be a prime divisor of $|T|$ and P a Sylow p-subgroup of T. We have to prove that P is a normal subgroup of T. Now, PD/D is a Sylow p-subgroup of T/D; hence, by the hypothesis a characteristic subgroup of T/D. It follows that PD is a normal subgroup of G. Moreover, P and all its conjugates are Sylow p-subgroups of PD. Hence, for any element g of G there exists an element x in PD such that $g^{-1}Pg = x^{-1}Px$. Put $x = yz$ where $y \in P$, $z \in D$. Since gx^{-1} belongs to the normalizer $\mathsf{N}(P)$ of P in G, we have

$$g \in \mathsf{N}(P)\, x = \mathsf{N}(P)\, yz \subseteq \mathsf{N}(P)\, D.$$

This means that $G = \mathsf{N}(P)D$. By $D \subseteq \Phi(G)$ and Theorem 9.3.2, it follows that $\mathsf{N}(P) = G$. Thus, P is a normal subgroup of T.

9.3.10 Wielandt's Theorem. *A finite group G is nilpotent if and only if $G' \subseteq \Phi(G)$.*

Proof. Assume that G is nilpotent and let M denote a maximal subgroup of G. By Theorem 9.2.2, the normalizer of M in G is larger than M so that M is a normal subgroup of G. Since M is maximal, the factor group G/M has no proper subgroup other than e. Thus, G/M is cyclic of prime order and hence certainly abelian. This implies that $G' \subseteq M$. As M was an arbitrary maximal subgroup, we conclude that $G' \subseteq \Phi(G)$.

Conversely, suppose that $G' \subseteq \Phi(G)$ and let P be a Sylow p-subgroup of G. If the normalizer $\mathsf{N}(P)$ of P in G is a proper subgroup of G, then it is contained in some maximal subgroup M of G. Then, we have $G' \subseteq \Phi(G) \subseteq M$. As M/G' is a subgroup of the abelian group G/G', we conclude that M is a normal subgroup of G. On the other hand, by Theorem 3.3.2, M coincides with its normalizer in G. Thus, the assumption that $\mathsf{N}(P) \subset G$ leads to a contradiction. Consequently, the normalizer of every Sylow subgroup of G coincides with G, i. e. G is nilpotent.

9.3.11 *The commutator subgroup G' of a finite group G is nilpotent if and only if $G'' \subseteq \Phi(G)$.*

Proof. Suppose that G' is nilpotent. Then, Theorem 9.3.10 gives $G'' \subseteq \Phi(G')$. By Theorem 9.3.7, this implies that $G'' \subseteq \Phi(G)$.

Conversely, assume that $G'' \subseteq \Phi(G)$. Then, we have $G'' \subseteq G' \cap \Phi(G)$. Since G'/G'' is nilpotent, in fact abelian, we can apply Theorem 9.3.9 to see that G' is nilpotent.

9.4 Supersoluble Groups

A group G is said to be *supersoluble* if it has a normal series whose factors are cyclic. Clearly, every supersoluble group is soluble. Let

$$G = G_0 \supset G_1 \supset \cdots \supset G_{r-1} \supset G_r = e \tag{9.26}$$

be a normal series with cyclic factors. If a_i is an element of G_i such that the natural homomorphism of G_i onto G_i/G_{i+1} maps a_i onto a generating element of G_i/G_{i+1}, then we have $G_i = \langle G_{i+1}, a_i \rangle$. Thus, $a_0, a_1, \ldots, a_{r-1}$ generate G. Hence, every supersoluble group is finitely generated.

9.4.1 *All subgroups and factor groups of supersoluble groups are supersoluble.*

Proof. Let U be a subgroup of the supersoluble group G. We form the intersections $U_i = U \cap G_i$ with the terms of the normal series (9.26). Then,

$$U = U_0 \supseteq U_1 \supseteq \cdots \supseteq U_{r-1} \supseteq U_r = e \tag{9.27}$$

is a normal series of U. By Theorem 2.7.1, the factors of (9.27) are isomorphic to subgroups of the corresponding factors of (9.26), and hence are cyclic. This shows that U is supersoluble.

Let N be a normal subgroup of G. The natural homomorphism of G onto G/N maps (9.26) onto the normal series

$$G/N = G_0/N \supseteq G_1N/N \supseteq \cdots \supseteq G_{r-1}N/N \supseteq N/N$$

of G/N. Every factor group $(G_iN/N)/(G_{i+1}N/N)$ is isomorphic to a subgroup of G_i/G_{i+1}, and hence cyclic. Thus, G/N is supersoluble. This completes the proof.

Suppose that the cyclic factor group $F = G_i/G_{i+1}$ of the series (9.26) is of finite order m. Let $m = p_1p_2 \ldots p_k$ be the decomposition of m into a product of not necessarily distinct prime numbers p_j. If m_1 is any divisor of m then, by Theorem 1.4.1, F contains one and only one subgroup of order m_1. Clearly, this subgroup is characteristic. Thus, F contains characteristic subgroups

$$F = F_k,\ F_{k-1}, \ldots, F_2, F_1, F_0 = e$$

whose respective orders are

$$p_1 \cdots p_k, p_1 \cdots p_{k-1}, \ldots, p_1p_2, p_1, 1.$$

Each F_j has the form

$$F_j = H_j/G_{i+1} \quad (j = k, k-1, \ldots, 1, 0),$$

where H_j is a normal subgroup of G. So, we obtain the sequence

$$G_i = H_k, H_{k-1}, \ldots, H_2, H_1, H_0 = G_{i+1}$$

of normal subgroups of G. By inserting these normal subgroups in (9.26) between G_i and G_{i+1}, we obtain a refinement of (9.26). By carrying out similar refinements in every finite factor of (9.26), we arrive at a normal series of G whose factors are cyclic of infinite or prime order. In particular, we obtain:

9.4.2 *A finite group is supersoluble if and only if its chief indices are prime numbers.*

In a finite supersoluble group, we can even construct a chief series whose indices occur in their natural order:

9.4.3 *Let G be a finite supersoluble group and*

$$|G| = p_1 p_2 \cdots p_n$$

the decomposition of its order into prime factors p_i where $p_1 \leqq p_2 \leqq \cdots \leqq p_n$. Then, there exists a chief series

$$G = H_0 \supset H_1 \supset H_2 \supset \cdots \supset H_{n-1} \supset H_n = e$$

such that H_{i-1}/H_i is of order p_i.

Proof. It is sufficient to prove the following assertion. Suppose that $A \supset B \supset C$ are three consecutive terms of a chief series of G such that $|A:B| = p > q = |B:C|$. Then, B can be replaced by another normal subgroup B^* of G such that $A \supset B^* \supset C$ and $|A:B^*| = q$, $|B^*:C| = p$. Starting with an arbitrary chief series, repeated application of this procedure will yield a chief series whose indices are ordered as required.

The factor group $F = A/C$ is of order pq. It is easy to see that F has a single Sylow p-subgroup. For the number of Sylow p-subgroups of F is $\equiv 1 \pmod p$, whereas this number cannot exceed q and $q < p$, so that the number in question is 1. Thus, the only Sylow p-subgroup P of F is a characteristic subgroup of F and can be written in the form $P = B^*/C$ where B^* is a normal subgroup of G. We have $|A:B^*| = q$, $|B^*:C| = p$ which completes the proof.

9.5 Maximal Subgroups

Nilpotent and supersoluble groups can also be characterized by properties of their maximal subgroups. In this section, we derive some pertinent results.

9.5.1 *A finite group is nilpotent if and only if all its maximal subgroups are normal.*

Proof. It follows immediately from 9.2.2 that in a nilpotent group every maximal subgroup is normal.

Conversely, assume that every maximal subgroup of the finite group G is a normal subgroup of G. By Theorem 3.3.2, every subgroup of G that contains the normalizer $\mathsf{N}(P)$ of a Sylow subgroup P of G coincides with its normalizer in G. It follows that $\mathsf{N}(P)$ is not contained in a maximal subgroup, which means that $\mathsf{N}(P) = G$. Thus, every Sylow subgroup of G is normal, hence G is nilpotent. This completes the proof.

A *maximal series* of G is a series

$$G = M_0 \supset M_1 \supset \cdots \supset M_{k-1} \supset M_k = e \tag{9.28}$$

in which each M_i is a maximal subgroup of M_{i-1}. If G is finite and supersoluble, then G possesses a chief series

$$G = H_0 \supset H_1 \supset \cdots \supset H_{n-1} \supset H_n = e \tag{9.29}$$

whose indices are prime numbers. Clearly, this is a maximal series of G. We shall prove that the indices in every maximal series of a finite supersoluble group are prime numbers.

Let (9.28) be any maximal series of G. Since M_1 is again supersoluble, it suffices to show that $|G: M_1|$ is a prime number. There is a unique index i in the chief series (9.29) such that

$$H_{i-1} \nsubseteq M_1, \ H_i \subseteq M_1.$$

Since $|H_{i-1}: H_i|$ is a prime number, there is no subgroup of G properly between H_{i-1} and H_i, hence $H_{i-1} \cap M_1 = H_i$. As M_1 is maximal, we have $H_{i-1}M_1 = G$. This gives

$$G/H_{i-1} = H_{i-1}M_1/H_{i-1} \cong M_1/(H_{i-1} \cap M_1) = M_1/H_i,$$

and hence

$$|G: M_1| = |H_{i-1}: H_i|.$$

Consequently, $|G: M_1|$ is a prime number. So, we obtain the following theorem:

9.5.2 *The index of every maximal subgroup of a finite supersoluble group G is a prime number. All maximal series of G have the same length, which is equal to the number of prime factors of $|G|$.*

It is a much deeper result that conversely a group is supersoluble if the index of every maximal subgroup is a prime number. As a preliminary to proving this theorem, we show that the groups in question are soluble:

9.5.3 *Let G be a finite group such that the index of every maximal subgroup of G is a prime number. Then, G is soluble.*

Proof. We proceed by induction on the order and assume that the theorem is true for all groups of order less than $|G|$.

Let M be a maximal subgroup of G. Then, the index $|G:M| = q$ is a prime number. We consider the permutational representation of G of degree q induced by M. Let M_1 denote the kernel of this representation. Evidently, G/M_1 satisfies the condition of our theorem. Hence, if $M_1 \neq e$, the factor group G/M_1 is soluble by the inductive hypothesis. For another maximal subgroup L of G, let L_1 denote the kernel of the permutational representation induced by L. As above, we conclude that G/L_1 is soluble provided that $L_1 \neq e$. We have

$$L_1 M_1/M_1 \cong L_1/(M_1 \cap L_1)$$

so that $G/(M_1 \cap L_1)$ is also soluble. If none of the permutational representations induced by the maximal subgroups is faithful and if

$$D = M_1 \cap L_1 \cap \cdots$$

denotes the intersection of their kernels, then it follows that G/D is soluble. Now D is contained in all maximal subgroups of G and hence in their intersection $\Phi(G)$. By Theorem 9.3.5, $\Phi(G)$ is nilpotent so that D is nilpotent and therefore certainly soluble. As G/D and D are soluble, it follows that G itself is soluble.

We now consider the case that G contains a maximal subgroup that induces a faithful permutational representation of G. The index of this maximal subgroup is a prime number p, and G is isomorphic to a subgroup of the symmetric group of degree p. It follows that $|G|$ is divisible by p but not by p^2. Let T be a minimal normal subgroup of G. By Theorem 4.2.2, T is simple or the direct product of isomorphic simple groups. By Theorem 7.1.10, T is transitive so that $|T|$ is divisible by the degree p of the permutational representation. Since $|T|$ is not divisible by p^2, we conclude that T is simple. Now, T contains a Sylow p-subgroup P of G, and as T is a normal subgroup, it contains all the Sylow p-subgroups of G. Moreover, the Sylow p-subgroups of G are conjugate in T. This shows that $G = \mathsf{N}(P)T$, where $\mathsf{N}(P)$ denotes the normalizer of P in G.

If at $\mathsf{N}(P) \neq G$, then $\mathsf{N}(P)$ is contained in a maximal subgroup M of G whose index is a prime number $q \neq p$. Since p is the largest prime divisor of $|G|$, we infer that $q < p$. We have $G = TM$ and $G/T \cong M/S$, where $S = T \cap M$. This gives $|T|:|S| = |G|:|M| = q$. Consequently, S induces a permutational representation of T of degree q. Since $q < p$, this representation is not faithful;

for, there exists an element of T of order p, whereas no permutation group of degree $q < p$ contains a permutation of order p. But, the existence of a non-faithful representation of T contradicts the fact that T is simple.

It follows that $\mathsf{N}(P) = G$. By the inductive hypothesis, G/P is soluble, and since P is certainly soluble, we conclude that G is soluble. This completes the proof.

As a corollary of Theorem 9.5.3, we obtain:

9.5.4 *If the index of every maximal subgroup of the finite group G is a prime number, then the commutator subgroup G' is nilpotent.*

Proof. By Theorem 9.5.3, G is soluble. Let M be a maximal subgroup of G of prime index q. Let M_1 denote the kernel of the permutational representation induced by M. By Theorem 7.1.11, this permutational representation is two-step metabelian. Thus, the second derived group of G/M_1 is the identity so that $G'' \subseteq M_1 \subseteq M$. We conclude that G'' is contained in every maximal subgroup and, hence, in their intersection $\Phi(G)$. By Theorem 9.3.11, $G'' \subseteq \Phi(G)$ implies that G' is nilpotent.

We now come to the converse of Theorem 9.5.2:

9.5.5 Huppert's Theorem. *If the index of every maximal subgroup of the finite group G is a prime number, then G is supersoluble.*

Proof. Suppose that the theorem is false and let G denote a group of smallest order that satisfies the condition of the theorem without being supersoluble. Note that every factor group G/N also satisfies the condition of our theorem.

By Theorem 9.5.3, G is soluble so that every minimal normal subgroup of G is elementary abelian. Let F be a minimal normal subgroup of G and $|F| = p^k$. Owing to the choice of G, G/F is supersoluble. It follows that F is not cyclic, i. e. $k > 1$, for, otherwise, G itself would be supersoluble. Moreover, F/e is the only non-cyclic chief factor of G. This implies that F is the only minimal normal subgroup of G. For suppose that F_1 is another minimal normal subgroup; then, G/F_1 would be supersoluble, which contradicts the fact that G/F_1 has a non-cyclic chief factor isomorphic to F.

Let H/F be a minimal normal subgroup of G/F. As G/F is supersoluble, the order of H/F is a prime number. We have to consider two cases:

$$\text{I.} \quad |H:F| = q \neq p; \qquad \text{II.} \quad |H:F| = p.$$

I. Let Q denote a Sylow q-subgroup of H. Then, we have $H = QF$. Since H is a normal subgroup, we obtain

$$g^{-1}Qg \subset H$$

for every $g \in G$. Hence, there is an element x of F such that $g^{-1} Qg = x^{-1}Qx$. Consequently, gx^{-1} belongs to the normalizer $\mathsf{N}(Q)$ of Q in G so that

$$G = F\mathsf{N}(Q). \tag{9.30}$$

Now, $F \cap \mathsf{N}(Q)$ is a normal subgroup of $\mathsf{N}(Q)$, but this intersection is also a normal subgroup of F, because F is abelian. Hence, (9.30) shows that $F \cap \mathsf{N}(Q)$ is a normal subgroup of G. Since F is a minimal normal subgroup, we have $F \cap \mathsf{N}(Q) = F$ or $F \cap \mathsf{N}(Q) = e$. The first possibility gives $F \subseteq \mathsf{N}(Q)$ and, hence, $G = \mathsf{N}(Q)$ by (9.30). But, then, Q would be a minimal normal subgroup of G distinct from F, which is impossible. So, we have $F \cap \mathsf{N}(Q) = e$ and, hence, $|G : \mathsf{N}(Q)| = p^k$ by (9.30). On the other hand, $\mathsf{N}(Q)$ is a maximal subgroup of G. For, $\mathsf{N}(Q) \subset U \subset G$ would imply that $e \subset U \cap F \subset F$; as above, we could infer that $U \cap F$ is a normal subgroup of G, which is impossible, because F is minimal. Hence, $\mathsf{N}(Q)$ is a maximal subgroup whose index p^k is not a prime number, contrary to the hypothesis.

II. In this case, H is abelian. For, otherwise, H' would be a proper subgroup of F, because, by Theorem 3.4.10, the index $|H : H'|$ is divisible by p^2. But, this is impossible, for H' is a characteristic subgroup of H and hence normal in G. Moreover, H does not contain an element of order p^2. For, otherwise, the p-th powers of the elements of H would form a characteristic subgroup of order p of H, so that G would contain a normal subgroup of order p, which contradicts the fact that F of order $p^k > p$ is the only minimal normal subgroup of G. It follows that H is an elementary abelian group of order p^{k+1}.

Transformation of H by the elements of G yields a representation of G by automorphisms of H. If we choose a basis of H, we obtain a representation Γ of G by $(k+1) \times (k+1)$ matrices whose elements belong to the Galois field $GF(p)$ with p elements. Let S denote the normal subgroup of G consisting of all those elements whose images in Γ are scalar multiples of the unit matrix. It is easily seen that S can also be characterized as follows: an element a belongs to S if and only if $a^{-1}ha = h^m$ for every $h \in H$, where $m = m(a)$ depends on a but not on h. Moreover, let $C = \mathsf{C}(H)$ denote the centralizer of H in G, so that Γ is a faithful representation of G/C. Clearly, S/C belongs to the centre of G/C. Moreover, S is a proper subgroup of G; for, every subgroup of H is a normal subgroup of S, so that $S = G$ would imply that a proper subgroup of F is a normal subgroup of G.

Let R/S be a minimal normal subgroup of G/S. Since G/S is supersoluble, $|R : S|$ is a prime number. We consider the two cases

$$\text{(a)} \quad |R : S| = p, \qquad \text{(b)} \quad |R : S| = q \neq p$$

separately.

(a) The order of the factor group S/C is relatively prime to p, because the order of every non-zero element of $GF(p)$ divides $p - 1$. Moreover, S/C belongs to the centre of G/C. This shows that

$$R/C = (S/C) \times (R_1/C),$$

where R_1/C is a cyclic group of order p, $R_1 = \langle C, a\rangle$ say. R_1/C is a characteristic subgroup of R/C and hence a normal subgroup of G/C; therefore, R_1 is a normal subgroup of G. Transformation of F by the elements of R_1 gives rise to a representation of R_1/C by $k \times k$ matrices with coefficients of $GF(p)$. To the element a of R_1, this representation assigns a matrix $[a]$ whose p-th power is the unit matrix. It follows that all the characteristic roots of $[a]$ are equal to the unit element of $GF(p)$. Consequently, there exists a triangular matrix similar to $[a]$ and with elements in $GF(p)$ such that all the elements in its principal diagonal are equal to the unit element of $GF(p)$. This shows that F contains at least one element other than e that commutes with a. Moreover, F commutes elementwise with C, so that the intersection of F and the centre $\mathbf{Z}(R_1)$ of R_1 is not trivial. As F is minimal, this gives $F \subseteq \mathbf{Z}(R_1)$. Let $H = \langle F, b\rangle$. Then, we obtain

$$[H, R_1] = [\langle F, b\rangle, \langle C, a\rangle] = \langle [a, b]\rangle.$$

We have $[a, b] \in H$ and $[a, b] \neq e$, since a does not belong to the centralizer C of H. It follows that $[a, b]$ is of order p. Since H and R_1 are normal subgroups of G, so is $[H, R_1] = \langle [a, b]\rangle$. So, we arrive at the contradiction that $\langle [a, b]\rangle$ is a minimal normal subgroup of G other than F. Hence, case (a) cannot occur.

(b) As we observed above, the order of S/C is relatively prime to p so that in the present case the order of R/C is not divisible by p. Consequently, H (regarded as a representation module of R/C) satisfies the condition of Theorem 13.1.2 so that we obtain $H = F \times P$, where $P = \langle b\rangle$ is of order p and a normal subgroup of R. All the subgroups conjugate to P in G are contained in H and their product T is a normal subgroup of G. Now, T is distinct from F, and F is the only minimal normal subgroup of G; this shows that $T = H$. Since P is not contained in F, no subgroup is conjugate to P. Let P_1 be conjugate to P and $P_1 \neq P$. Then, the intersection $D = PP_1 \cap F$ is of order p, since $PP_1 \subset H$, $|PP_1| = p^2$, $|H:F| = p$. We can choose a generating element c of P_1 such that $D = \langle cb\rangle$. Since P, P_1, and D are normal subgroups of R, transformation by an element a of R leads to the relations

$$a^{-1}ba = b^m, \quad a^{-1}ca = c^s, \quad a^{-1}(cb)a = (cb)^t.$$

This shows that

$$(cb)^t = c^s b^m,$$

from which it follows that $m = s = t$; for, b and c commute, because P and P_1 are normal subgroups with trivial intersection. Since P_1 is an arbitrary conjugate of P other than P and the product of all conjugates of P coincides with H, we conclude that

$$a^{-1}ha = h^m$$

for every $h \in H$ and every $a \in R$. Here, $m = m(a)$ does not depend on h. So, we have $R \subseteq S$, and this contradiction to $|R:S| = q$ completes the proof.

Exercises

1. Prove that

$$[[x, y], z^x]\ [[z, x], y^z]\ [[y, z], x^y] = e.$$

2. Show that the alternating group A_4 is soluble but not supersoluble.

3. Prove that every finite nilpotent group is supersoluble. Give an example of a supersoluble group that is not nilpotent.

4. Determine the Frattini subgroup of a finite abelian group.

5. The definition of the Frattini subgroup $\Phi(G)$ can be extended to an infinite group G provided that G has maximal subgroups. In case G has no maximal subgroup, we define $\Phi(G) = G$. Prove that a finitely generated abelian group G is finite if and only if $G/\Phi(G)$ is finite.

6. Let T_n denote the group of all $n \times n$ triangular matrices whose elements belong to a field and all of whose characteristic roots are equal to each other.

(a) Prove that T_n is nilpotent and determine the class of T_n.

(b) Determine the upper and the lower central series of T_3.

7. Determine all commutator polynomials of weight 5.

8. Determine all finite groups in which every subgroup is subnormal.

CHAPTER 10

FINITE p-GROUPS

10.1 Elementary Properties

To begin with, we list some previous results on finite p-groups:

Theorem 3.4.2. Every finite p-group has a non-trivial centre.

Theorem 3.4.5. All chief indices of a finite p-group are equal to p. In particular, every finite p-group is soluble.

Theorem 3.4.6. If U is a proper subgroup of a finite p-group, then U is properly contained in its normalizer.

Theorem 3.4.7. Every maximal subgroup of a finite p-group is a normal subgroup and has index p.

Theorem 3.4.8. Every normal subgroup of order p in a finite p-group lies in the centre.

Theorem 3.4.8. The index of the centre of a finite non-abelian p-group is divisible by p^2.

Theorem 3.4.10. Every group of order p^2 is abelian. Every normal subgroup of index p^2 contains the commutator subgroup.

Theorem 9.2.3. Every finite p-group is nilpotent.

These results will be used in what follows without further reference.

Let G be a finite p-group and M an arbitrary maximal subgroup of G. Then, M contains the commutator subgroup G'. Moreover, if x is any element of G, then x^p belongs to M. Since this holds for every maximal subgroup, we conclude that the Frattini subgroup $\Phi(G)$ contains G' and that every element of $G/\Phi(G)$ other than the identity is of order p. This shows that $G/\Phi(G)$ is an elementary abelian group of order p^d say.

If $a_1, \ldots, a_r$ generate G, then the cosets $\Phi(G)a_1, \ldots, \Phi(G)a_r$ generate $G/\Phi(G)$. From these cosets, we can therefore choose a basis of the elementary abelian group $G/\Phi(G)$, say $\Phi(G)a_1, \ldots, \Phi(G)a_d$. It follows that $a_1, \ldots, a_d$ generate G. For, otherwise, there exists a maximal subgroup M of G such that $\langle a_1, \ldots, a_d \rangle \subseteq M$.

Since $\Phi(G) \subseteq M$, this would imply that $\langle \Phi(G)a_1, \ldots, \Phi(G)a_d \rangle \subseteq M/\Phi(G)$, which contradicts the choice of $a_1, \ldots, a_d$. The group cannot be generated by fewer than d elements. For, the natural homomorphism of G onto $G/\Phi(G)$ carries every system of generators of G into a system of generators of $G/\Phi(G)$, and the smallest number of generators of the elementary abelian group $G/\Phi(G)$ of order p^d is d.

Any system of d elements that generates G is called a minimal basis of G.

We summarize our results:

10.1.1 Burnside's Basis Theorem. *Let G be a finite p-group. Then, $G/\Phi(G)$ is an elementary abelian group. If $|G:\Phi(G)| = p^d$, then G contains generating systems of d elements, but cannot be generated by fewer than d elements. Every system of d elements that generates G is called a minimal basis of G. Every system of generators of G contains a minimal basis. A system of d elements is a minimal basis if and only if it is carried into a basis of $G/\Phi(G)$ under the natural homomorphism of G onto $G/\Phi(G)$.*

Let A_r be an elementary abelian group of order p^r. Clearly, every subgroup of A_r is also elementary abelian. The number of distinct bases of A_r is

$$(p^r - 1)(p^r - p) \cdots (p^r - p^{r-1}). \tag{10.1}$$

For, if we wish to choose a basis $x_1, \ldots, x_r$ of A_r, then x_1 may be any element other than e so that there are $p^r - 1$ choices for x_1. After that, x_2 may be chosen arbitrarily from among the $p^r - p$ elements of A_r not contained in $\langle x_1 \rangle$. Generally, we may take x_i as any one of the $p^r - p^{i-1}$ elements that do not belong to $\langle x_1, \ldots, x_{i-1} \rangle = \langle x_1 \rangle \times \cdots \times \langle x_{i-1} \rangle$. This shows that the number of distinct ways of choosing a basis of A_r is given by (10.1). Moreover, the first k elements of a basis may be chosen in

$$(p^r - 1)(p^r - p) \cdots (p^r - p^{k-1})$$

ways.

If $x_1, \ldots, x_k$ are the first k elements of a basis of A_r, then they form the basis of a subgroup of order p^k. By Theorem 4.2.3, every basis of a subgroup of order p^k can be extended to a basis of A_r. Consequently, the number $\varphi_{r,k}$ of subgroups of order p^k of A_r is given by

$$\begin{aligned} \varphi_{r,k} &= \frac{(p^r - 1)(p^r - p) \cdots (p^r - p^{k-1})}{(p^k - 1)(p^k - p) \cdots (p^k - p^{k-1})} \\ &= \frac{(p^r - 1)(p^{r-1} - 1) \cdots (p^{r-k+1} - 1)}{(p^k - 1)(p^{k-1} - 1) \cdots (p - 1)}. \end{aligned} \qquad (1 \leqq k \leqq r)$$

One easily verifies that

$$\varphi_{r,k} \equiv 1 \pmod{p} \tag{10.2}$$

and that

$$\varphi_{r,k} = \varphi_{r,r-k}.$$

10.1.2 *The number of normal subgroups of a given order p^k in a finite p-group is $\equiv 1$ (mod p).*

Proof. Let G be a finite p-group. Every normal subgroup of order p lies in the centre of G. Hence, all normal subgroups of order p generate an elementary abelian subgroup of G, of order p^r say. Thus, the number of normal subgroups of order p in G is $\varphi_{r,1}$, and (10.2) shows that in this special case our theorem is true.

The assertion is trivial if G is of order p. We proceed by induction on the order and assume that our theorem holds for all p-groups of order less than $|G|$.

Every normal subgroup of order p^k occurs in some chief series of G, and hence it contains a normal subgroup P of order p. By the inductive hypothesis, the number of normal subgroups of order p^{k-1} of G/P is $\equiv 1$ (mod p). Thus, the number of normal subgroups of order p^k of G that contain P is $\equiv 1$ (mod p). To every normal subgroup of order p, there belongs such a system of normal subgroups of order p^k and, as the number of subgroups of order p is $\equiv 1$ (mod p), it follows that the total number of normal subgroups of order p^k in all these systems is also $\equiv 1$ (mod p). A fixed normal subgroup N of order p^k may belong to several systems so that it is counted several times. Clearly, the number m_N of systems containing N is equal to the number of normal subgroups of order p that are contained in N. We saw above that $m_N \equiv 1$ (mod p). Hence, if for any N the corresponding multiplicity m_N is replaced by 1, then the total number remains $\equiv 1$ (mod p). This completes the proof.

As a corollary of Theorem 10.1.2, we obtain:

10.1.3 *The number of subgroups of a given order p^k in a finite p-group is $\equiv 1$ (mod p).*

Proof. The number of normal subgroups of order p^k is $\equiv 1$ (mod p). The non-normal subgroups of order p^k occur in full classes of conjugates, and the number of subgroups in each class is divisible by p.

10.2 Special p-Groups

From the results in section 6.4, it follows that two elements a and b subject to the relations

$$a^4 = e, \; a^2 = b^2, \; b^{-1}ab = a^{-1}$$

generate a group of order 8. This group is called the *quaternion group*. Writing

$$e = 1, \; a = i, \; a^2 = -1, \; b = j, \; ab = k,$$

we obtain the familiar multiplication rule of the quaternions:

$$i^2 = j^2 = k^2 = -1,$$
$$ij = -ji = k, \; jk = -kj = i, \; ki = -ik = j.$$

The quaternion group contains one subgroup of order 2, namely $\langle a^2 \rangle$, and three subgroups of order 4, namely $\langle a \rangle$, $\langle b \rangle$, $\langle ab \rangle$. By Theorem 3.4.10, $\langle a^2 \rangle$ is the centre and the commutator subgroup.

For $n > 3$, the so-called *generalized quaternion groups* are defined by

$$a^{2^{n-1}} = e, \; b^2 = a^{2^{n-2}}, \; b^{-1}ab = a^{-1}. \tag{10.3}$$

Its order is 2^n.

10.2.1 *All the groups of order p^n that contain a cyclic subgroup of order p^{n-1} belong to the following types:*

Abelian groups

(I) $n \geqq 1, \; a^{p^n} = e,$

(II) $n \geqq 2, \; a^{p^{n-1}} = e, \; r^p = e, \; r^{-1}ar = a.$

Non-abelian groups

(III) p odd, $n \geqq 3, \; a^{p^{n-1}} = e, \; r^p = e, \; r^{-1}ar = a^{1+p^{n-2}};$

(IV) $p = 2, \; n \geqq 3, \; a^{2^{n-1}} = e, \; r^2 = a^{2^{n-2}}, \; r^{-1}ar = a^{-1}$
(*Quaternion group or generalized quaternion group*)

(V) $p = 2, \; n \geqq 3, \; a^{2^{n-1}} = e, \; r^2 = e, \; r^{-1}ar = a^{-1}$
(*Dihedral group*)

(VI) $p = 2, \; n \geqq 4, \; a^{2^{n-1}} = e, \; r^2 = e, \; r^{-1}ar = a^{1+2^{n-2}}$

(VII) $p = 2, \; n \geqq 4, \; a^{2^{n-1}} = e, \; r^2 = e, \; r^{-1}ar = a^{-1+2^{n-2}}$

Proof. The abelian case is obvious.

The cyclic subgroup of order p^{n-1} is normal, so that all the groups are extensions of this normal subgroup by a cyclic group of order p. From the results of section 6.4, it follows that such groups can be generated by two elements a and r with the defining relations

$$a^{p^{n-1}} = e, \; r^p = a^k, \; r^{-1}ar = a^g,$$

where the natural numbers g and k satisfy the conditions

$$g^p \equiv 1 \quad (\text{mod } p^{n-1}), \tag{10.4}$$

$$kg \equiv k \quad (\text{mod } p^{n-1}). \tag{10.5}$$

As the remaining groups are non-abelian, we may assume that $g \not\equiv 1 \pmod{p^{n-1}}$.

Let us first assume that p is odd. By (10.4), we have $g = 1 + cp^{n-2}$ and $(c, p) = 1$. We may replace r by $r_1 = r^x a^y$ with $(x, p) = 1$. This shows that

$$r_1^{-1} a r_1 = a^{g_1},$$

where

$$g_1 \equiv g^x = (1 + cp^{n-2})^x \equiv 1 + cxp^{n-2} \equiv 1 + p^{n-2} \qquad \pmod{p^{n-1}}$$

provided that we choose x such that

$$cx \equiv 1 \qquad \pmod{p}.$$

Next, we obtain

$$r_1^p = r^{px} a^z = a^{kx+z}, \tag{10.6}$$

where

$$\begin{aligned} z &\equiv y(g_1^{p-1} + g_1^{p-2} + \cdots + g_1 + 1) \\ &\equiv y[p + p^{n-2}(p - 1 + p - 2 + \cdots + 1)] \\ &\equiv y\left[p + p^{n-1}\frac{p-1}{2}\right] \equiv yp \qquad \pmod{p^{n-1}}. \end{aligned}$$

Since r_1 is not of order p^n, we conclude that p divides k, $k = pk'$ say. Thus, the exponent on the right-hand side of (10.6) satisfies the congruence

$$kx + z \equiv p(k'x + y) \qquad \pmod{p^{n-1}}.$$

If we choose k such that

$$y \equiv -k'x \qquad \pmod{p^{n-2}},$$

we find that $r_1^p = e$. Hence, the generating elements a and r_1 satisfy the relations of type (III).

Now, assume that $p = 2$ and $a^{2^{n-1}} = e$. We have $r^{-1}ar = a^g$, where

$$g^2 \equiv 1 \pmod{2^{n-1}}, \qquad g \not\equiv 1 \pmod{2^{n-1}}.$$

Since only the residue class mod 2^{n-1} of g is relevant, there are three possible values of g:

$$g = -1, \quad g = 1 + 2^{n-2}, \quad g = -1 + 2^{n-2}.$$

Furthermore, $r^2 = a^k$, where $kg \equiv k \pmod{2^{n-1}}$. For $g = -1$, this leads to $-k \equiv k \pmod{2^{n-1}}$ or $2k \equiv 0 \pmod{2^{n-1}}$. So, we obtain $k = 0$ or $k = 2^{n-2}$. These are the types (IV) and (V).

If $n = 3$, the two other values of g do not lead to new types. For $g = 1 + 2$, we have $g \equiv -1 \pmod 4$, which is the case just considered. For $g = -1 + 2^{3-2} = 1$, we obtain an abelian group of type (II).

Now, let $n \geqq 4$ and $g = 1 + 2^{n-2}$. From $kg \equiv k \pmod{2^{n-1}}$, we conclude that $2^{n-2}k \equiv 0 \pmod{2^{n-1}}$ so that k is even, $k = 2k_1$ say. If we choose y such that

$$y(1 + 2^{n-3}) + k_1 \equiv 0 \qquad \pmod{2^{n-2}}$$

and replace r by $r_1 = ra^y$, we obtain

$$r_1^2 = r^2 a^{y(1+2^{n-2})+y} = a^{2k_1+y(2+2^{n-2})} = e\,.$$

This is type (VI).

Finally, suppose that $n \geqq 4$ and take $g = -1 + 2^{n-2}$. The condition $kg \equiv k \pmod{2^{n-1}}$ yields

$$k(2 - 2^{n-2}) \equiv 0 \qquad \pmod{2^{n-1}}$$

or

$$k \equiv 0 \qquad \pmod{2^{n-2}}.$$

So, we have $r^2 = e$ or $r^2 = a^{2^{n-2}}$. The first case is type (VII), but the second case also leads to the same type. For, if we replace r by $r_1 = ra$, we obtain

$$r_1^2 = rara = r^2 a^{-1+2^{n-2}} a = e\,.$$

This completes the proof.

It is now easy to give a survey of all groups of order p^3.

10.2.2 *There are the following types of groups of order* p^3:

Abelian groups

(I) $a^{p^3} = e;$

(II) $a^{p^2} = e,\ b^p = e,\ ab = ba;$

(III) $a^p = b^p = c^p = e,\ ab = ba,\ ac = ca,\ bc = cb;$

Non-abelian groups
p odd

(IV) $a^{p^2} = e,\ b^p = e,\ b^{-1}ab = a^{1+p};$

(V) $a^p = b^p = c^p = e,\ a^{-1}b^{-1}ab = c,\ ac = ca,\ bc = cb;$

$p = 2$

(IV) $a^4 = e,\ b^2 = e,\ b^{-1}ab = a^{-1}$ (*Dihedral group*)

(VI) $a^4 = e,\ b^2 = a^2,\ b^{-1}ab = a^{-1}$ (*Quaternion group*)

(I), (II), (III) are clearly all the possible types of abelian groups of order p^3.

We now turn to non-abelian groups. Suppose first that p is odd. If there exists an element of order p^2, then we can apply Theorem 10.2.1. Accordingly, there is only type (III) of Theorem 10.2.1, which is type (IV) of the present theorem. Next, we have to consider groups of exponent p. By Theorem 3.4.10, the commutator subgroup and the centre coincide and are of order p. In our case, the factor group with respect to the commutator subgroup is elementary abelian. Let a and b be representatives of two generating cosets of this factor group and let c be a generator of the commutator subgroup. Evidently, these elements satisfy the relations of type (V).

For $p = 2$, there exists an element of order 4 so that Theorem 10.2.1 gives the two above types.

Thus, our list contains all the possible types.

10.2.3 *A group of order p^n that contains only one subgroup of order p is cyclic, a quaternion group, or a generalized quaternion group.*

Proof. Our assertion is obvious for $n = 1$. Let us assume that our theorem is true for all p-groups of order less than p^n.

Suppose first that p is odd. By the inductive hypothesis, every subgroup of index p is cyclic. It follows that the group belongs to type (I), (II), or (III) of Theorem 10.2.1. As the groups of type (II) or (III) contain more than one subgroup of order p, there only remain the cyclic groups.

Now, let $p = 2$. If the group contains a cyclic subgroup of order 2^{n-1}, then Theorem 10.2.1 shows that it is cyclic, a quaternion group, or a generalized quaternion group.

It remains to consider the cases that are not covered by Theorem 10.2.1. We may assume that every subgroup of index 2 is a quaternion group or a generalized quaternion group. We shall show that these cases cannot occur. Of course, we have to deal only with exponents $n \geqq 4$.

First, let $n = 4$. A subgroup Q of index 2 is a quaternion group so that Q is generated by two elements a and b satisfying the relations $a^4 = e$, $a^2 = b^2$, $b^{-1}ab = a^{-1}$. With an element c that does not belong to Q, we obtain the coset decomposition $G = Q \cup Qc$. Now, Q contains three subgroups of order 4, namely $\langle a \rangle$, $\langle b \rangle$, $\langle ab \rangle$. As the order of c is a power of 2, transformation by c and its powers cannot induce a transitive permutation group on these three subgroups. It follows that at least one of them is invariant under transformation by c. We may assume that the notation is such that $c^{-1}\langle a \rangle c = \langle a \rangle$. Then, we have $c^{-1}ac = a$ or $c^{-1}ac = a^{-1}$. In the first case, $\langle a, c \rangle$ is an abelian subgroup of index 2, which contradicts our assumption. In the second case, we have $(cb)^{-1}a(cb) = a$ so that now $\langle a, cb \rangle$ would be an abelian subgroup of index 2, again providing a contradiction.

Finally, let $n \geqq 5$ and let Q_1 be a generalized quaternion group of index 2. We may write $Q_1 = \langle a, b \rangle$, where $a^{2^{n-2}} = e$, $b^2 = a^{2^{n-3}}$, $b^{-1}ab = a^{-1}$. We have $G = Q_1 \cup Q_1 c$, where c is any element not contained in Q_1. The only subgroup of order 2^{n-2} of Q_1 is $\langle a \rangle$. Consequently, $\langle a \rangle$ is invariant under transformation by c, so that $c^{-1}ac = a^r$ for some integer r. Moreover, $c^2 = a^k b$ or $c^2 = a^k$ for some k. In the first case, $c^{-2}ac^2 = a^{-1}$ and hence $r^2 \equiv -1 \pmod{2^{n-2}}$, which is impossible. But, if $c^2 = a^k$, then $\langle a, c\rangle$ is a subgroup of index 2 and hence a generalized quaternion group. This gives $c^{-1}ac = a^{-1}$ which implies that $(cb)^{-1}a(cb) = a$. Thus, $\langle cb, a\rangle$ would be an abelian subgroup of index 2, which contradicts our assumption. This completes the proof.

The analogue to Theorem 10.2.3 for subgroups of order greater than p does not involve the exceptional quaternion groups and hence gives a characteristic property of cyclic p-groups.

10.2.4 *Let G be a group of order p^n and m a fixed natural number, $1 < m < n$. If G contains only one subgroup of order p^m, then G is cyclic.*

Proof. The condition of our theorem can be satisfied only for $n \geqq 3$.

First, let $m = n - 1$. If a is an element that does not belong to the only subgroup of order p^{n-1}, then we have necessarily $G = \langle a\rangle$, so that G is cyclic. Thus, our theorem is proved for $n = 3$ and for arbitrary n in case $m = n - 1$. We proceed by induction on n and may assume that $m < n - 1$.

Let P denote the only subgroup of order p^m of G. Now, P is contained in some maximal subgroup M. As $|M| = p^{n-1}$ and $1 < m < n - 1$, it follows from the inductive hypothesis that M is cyclic. Hence, P is also cyclic. Every subgroup of G of order p or p^2 is contained in a subgroup of order p^m; and, since P is the only subgroup of order p^m, we conclude that P contains every subgroup of order p or p^2. Being a cyclic group, P contains a single subgroup of order p and a single subgroup of order p^2. Hence, it follows from Theorem 10.2.3 that G is cyclic, a quaternion group, or a generalized quaternion group. But, since the quaternion group and the generalized quaternion groups contain more than one subgroup of order 4, we conclude that G is cyclic. Thus, our theorem is proved.

In an abelian group, every subgroup is normal. But, there are also non-abelian groups all of whose subgroups are normal. Such groups are called *hamiltonian*. The next theorem gives a complete survey of all hamiltonian groups.

10.2.5 *Let Q denote a quaternion group, U an abelian group in which every element is of odd order, and Z an abelian group of exponent 2. Then, every hamiltonian group belongs to one of the following types: Q, $Q \times U$, $Q \times Z$, $Q \times U \times Z$.*

Proof. We begin with a proof of the following formula: If $[x, y]$ belongs to the centre of $\langle x, y\rangle$, then for every natural number n

$$(xy)^n = x^n y^n [y, x]^{n(n-1)/2}. \tag{10.7}$$

This is trivial for $n = 1$. Assume that

$$(xy)^{n-1} = x^{n-1}y^{n-1}[y, x]^{(n-1)(n-2)/2}.$$

Then, we have

$$\begin{aligned}(xy)^n = (xy)^{n-1}xy &= x^{n-1}y^{n-1}[y, x]^{(n-1)(n-2)/2}\,xy \\ &= x^{n-1}y^{n-1}xy[y, x]^{(n-1)(n-2)/2} \\ &= x^n y^{n-1}[y^{n-1}, x]\; y[y, x]^{(n-1)(n-2)/2}.\end{aligned}$$

By (9.3), we obtain

$$(xy)^n = x^n y^{n-1}[y, x]^{n-1}y[y, x]^{(n-1)(n-2)/2} = x^n y^n [y, x]^{n(n-1)/2}.$$

Thus, (10.7) holds for all natural numbers n.

Let H be a hamiltonian group. Since H is not abelian, there are two elements a and b in H that do not commute. As $\langle a\rangle$ and $\langle b\rangle$ are normal subgroups of H, we have

$$c = [a, b] \in \langle a\rangle \cap \langle b\rangle \tag{10.8}$$

so that c belongs to the centre of $Q = \langle a, b\rangle$. The commutator subgroup Q' is generated by c and is a proper subgroup of $\langle a\rangle$ as well as of $\langle b\rangle$.

By (10.8), $c = a^r = b^s$, where $rs \neq 0$ since $c \neq e$. Since c belongs to the centre of $\langle a, b\rangle$, it follows from (9.4) that

$$c^s = [a, b]^s = [a, b^s] = e.$$

Hence, a and b are of finite order.

We conclude that every element of H that does not belong to the centre is of finite order. If the element h of H is permutable with a and b, then ah is not permutable with b so that ah is of finite order. As h and a commute, we conclude that h is of finite order. This shows that every element of H has finite order.

The elements a and b are only subject to the condition that they do not commute. We now choose a and b subject to this condition such that their respective orders m and n are as small as possible. For a prime divisor p of m, we then obtain $[a^p, b] = e$ and hence $c^p = [a, b]^p = [a^p, b] = e$. Similarly, we conclude that $c^q = e$ for any prime divisor q of n. Hence, $c \neq e$ implies that $p = q$. Thus, the orders of a and b are powers of one and the same prime number p. The orders of a and b are divisible by p^2, because $\langle c\rangle$ is a proper subgroup of $\langle a\rangle$ and of $\langle b\rangle$. Moreover, a^p and b^p are contained in the centre of Q.

Suppose that $a^{p^k} = c^t$, $b^{p^l} = c^u$, where t and u are not divisible by p. Then, we may replace a by a^t and b by b^u. Hence, it is no loss of generality to assume that

$$a^{p^k} = b^{p^l} = [a, b] = c$$

and that $k \geqq l > 0$.

The elements a and $b_1 = a^{-p^{k-l}}b$ generate Q. Hence, the order of b_1 cannot be smaller than the order of b. By (10.7), we obtain

$$b_1^p = a^{-p^{k-l+1}}b^p[b, a]^{\frac{1}{2}(p-1)p^{k-l+1}} = a^{-p^{k-l+1}}b^p c^{-\frac{1}{2}(p-1)p^{k-l+1}},$$

which gives

$$b_1^{p^l} = c^{-\frac{1}{2}(p-1)p^k}.$$

Now, $b_1^{p^l} \neq e$, but $c^p = e$ so that $p = 2$, $k = 1$. This implies that $l = 1$. So, we have $a^2 = b^2 = c$, $c^2 = e$. It is easy to see that these relations too define the quaternion group. Thus, $Q = \langle a, b\rangle$ is a quaternion group.

We now prove that $H = QC$, where C denotes the centralizer of Q in H. Suppose that an element x of H does not belong to C, because it does not commute with a. Then, we have $x^{-1}ax = a^{-1}$, which implies that xb commutes with a. Similarly, if x is not permutable with b, then xa commutes with b. If x commutes neither with a nor with b, then xab commutes with both a and b. Thus, one of the elements x, xa, xb, xab belongs to C. This means that $H = QC$.

Next, C does not contain an element of order 4. Indeed, for $x \in C$ we have $[a, bx] \neq e$, so that from $(bx)^4 = e$ it follows that $a^{-1}(bx)\,a = (bx)^{-1}$ or $a^{-1}bax = b^{-1}x^{-1}$; hence, $x^2 = e$. As C contains no element of order 4, we conclude, in particular, that C does contain not a quaternion group as a subgroup. It follows from our above arguments that any two elements of C commute, i. e. C is abelian.

We conclude that C is the direct product of its subgroup U of all elements of odd order and its subgroup Z_1 of all elements whose square is the unit element. The element $c = [a, b]$ belongs to Z_1. Let Z denote a subgroup of Z_1 that is maximal, subject to the condition that it does not contain c. Then, $c \in \langle Z, x\rangle$ for every element x of Z_1 that is not contained in Z. Now, $x^2 = e$ implies that $|\langle Z, x\rangle : Z| = 2$. Moreover, $|\langle Z, c\rangle : Z| = 2$. This shows that $\langle Z, x\rangle = \langle Z, c\rangle$ and hence $\langle Z, c\rangle = Z_1$, $Z \cap \langle c\rangle = e$. So, we have $C = U \times Z \times \langle c\rangle$. Since of $Q \cap C = \langle c\rangle$, we find that $Q \cap (U \times Z) = e$ and $Q(U \times Z) = H$, hence $H = Q \times U \times Z$. Of course, U or Z or both of them may reduce to the identity. This proves our theorem.

Conversely, every group of one of the types in question is hamiltonian. It is not abelian, because Q is not. It remains to show that every cyclic subgroup $\langle quz\rangle$ where $q \in Q$, $u \in U$, $z \in Z$ is normal. As U and Z belong to the centre, it suffices to show that $\langle quz\rangle$ is invariant under transformation by a and b. We have $a^{-1}quza = q^k uz$, where $k = 1$ or 3. The order of u is an odd number n, and the order of z is 1 or 2. The congruences $x \equiv k \pmod 4$, $x \equiv 1 \pmod n$ have a solution x so that $a^{-1}quza = (quz)^x$. A similar argument applies to transformation by b.

Exercises

1. Prove that two elements a and b subject to the relations $a^2 = b^2 = (ab)^2$ define the quaternion group.
2. Prove that

$$\varphi_{r+1,k} = \varphi_{r,k} + p^{r-k+1}\,\varphi_{r,k-1},$$

where $\varphi_{r,k}$ is defined as in section 10.1.
3. Determine the commutator subgroup of a generalized quaternion group.
4. Determine the Frattini subgroups of the groups of order p^3.
5. Let G be the group of all matrices

$$\begin{bmatrix} 1 & 0 & 0 \\ a & 1 & 0 \\ b & c & 1 \end{bmatrix}$$

whose elements belong to the Galois field $GF(p)$. Find the type in Theorem 10.2.2 to which G belongs.

CHAPTER 11

FINITE SOLUBLE GROUPS

11.1 Hall Subgroups of Soluble Groups

A subgroup H of a finite group G is called a *Hall subgroup* if its order $|H|$ is relatively prime to its index $|G:H|$. For example, all Sylow subgroups are Hall subgroups. Though Sylow's Theorem 3.2.5 cannot be extended to Hall subgroups of an arbitrary finite group G, an important result due to P. Hall says that such a generalization is possible provided that G is soluble. This result should be viewed in the light of the Feit–Thompson Theorem 2.8.1.

11.1.1 P. Hall's Theorem. *Let G be a finite soluble group.*

(a) *If $|G| = mn$ is any factorization with $(m, n) = 1$, then G contains at least one Hall subgroup of order m.*

(b) *Any two Hall subgroups of order m are conjugate.*

(c) *Every subgroup of G whose order divides m is contained in some Hall subgroup of order m.*

(d) *The number h_m of Hall subgroups of order m of G can be represented as a product of factors a_i such that*

(1) $a_i \equiv 1 \pmod{p_i}$, *where p_i is a prime divisor of m.*

(2) *a_i is a prime power and divides some chief index of G.*

Here, (a), (b) and (c) are analogues to the corresponding parts of Theorem 3.2.5, while (d) is a stronger statement than its analogue for Sylow subgroups.

Proof. For $|G| = 1$, there is nothing to prove. We proceed by induction on the group order and assume that our theorem is true for all soluble groups of order less than $|G|$. We consider two cases.

First case: Suppose that G contains a proper normal subgroup T whose order is not divisible by n. Putting $|T| = m_1 n_1$ and $|G:T| = m_2 n_2$, where $m_1 m_2 = m$, $n_1 n_2 = n$, we have $n_1 < n$.

By the inductive hypothesis, G/T contains a subgroup D/T of order m_2. The order mn_1 of D is less than $|G|$, so that by the inductive hypothesis D contains a subgroup H of order m. Thus, (a) is proved.

Let H and H^* be two subgroups of order m. The order of $D = TH$ is a divisor of $|T||H| = m_1n_1m_1m_2$. But, this order divides mn_1 since it is a divisor of $|G| = mn$. On the other hand, the order of TH is divisible by $|T| = m_1n_1$ as well as by $|H| = m$. Consequently, we have $|D| = |TH| = mn_1$. In the same way, we see that $D^* = TH^*$ is of order mn_1. Consequently, D/T and D^*/T are two Hall subgroups of order m_2 of G/T. It follows from the inductive hypothesis that D/T and D^*/T are conjugate in G/T. This means that $D = x^{-1}D^*x$ for some $x \in G$, and hence $x^{-1}H^*x \subset D$. Again, by induction, H and $x^{-1}H^*x$ are conjugate in D so that H and H^* are conjugate in G. This completes the proof of (b) for G.

Now, let U be a subgroup of G whose order divides m. As (c) is true for G/T, the factor group UT/T is contained in a subgroup D/T of G/T. This gives $UT \subseteq D$ and hence $U \subset D$. By the inductive hypothesis, U is contained in a subgroup of order m of D, which proves (c) for G.

By (b), the number h_m is equal to the number of conjugates of a Hall subgroup H of order m. Now, this number is equal to the product of the number of conjugates of H contained in $D = TH$, and the number h_{m_2} of subgroups of order m_2 of G/T. The chief indices of D are divisors of certain chief indices of G, and the chief indices of G/T form a subset of those of G. By the inductive hypothesis, h_m is therefore the product of two natural numbers both of which satisfy the conditions (1) and (2) of (d). Consequently, h_m itself satisfies those conditions.

Second case: We have now to consider the case when the order of every proper normal subgroup of G is divisible by n. As G is soluble, the order of every minimal normal subgroup is a power of a prime number. Let K be a minimal normal subgroup of order p^a. Since all other situations are covered by the first case, we may assume that $n = p^a$ and that every minimal normal subgroup of G is of order p^a. Thus, the minimal normal subgroups of G are Sylow p-subgroups so that K is the only minimal normal subgroup of G. By Theorem 4.2.2, K is elementary abelian.

Let L/K be a minimal normal subgroup of G/K. The order of L/K is a power q^b of a prime number $q \neq p$. If Q denotes a Sylow q-subgroup of L, then we have $|Q| = q^b$ and $L = QK$. We consider the intersection $R = \mathsf{N}(Q) \cap K$, where $\mathsf{N}(Q)$ denotes the normalizer of Q in G. As a subgroup of K, R is elementary abelian, and clearly R is a normal subgroup of $\mathsf{N}(Q)$. The normal subgroups R and Q of $\mathsf{N}(Q)$ intersect trivially so that they commute elementwise. This means that R belongs to the centre C of L. Now, C is normal in G, since it is a characteristic subgroup of the normal subgroup L of G. As K is minimal, the intersection $K \cap C$ coincides with K or is equal to the unit element. The case $K \cap C = K$ can readily be excluded; for, in that case, we would have

$L = Q \times K$ so that Q would be characteristic in L and hence normal in G, but this contradicts our above assumption on K. So, we have $K \cap C = e$ and, since $R \subseteq K \cap C$, this implies that $R = e$. Consequently, Q coincides with its normalizer in L so that Q has $|L:Q| = p^a$ distinct conjugates in L. As L is normal in G, we conclude that every subgroup conjugate to Q in G is contained in L, and hence the total number of conjugates of Q in G is p^a. This shows that $|G:\mathsf{N}(Q)| = p^a = n$. Consequently, $\mathsf{N}(Q)$ is a subgroup of order m of G, and (a) is proved.

The normalizers of the p^a conjugates of Q in G are distinct from each other and form a complete system of conjugate subgroups of G. For the Sylow q-subgroup Q of L is the intersection of L and some Sylow q-subgroup $\bar{Q}$ of G so that the normalizer of $\bar{Q}$ in G is contained in $\mathsf{N}(Q)$; by Theorem 3.3.2, $\mathsf{N}(Q)$ coincides with its normalizer in G, and hence G contains exactly p^a subgroups conjugate to $\mathsf{N}(Q)$. Moreover, we have $p^a \equiv 1 \pmod{q}$, since p^a is the number of Sylow q-subgroups of L. Let M be any subgroup of G of order m. Then, the order of ML is divisible by m and by $n = p^a$ and hence $ML = G$. From

$$G/L = ML/L \cong M/M \cap L,$$

it follows that the order of $M \cap L$ is q^b, which implies that $M \cap L$ is conjugate to Q in L. Moreover, $M \cap L$ is a normal subgroup of M so that M is the normalizer of a conjugate of Q. Thus, the conjugates of $\mathsf{N}(Q)$ exhaust all the subgroups of order m of G. This proves (b) and (d).

Finally, let U be a subgroup of G whose order divides m. We take an arbitrary subgroup M of order m and put $U^* = M \cap UK$. Then, $\langle M, UK \rangle = G$, and hence Theorem 1.5.4 shows that $|U^*| = |U|$. Using (b) for UK we conclude that U^* is conjugate to U. Thus, U is contained in a conjugate of M. This proves (c).

The proof of Theorem 11.1.1 is now complete.

There are further theorems on Sylow subgroups that can be extended to Hall subgroups of soluble groups.

11.1.2 *Let G be a soluble group of order mn with $(m, n) = 1$, and let U denote a subgroup of order u, where $(u, m) = u_1$. Then, two distinct Hall subgroups H_1 and H_2 of order u_1 of U are not contained in the same Hall subgroup of order m of G.*

Proof. Assume that H_1 and H_2 are contained in the same Hall subgroup H of order m. Then, $\langle H_1, H_2 \rangle$ is a subgroup of H so that $|\langle H_1, H_2 \rangle|$ divides m. But, since $\langle H_1, H_2 \rangle$ belongs to U, the order $|\langle H_1, H_2 \rangle|$ also divides u. Thus, $|\langle H_1, H_2 \rangle|$ is a divisor of $(u, m) = u_1$ which implies that $\langle H_1, H_2 \rangle = H_1 = H_2$.

11.1.3 *If a subgroup A of the soluble group G contains the normalizer $\mathsf{N}(H)$ of a Hall subgroup H of G, then A coincides with its normalizer in G.*

Proof. Suppose that $x^{-1}Ax = A$. Then, H and $x^{-1}Hx$ are two Hall subgroups of A. By Theorem 11.1.1, there exists an element y of A such that $y^{-1}x^{-1}Hxy = H$. Therefore, $xy \in \mathsf{N}(H)$ and hence $xy \in A$, which shows that $x \in A$.

11.1.4 *Suppose that the complexes K and L of the soluble group G are conjugate in G. Suppose also that*

$$a^{-1}Ka = K, \quad a^{-1}La = L$$

for every element a of a Hall subgroup H of G. Then, the normalizer $\mathsf{N}(H)$ of H in G contains an element s such that $s^{-1}Ks = L$.

Proof. Let $\mathsf{N}(K)$ denote the normalizer of K in G. By the hypothesis of our theorem, $H \subseteq \mathsf{N}(K)$. From $x^{-1}Kx = L$ we conclude that $\mathsf{N}(L) = x^{-1}\mathsf{N}(K)\,x$. This shows that $x^{-1}Hx \subseteq \mathsf{N}(L)$. Moreover, $H \subseteq \mathsf{N}(L)$ by the hypothesis. Thus, H and $x^{-1}Hx$ are two Hall subgroups of $\mathsf{N}(L)$. By Theorem 11.1.1, there is an element y of $\mathsf{N}(L)$ such that $H = y^{-1}x^{-1}Hxy$. Therefore, $s = xy$ is contained in $\mathsf{N}(H)$, and we have

$$s^{-1}Ks = y^{-1}x^{-1}Kxy = y^{-1}Ly = L.$$

This completes the proof.

We now turn to another remarkable theorem of P. Hall. It states that the solubility of G is not only sufficient, but also necessary, for statement (a) of 11.1.1 to be true.

11.1.5 *Suppose that a finite group G satisfies the following condition: For every factorization $|G| = mn$ with $(m, n) = 1$, there is at least one subgroup of order m of G. Then, G is soluble.*

Let p^a denote the highest power of the prime number p that divides the order of the group G. Then, any subgroup of index p^a of G is called a *Sylow p-complement*.

Clearly, Theorem 11.1.5 is equivalent to the following theorem:

11.1.5′ *If a group G contains at least one Sylow p-complement for every prime divisor p of $|G|$, then G is soluble.*

The proof requires a lemma whose main assertion will only be proved in the last chapter.

Lemma. Let G be a group of order $p^a q^b m$, where p and q are distinct primes, $a > 0$, $b > 0$, and $(p, m) = (q, m) = 1$. Suppose that G contains a subgroup H of order $p^a q^b$ and two proper subgroups whose indices are powers of p and q, respectively. Then, G is not simple.

Proof of the Lemma. Theorem 13.6.2, which we use here, states that H is soluble. Hence, H contains a minimal normal subgroup A whose order is a power of a prime number. We may assume that the notation is such that $|A|$ is a power of p. By the hypothesis, G contains a subgroup B whose index is a power of q. Hence, B contains a Sylow p-subgroup of G and, on the other hand, A is contained in some Sylow p-subgroup of G. Therefore, we may assume that $A \subseteq B$. For, if necessary, we can replace B by a conjugate subgroup containing A. Since $|G:H|$ and $|G:B|$ are relatively prime, we have $G = BH$. It follows that all conjugates of B in G are obtained by transforming B by the elements of H. Since A is contained in B and a normal subgroup of H, we conclude that A lies in all conjugates of B, and hence that $A \subseteq D$, where D denotes the intersection of all conjugates of B. Therefore, $e \neq A \subseteq D \subseteq B \neq G$ so that D is a non-trivial normal subgroup of G.

Proof of Theorem 11.1.5′. By Theorem 13.6.2, a group is soluble if its order has only two prime divisors. Therefore, we may assume that the order of G has the factorization

$$|G| = p_1^{a_1} p_2^{a_2} \cdots p_r^{a_r}$$

into powers of distinct prime numbers p_i with $r > 2$. Furthermore, we assume that the theorem is true for all groups whose order is less than $|G|$. Let K_i denote a Sylow p_i-complement of G ($i = 1, \ldots, r$). By Theorem 1.5.5, the intersection

$$H = K_3 \cap \cdots \cap K_r$$

is of order $p_1^{a_1} p_2^{a_2}$. Now, G contains subgroups whose indices are $p_1^{a_1}$ and $p_2^{a_2}$, respectively. So, we can apply the Lemma which guarantees the existence of a non-trivial normal subgroup D of G. Just as in the proof of Theorem 3.2.3, it follows that $D \cap K_i$ is a Sylow p_i-complement of D and that $K_i D/D$ is a Sylow p_i-complement of G/D. Consequently, both D and G/D contain Sylow complements for all prime divisors of their orders so that they are soluble by the inductive hypothesis. Thus, G itself is soluble.

11.2 Sylow Systems

Let G be a finite soluble group and let

$$|G| = p_1^{a_1} p_2^{a_2} \cdots p_r^{a_r}$$

be the factorization of its order into powers of distinct prime numbers p_i. By Theorem 11.1.1, there exists at least one Sylow p_i-complement for every p_i. Any set

$$K_1, K_2, \ldots, K_r \tag{11.1}$$

of Sylow p_i-complements corresponding to the r prime divisors p_i of G is called a *complete system of Sylow complements*.

For a non-empty subset α of $\{1, 2, \ldots, r\}$, we define

$$K_\alpha = \bigcap_{i \in \alpha} K_i . \tag{11.2}$$

If α is empty, we put $K_\alpha = G$. In this way, we obtain a system of 2^r subgroups K_α, a so-called *Sylow system of G*.

By Theorem 1.5.5, we have

$$|G:K_\alpha| = \prod_{i \in \alpha} |G:K_i| = \prod_{i \in \alpha} p_i^{a_i}.$$

Hence,

$$|K_\alpha| = \prod_{j \in \alpha'} p_j^{a_j}, \tag{11.3}$$

where α' denotes the complement of α in $\{1, 2, \ldots, r\}$. Thus, all the K_α are Hall subgroups of G. But, the orders of the K_α are precisely the possible orders of Hall subgroups of G, where we agree to count e and G among the Hall subgroups. By Theorem 11.1.1, a suitable conjugate of every Hall subgroup is some K_α.

For another subset β of $\{1, 2, \ldots, r\}$, it follows immediately from the definition (11.2) that

$$K_{\alpha \cup \beta} = K_\alpha \cap K_\beta . \tag{11.4}$$

On the other hand,

$$K_{\alpha \cap \beta} = K_\alpha K_\beta . \tag{11.5}$$

Obviously, both K_α and K_β are contained in $K_{\alpha \cap \beta}$. To prove (11.5), we have to show that $K_\alpha K_\beta$ consists of $|K_{\alpha \cap \beta}|$ distinct elements. By Theorem 1.3.5, the number of distinct elements in $K_\alpha K_\beta$ is

$$|K_\alpha||K_\beta||K_\alpha \cap K_\beta|^{-1}$$

and by (11.4), this number is equal to $|K_\alpha||K_\beta||K_{\alpha \cup \beta}|^{-1}$. Using (11.3), we obtain

$$|K_\alpha||K_\beta||K_{\alpha \cup \beta}|^{-1} = \prod_{j \in \alpha'} p_j^{a_j} \prod_{k \in \beta'} p_k^{a_k} \Big(\prod_{l \in (\alpha \cup \beta)'} p_l^{a_l} \Big)^{-1}$$

$$= \prod_{j \in \alpha'} p_j^{a_j} \prod_{k \in \beta'} p_k^{a_k} \Big(\prod_{l \in \alpha' \cap \beta'} p_l^{a_l} \Big)^{-1} = \prod_{i \in \alpha' \cup \beta'} p_i^{a_i}$$

$$= \prod_{i \in (\alpha \cap \beta)'} p_i^{a_i} = |K_{\alpha \cap \beta}| .$$

This proves (11.5). In particular, we conclude that

$$K_\alpha K_\beta = K_\beta K_\alpha . \tag{11.6}$$

By a *complete system of Sylow subgroups,* we mean any system of r Sylow p_i-subgroups for $i = 1, 2, \ldots, r$. Every Sylow system contains a complete system of Sylow subgroups. By (11.6), these Sylow subgroups are even permutable in pairs. This is not true in general for an arbitrary complete system of Sylow subgroups. If

$$P_1, P_2, \ldots, P_r$$

denotes the complete system of Sylow subgroups that belongs to our Sylow system, then we obviously have

$$K_\alpha = \prod_{i \in \alpha'} P_i \, .$$

Conversely, if a group contains a complete system of pairwise permutable Sylow subgroups, then it is soluble by Theorem 11.1.5. So, we have the following theorem:

11.2.1 *A finite group is soluble if and only if it is the product of pairwise permutable subgroups of prime power order.*

This theorem can be generalized. The most far-reaching generalization known states that a finite group is soluble if and only if can be expressed as a product of pairwise permutable nilpotent subgroups. For the proof, we refer to [39].

11.2.2 *Every system $\mathfrak{T}$ of pairwise permutable Hall subgroups of a soluble group G is contained in some Sylow system $\mathfrak{S}$ of G.*

Proof. Let V_i denote the product of all those Hall subgroups in $\mathfrak{T}$ whose orders are relatively prime to p_i. If there is no Hall subgroup in $\mathfrak{T}$ whose order is relatively prime to p_i, then we put $V_i = e$. Clearly, $|V_i|$ is relatively prime to p_i. It follows from Theorem 11.1.1 that, for $i = 1, 2, \ldots, r$, there exists a Sylow p_i-complement K_i of G containing V_i. Let $\mathfrak{S}$ denote the Sylow system of G that is obtained from these Sylow complements K_i.

Now, let H be an arbitrary Hall subgroup belonging to $\mathfrak{T}$. Let $p_{i_1}, \ldots, p_{i_t}$ be the prime divisors of the index $|G:H|$ and denote the set $\{i_1, \ldots, i_t\}$ by τ. We show that $H = K_\tau$. Since H is a Hall subgroup, its order is relatively prime to $p_{i_1}, \ldots, p_{i_t}$. So, we have

$$H \subseteq V_{i_j}, \quad j = 1, \ldots, t$$

and hence

$$H \subseteq K_{i_j}, \quad j = 1, \ldots, t,$$

and, therefore, $H \subseteq K_\tau$. But, since H is a Hall subgroup, its order must be the same as that of K_τ. Therefore, $H = K_\tau$. This completes the proof.

11.2.3 *Any two Sylow systems of a soluble group are conjugate.*

Proof. Let $\mathfrak{S}$ and $\mathfrak{S}^*$ be two Sylow systems of G. They arise from complete systems of Sylow complements

$$\mathfrak{K}: K_1, \ldots, K_r$$

and

$$\mathfrak{K}^*: K_1^*, \ldots, K_r^*,$$

respectively. Suppose that $\mathfrak{K}$ and $\mathfrak{K}^*$ have s groups in common. When $s = r$, then $\mathfrak{S} = \mathfrak{S}^*$ and the theorem is true. So, we have only to consider the case $s < r$. It is sufficient to prove the following proposition: There is an element x in G such that $x^{-1}\mathfrak{K}x$ and $\mathfrak{K}^*$ have $s + 1$ groups in common. For, starting with $\mathfrak{K}$ we arrive, after $r - s$ steps, at a conjugate system that coincides with $\mathfrak{K}^*$.

Assume that $K_i \neq K_i^*$. Let P_i denote the Sylow p_i-subgroup belonging to $\mathfrak{S}$. Then, $G = K_i P_i$ so that transformation of K_i by the elements of P_i gives all the conjugates of K_i in G. By Theorem 11.1.1, K_i and K_i^* are conjugate. Consequently, there is an element x of P_i such that $x^{-1}K_i x = K_i^*$. We know that P_i is the intersection of all the groups K_j of $\mathfrak{K}$ other than K_i. This means that $x \in K_j$ for $j \neq i$, and hence $x^{-1}K_j\, x = K_j$ for $j \neq i$. Therefore, $x^{-1}\mathfrak{K}x$ and $\mathfrak{K}^*$ have $s + 1$ groups in common, namely the s groups common to $\mathfrak{K}$ and $\mathfrak{K}^*$, and K_i^*.

This completes the proof.

The following theorem states a simple relationship between the Sylow system of G and those of subgroups of G.

11.2.4 *Every Sylow system of a subgroup U of G is the intersection of U with the groups of some Sylow system of G.*

Proof. Let us assume that the prime divisors $p_1, \ldots, p_r$ of $|G|$ are numbered so that $p_1, \ldots, p_s$ divide $|U|$. Of course, the case $s = r$ is possible. Let $\mathfrak{W}$ be a Sylow system of U. Then, $\mathfrak{W}$ contains a complete system of Sylow complements

$$L_1, \ldots, L_s,$$

where $|U:L_i|$ is a power of p_i $(i = 1, \ldots, s)$. By Theorem 11.1.1, for each of the prime numbers $p_1, \ldots, p_s$, there exists a Sylow p_i-complement K_i of G containing L_i. The same theorem asserts that, for each of the prime numbers $p_{s+1}, \ldots, p_r$, we can find a Sylow p_j-complement K_j of G containing U. So, we obtain a complete system of Sylow complements

$$K_1, \ldots, K_s, K_{s+1}, \ldots K_r$$

of G.

Now, let σ be a subset of $\{1, \ldots, s\}$ and ϱ a subset of $\{s+1, \ldots, r\}$. We show that

$$L_\sigma = U \cap K_{\sigma \cup \varrho} \tag{11.7}$$

and this will prove our theorem.

In accordance with the choice of the K_i, the right-hand side of (11.7) contains L_σ. On the other hand, the order of the right-hand side is equal to the greatest common divisor of $|U|$ and $|K_{\sigma \cup \varrho}|$. But, since $|L_\sigma|$ is the greatest common divisor of $|U|$ and $|K_{\sigma \cup \varrho}|$, we obtain (11.7).

11.3 System Normalizers

For a Sylow system $\mathfrak{S}$ of the soluble group G, let

$$K_1, \ldots, K_r$$

denote the Sylow p_i-complements and

$$P_1, \ldots, P_r$$

the Sylow p_i-subgroups that belong to $\mathfrak{S}$.

By the *system normalizer* $\mathsf{N}(\mathfrak{S})$ of $\mathfrak{S}$, we mean the largest subgroup of G in which each group of $\mathfrak{S}$ is normal. Since all the groups of $\mathfrak{S}$ are obtained by taking intersections of certain Sylow complements, an element x of G belongs to $\mathsf{N}(\mathfrak{S})$ if and only if

$$x^{-1}K_i x = K_i \quad \text{for} \quad i = 1, \ldots, r.$$

On the other hand, the groups of $\mathfrak{S}$ can also be obtained as products of certain Sylow subgroups. So, we have $x \in \mathsf{N}(\mathfrak{S})$ if and only if

$$x^{-1}P_i x = P_i \quad \text{for} \quad i = 1, \ldots, r.$$

This shows that

$$\mathsf{N}(\mathfrak{S}) = \bigcap_{i=1}^{r} \mathsf{N}(K_i) = \bigcap_{i=1}^{r} \mathsf{N}(P_i).$$

It follows from Theorem 11.2.3 that all system normalizers of G are conjugate. Thus, any automorphism of G permutes the system normalizers of G.

The next theorem states that the system normalizers are nilpotent.

11.3.1 *The system normalizer* $\mathsf{N}(\mathfrak{S})$ *is the direct product of its Sylow subgroups* N_i, *where*

$$N_i = P_i \cap \mathsf{N}(K_i) \quad (i = 1, \ldots, r). \tag{11.8}$$

Proof. We define the subgroup N_i by (11.8) and then prove that N_i is the only Sylow p_i-subgroup of $\mathsf{N}(S)$.

From $N_i \subseteq P_i$, we conclude that $N_i \subseteq K_j$ for $j \neq i$. Consequently, N_i is contained in the normalizer of each K_j, $j \neq i$. By the definition of N_i, it is also contained in the normalizer of K_i. Therefore, $N_i \subseteq \mathsf{N}(\mathfrak{S})$.

Now, let N_i^* be any Sylow p_i-subgroup of $\mathsf{N}(\mathfrak{S})$. Then, N_i^* is contained, in particular, in $\mathsf{N}(P_i)$. But, P_i is the only Sylow p_i-subgroup of $\mathsf{N}(P_i)$ and $|N_i^*|$ is a power of p_i so that $N_i^* \subseteq P_i$. Moreover, N_i^* is contained in $\mathsf{N}(K_i)$, and hence $N_i^* \subseteq N_i$. On the other hand, from $N_i \subseteq \mathsf{N}(\mathfrak{S})$ and from the fact that $|N_i|$ is a power of p_i, it follows that $N_i^* = N_i$. This shows that N_i is the only Sylow p_i-subgroup of $\mathsf{N}(\mathfrak{S})$.

11.3.2 *The system normalizers of G generate G. No proper normal subgroup of G contains a system normalizer.*

Proof. Since all system normalizers are conjugate, they generate a normal subgroup of G. Thus, it is sufficient to show that no system normalizer $\mathsf{N}(\mathfrak{S})$ is contained in a given maximal normal subgroup M of G.

The index of M in G is a prime number, $|G:M| = p_i$ say. Then, all the conjugates of K_i are contained in M so that $G = M\mathsf{N}(K_i)$. Since

$$G/M = M\mathsf{N}(K_i)/M = \mathsf{N}(K_i)/M \cap \mathsf{N}(K_i),$$

there is at least one coset of xK_i that belongs to $\mathsf{N}(K_i)$ and contains no element of M. Every coset of K_i contains one and only one element of P_i, so that we may assume that $x \in P_i$. By Theorem 11.3.1, $x \in P_i \cap \mathsf{N}(K_i)$ shows that x is contained in a Sylow p_i-subgroup N_i of $\mathsf{N}(\mathfrak{S})$. But, $x \notin M$, and hence $\mathsf{N}(\mathfrak{S})$ is not contained in M.

In contrast to Theorem 11.3.2, we have:

11.3.3 *Every maximal non-normal subgroup of G contains at least one system normalizer.*

Proof. As we saw in section 7.1 and section 7.2, a maximal subgroup H of G induces a primitive permutational representation of degree $|G:H|$ of G. Hence, it follows from Theorem 7.1.11 that $|G:H|$ is a power of a prime number, $|G:H| = p_i^b$ say. Consequently, every Sylow p_i-complement K_i of H is at the same time a Sylow p_i-complement of G. It suffices to show that H contains the normalizer $\mathsf{N}(K_i)$ of K_i in G.

Let D denote the intersection of all subgroups conjugate to H. The permutational representation of G that is induced by H is isomorphic to G/D. By Theorem 7.1.11, G/D contains a single minimal normal subgroup R/D, and the order of R/D is equal to p_i^b. The permutations corresponding to R/D form a regular abelian permutation group. Hence, it follows from Theorem 7.1.8 that R/D coincides with its centralizer in G/D.

We now show that the order of any minimal normal subgroup Q/R of G/R is a power of a prime number other than p_i. Since G/R is soluble, the order $|Q/R|$ is a prime power. Suppose that $|Q/R|$ is a power of p_i. Then, Q/D is a p_i-group, and hence has a non-trivial centre Z/D. Now, Z/D is a characteristic subgroup of the normal subgroup Q/D of G/D, so that Z/D is a normal subgroup of G/D. Since R/D is the only minimal normal subgroup of G/D, it follows that $R/D \subseteq Z/D$. Thus, R/D belongs to the centre of Q/D. But, this contradicts the fact that R/D is its own centralizer in G/D. Hence, we have $|Q/R| = p_j^c$, where $p_j \neq p_i$. It follows that $HQ = G$.

Let $M = H \cap Q$. Then, it follows from Theorem 1.5.4 that $|M/D| = p_j^c$. Since every p_j-subgroup of H/D is contained in some Sylow p_j-subgroup of H/D, we conclude that $M = K_iD \cap Q$. Since both D and Q are normal in G, it follows that $\mathsf{N}(K_i)$ is contained in the normalizer of M, namely H. This completes the proof.

11.4 System Normalizers and Factor Groups

When we talk of the Sylow structure or the *arithmetical structure* of a group, we have in mind those properties that are related to its Sylow subgroups or, more generally, to subgroups satisfying certain arithmetical conditions. On the other hand, the properties of a group that refer to its normal and subnormal subgroups reflect its *normal structure*. The definition of solubility in section 2.8 and Theorem 2.8.4, for instance, deals with the normal structure, whereas Theorems 11.1.1 and 11.1.5 describe the Sylow structure of soluble groups. Many important theorems establish links between these two aspects of group structure. In this section, we derive two theorems that exhibit a relation between the system normalizers of a finite soluble group and its normal structure.

For two normal subgroups A and B with $A \subset B$, the factor group B/A is called a *factor* of G. If there is no normal subgroup D of G such that $A \subset D \subset B$ with proper inclusions, then B/A is said to be a *chief factor*; otherwise, B/A is *composite*.

When B/A is composite, there is a chain

$$B = B_0 \supset B_1 \supset \cdots \supset B_l = A$$

of normal subgroups B_i of G. We say that the factor B/A is decomposed into the component factors B_{i-1}/B_i $(i = 1, \ldots, l)$. Thus, every composite factor can be decomposed into components that are chief factors.

Let C/A be the centralizer of B/A in G/A. Then, C is called the *centralizer* of B/A. Clearly, C consists of those elements c of G for which $c^{-1}xAc = xA$

for every coset xA contained in B; in other words, C is the largest subgroup of G such that $A \subseteq C$ and $[C,B] \subseteq A$.

A factor B/A is called *central* if the centralizer of B/A is the whole group G. A factor is said to be *hypercentral* if it can be decomposed into components all of which are central.

A chief factor of G that is not central is called *eccentric*. A *hypereccentric* factor is one that can be decomposed into eccentric chief factors.

The property of being hypercentral or hypereccentric does not depend on a particular decomposition into component factors. This follows from Theorem 2.7.3, when the operator domain consists of all the inner automorphisms of G.

We say that a subgroup U of G *covers* the factor B/A if U contains at least one element of every coset of A that belongs to B. Thus, U covers B/A if and only if $A(U \cap B) = B$.

On the other hand, U is said to *avoid* the factor B/A if all elements of $U \cap B$ are contained in A or, what is the same, if $A(U \cap B) = A$.

Let

$$G = G_0 \supset G_1 \supset \cdots \supset G_{l-1} \supset G_l = e \tag{11.9}$$

be a normal series of G. For a subgroup U of G, we put

$$U_i = U \cap G_i \quad (i = 0, \ldots, l).$$

Then, the U_i are normal in U, and we have

$$|U| = \prod_{i=0}^{l-1} |U_i/U_{i+1}|\,.$$

If U covers the factor G_i/G_{i+1}, we obtain therefore,

$$G_{i+1}(U \cap G_i) = G_{i+1} U_i = G_i\,,$$

$$G_i/G_{i+1} = G_{i+1} U_i/G_{i+1} \cong U_i/U_i \cap G_{i+1} = U_i/U_{i+1}\,.$$

But, if U avoids the factor G_i/G_{i+1}, then we have $G_{i+1}(U \cap G_i) = G_{i+1}$ and $U_i = U_{i+1}$.

If U either covers or avoids every factor of the normal series (11.9), the order of U is equal to the product of the orders of those factors that are covered by U.

11.4.1 *Every system normalizer of a finite soluble group G avoids every hypereccentric factor of G. In particular, it avoids every eccentric chief factor.*

Proof. Let $\mathfrak{S}$ be a Sylow system of G and $N = \mathsf{N}(\mathfrak{S})$. It suffices to show that N avoids every eccentric chief factor. For, if B/A is hypereccentric, then it can be decomposed into eccentric chief factors. If N avoids all these components, then it also avoids B/A.

Suppose that B/A is an eccentric chief factor of G. The centralizer C of B/A is a proper subgroup of G. Hence, we can choose a normal subgroup C^* of G such that C^*/C is a chief factor of G. As G is soluble, both $|B/A|$ and $|C^*/C|$ are powers of prime numbers.

First, we show that $|B/A|$ and $|C^*/C|$ are powers of distinct primes. Let us assume that, on the contrary, both orders are powers of one and the same prime number p. Since B/A is abelian, we have

$$A \subset B \subseteq C \subset C^*.$$

Let P^*/A be a Sylow p-subgroup of C^*/A and let Z^*/A denote the centre of P^*/A. The elementary abelian group B/A is contained in P^*/A. Thus, there is a chief series

$$A/A \subset A_1/A \subset \cdots$$

of P^*/A containing B/A. Then, A_1/A is a normal subgroup of P^*/A contained in B/A. But, as A_1/A is of order p, it belongs to Z^*/A. This shows that B/A and Z^*/A have a non-trivial intersection. On the other hand, B/A is not contained in Z^*/A for, otherwise, the centralizer of B/A would be larger than C. Therefore,

$$(Z^*/A) \cap (B/A) = B_0/A,$$

where $A \subset B_0 \subset B$. Clearly, $C^* = CP^*$. Since C is the centralizer of B/A, the group B_0/A can be described as the largest subgroup of B/A that commutes elementwise with C^*/A. Now, every inner automorphism of G leaves B/A and C^*/A fixed. Thus, B_0/A also remains fixed under every inner automorphism of G, so that B_0 is a normal subgroup of G. But, this is impossible, because B/A is a chief factor.

Consequently, we have

$$|B/A| = p_i^l, \ |C^*/C| = p_j^m, \ p_i \neq p_j.$$

Let N_j denote the normalizer in G of the Sylow p_j-subgroup P_j of G that belongs to $\mathfrak{S}$. Since $N \subseteq N_j$, it is sufficient to prove that N_j avoids B/A. Let us assume the contrary. Then, there exists an element x of N_j such that $x \in B$, $x \notin A$. Since $|C^*/C|$ is a power of p_j, every element of C^* can be expressed as a product yc, where $y \in P_j$, $c \in C$. Then, $x^{-1}y^{-1}xy \in P_j$, because $x \in N_j$, and $x^{-1}y^{-1}xy \in B$ because $x \in B$. Since $|P_j|$ and $|B/A|$ are relatively prime, it follows that $x^{-1}y^{-1}xy \in A$. Next, we have $x^{-1}c^{-1}xc \in A$, because $c \in C$, $x \in B$. Consequently, the commutator of x and any element of C^* belongs to A. The elements of B whose commutators with every element of C^* belong to A form a normal subgroup B_1 of G containing A. From the fact that $x \in B$, $x \notin A$, it follows that A is properly contained in B_1. But, this implies that $B_1 = B$, so that C^* belongs to the centralizer of B/A, and this contradicts the definition of C^*.

This completes the proof of Theorem 11.4.1. As a companion piece, we have the following theorem:

11.4.2 *Every system normalizer of a finite soluble group G covers every hypercentral factor of G. In particular, it covers every central factor of G.*

This theorem can be obtained as a corollary to a theorem on the behaviour of the system normalizers under homomorphisms of G.

Let $\mathfrak{S}$ denote an arbitrary system of pairwise permutable Hall subgroups of G. Thus, $\mathfrak{S}$ may, but need not be, a Sylow system of G. The image of $\mathfrak{S}$ under an arbitrary homomorphism σ of G is denoted by $\mathfrak{S}\sigma$. Obviously, $\mathfrak{S}\sigma$ is a system of pairwise permutable Hall subgroups of $G\sigma$.

11.4.3 *If N is the normalizer of $\mathfrak{S}$ in G, then $N\sigma$ is the normalizer of $\mathfrak{S}\sigma$ in $G\sigma$.*

Proof. Let G_0 denote the kernel of σ. We first consider the special case when $|G_0|$ is a power of a prime number p. Let S denote the product of all those groups in $\mathfrak{S}$ whose orders are relatively prime to p; of course, it may happen that $S = e$. By $G_0\mathfrak{S}$, we mean the system of all products of G_0 and the groups in $\mathfrak{S}$. Let x be an element of the normalizer of $G_0\mathfrak{S}$ in G. Such an element x can be characterized by the property that $x\sigma$ belongs to the normalizer of $\mathfrak{S}\sigma$ in $G\sigma$. Every group of $\mathfrak{S}$ whose order is divisible by p contains G_0 because it is a Hall subgroup. Thus, such a group remains fixed under transformation by x. On the other hand, $x^{-1}Sx = S^*$ implies that $G_0S = G_0S^*$. Evidently, S and S^* are Hall subgroups of G_0S. By Theorem 11.1.1, S and S^* are conjugate in G_0S. Consequently, there is an element y of G_0 such that $y^{-1}Sy = S^*$. Then, both $G_0\mathfrak{S}$ and S remain fixed under transformation by xy^{-1}. Thus, the system of all intersections $G_0\mathfrak{S} \cap S$ is invariant under transformation by xy^{-1}. Now, $G_0\mathfrak{S} \cap S$ consists precisely of those groups in $\mathfrak{S}$ whose orders are relatively prime to p. Clearly, all groups in $\mathfrak{S}$ that contain G_0 remain fixed under transformation by y so that transformation by xy^{-1} maps all the groups in $\mathfrak{S}$ onto themselves; hence, $xy^{-1} \in N$. It follows that $x \in G_0N$, and therefore, $x\sigma \in N\sigma$. Consequently, the normalizer of $\mathfrak{S}\sigma$ in $G\sigma$ is contained in $N\sigma$. On the other hand, every element of $N\sigma$ obviously belongs to the normalizer of $\mathfrak{S}\sigma$.

Thus, our theorem is proved when the kernel of σ is a p-group.

In the case of an arbitrary kernel G_0, we form a characteristic series

$$G_0 \supset G_1 \supset \cdots \supset G_{m-1} \supset G_m = e$$

in which the factor groups G_i/G_{i+1} are of prime power order. This is possible, because G_0 is soluble. The natural homomorphism of G onto G/G_0. can be decomposed into a sequence of natural homomorphisms, viz. of G onto G/G_{m-1}, of G/G_{m-1} onto G/G_{m-2}, etc. The kernels of all these homomorphisms are of prime power order, so that our argument above is applicable. This completes the proof.

Proof of Theorem 11.4.2. It is sufficient to show that any central factor B/A of G is covered by an arbitrary system normalizer $N = \mathsf{N}(\mathfrak{S})$. Let σ denote the natural homomorphism of G onto G/A. By Theorem 11.4.3, $N\sigma$ is the normalizer of $\mathfrak{S}\sigma$ in $G\sigma = G/A$. Now, B/A belongs to the centre of G/A, so that $B\sigma = B/A$ belongs to the normalizer of every Sylow system of $G\sigma$, and, in particular, $B\sigma \subseteq N\sigma$. But, $N\sigma = NA/A$, so that B is contained in NA or, in other words, N covers B/A.

11.5 The Hypercentre

According to section 9.2, the hypercentre Z^* of a finite group G is defined as the term at which the upper central chain of G becomes stationary. As in section 9.2, $Z^i = Z^i(G)$ denotes the i-th term of the upper central chain of G, so that $Z^* = Z^k = Z^{k+1} = \cdots$ for some k.

Lemma. For a normal subgroup L of G, L/e is a hypercentral factor of G if and only if $L \subseteq Z^*$.

Proof. If L/e is hypercentral, then there exists a chain

$$e = L_0 \subset L_1 \subset L_2 \subset \cdots \subset L_{m-1} \subset L_m = L \tag{11.10}$$

in which each L_i/L_{i-1} is a central chief factor of G. So, we have $[L_i, G] \subseteq L_{i-1}$. As the terms of the upper central chain are maximal in the sense indicated in section 9.2, we conclude that $L_i \subseteq Z^i$ for $i = 0, 1, \ldots, m$; therefore, $L \subseteq Z^*$. Conversely, suppose that $L \subseteq Z^*$. Putting $L_i = L \cap Z^i$, we have $[L_i, G] \subseteq [Z^i, G] \subseteq Z^{i-1}$ and $[L, G] \subseteq L$; hence, $[L_i, G] \subseteq L_{i-1}$. So, we obtain a chain of the form (11.10) in which each factor group L_i/L_{i-1} is a central factor of G. This proves the lemma.

In this and the next section, we frequently deal with subgroups of G that coincide with their normalizers in G. Such subgroups are called *self-normalizing*.

We now derive another characterization of the hypercentre:

11.5.1 *The hypercentre of an arbitrary finite group G is*

(a) *the intersection of the normalizers of the Sylow subgroups of G,*

(b) *the intersection of the self-normalizing subgroups of G.*

Proof. Let U be any subgroup of G that does not contain the hypercentre Z^* of G. If Z^i is the last term of the upper central chain of G to be contained in U, then Z^i is obviously a proper subgroup of Z^{i+1}. We have $[G, Z^{i+1}] \subseteq Z^i$ and, therefore, $[G, Z^{i+1}] \subseteq U$. This gives $u^{-1} z^{-1} uz \in U$ for arbitrary elements u, z with $u \in U$, $z \in Z^{i+1}$. Hence, Z^{i+1} belongs to the normalizer of U, but not to U itself. This shows that all self-normalizing subgroups of G contain Z^*. By Theorem 3.3.2, the normalizers of the Sylow subgroups of G are self-normalizing. Thus, (b) is a consequence of (a).

Let H denote the intersection of the normalizers of the Sylow subgroups of G. We have just shown that $Z^* \subseteq H$. To obtain the converse inclusion, we first prove that H is the direct product of its Sylow subgroups. If $|H|$ is a power of a prime number, then there is nothing to prove. Otherwise, let h_i and h_j be two elements of H whose orders are powers of distinct prime numbers p_i and p_j, respectively. Then, by the definition of H, $h_i^{-1} h_j^{-1} h_i h_j$ belongs to a Sylow p_i-subgroup as well as to a Sylow p_j-subgroup of G, so that $h_i^{-1} h_j^{-1} h_i h_j = e$. This shows that H is indeed the direct product

$$H = H_1 \times H_2 \times \cdots \times H_r$$

of its Sylow p_i-subgroups H_i.

If h_i is any element of H_i and x_j any element of G of order a power of p_j, $p_j \neq p_i$, then $c = x_j^{-1} h_i^{-1} x_j h_i$ is contained in H_i, since H_i is a characteristic subgroup of the characteristic subgroup H of G and hence certainly normal in G. But, c also lies in every Sylow p_j-subgroup of G that contains x_j. So, we have $c = e$, and, therefore, every element of H_i commutes with every element of G whose order is relatively prime to p_i. Hence, the index of the centralizer $\mathsf{C}(H_i)$ of H_i in G is a power of p_i. Now, transformation of the p_i-group H_i by the elements of G induces a group of automorphisms of H_i, which is isomorphic to $G/\mathsf{C}(H_i)$ and hence also a p_i-group. From this, it follows that H_i/e is a hypercentral factor of G. For, let B/A be a chief factor of G that is a component of H_i/e. Then, B/A is an elementary abelian p_i-group, and transformation by the elements of G induces a p_i-group of automorphisms of B/A. We consider the orbits under these automorphisms. The numbers of elements in these orbits are powers of p_i. Now, one of the orbits consists of the unit element of B/A alone, and the order of B/A is divisible by p_i. Therefore, there must be other orbits consisting of a single element. This means that there is a normal subgroup A_1 of G with $A \subset A_1 \subseteq B$, such that the elements of A_1/A remain fixed under transformation by the elements of G. Clearly, A_1 is a normal subgroup of G and, since B/A is a chief factor, we must have $A_1 = B$. Thus, G is the centralizer of every component chief factor of H_i/e, i. e. H_i/e is a hypercentral factor of G. Applying the lemma, we obtain $H_i \subseteq Z^*$. Since this holds for every Sylow subgroup of H, we see that $H \subseteq Z^*$, and this proves our theorem.

11.5.2 *The hypercentre of a finite soluble group G is the intersection of the system normalizers of G.*

Proof. Every system normalizer of G is the intersection of the normalizers of certain Sylow subgroups of G. Thus, by (a) of the previous theorem, every system normalizer contains the hypercentre. On the other hand, since the Sylow p-subgroups of G are conjugate, every Sylow subgroup of G occurs in some Sylow system of G, so that the intersection of the system normalizers is contained in the normalizer of every Sylow subgroup. Hence, our theorem follows from (a) of Theorem 11.5.1.

11.6 Carter Subgroups

Apart from the system normalizers, a finite soluble group contains further remarkable classes of conjugate subgroups. By way of example, we deal in this section with one such class.

A subgroup U of an arbitrary group G is called *abnormal* if $x \in \langle U, x^{-1}Ux \rangle$ for every $x \in G$.

Lemma. The subgroup U is abnormal in G if and only if it satisfies the following conditions:

(a) Every subgroup of G containing U is its own normalizer in G.

(b) U is not contained in two distinct conjugate subgroups of G.

Proof. Suppose that U is abnormal and let V be a subgroup of G that contains U. If x belongs to the normalizer of V, then

$$x \in \langle U, x^{-1}Ux \rangle \subseteq \langle V, x^{-1}Vx \rangle \subseteq V,$$

which shows that V is self-normalizing.

Let $U \subseteq V \cap yVy^{-1}$ for some subgroup V of G. Then, we have $y \in \langle U, y^{-1}Uy \rangle \subseteq V$ so that $yVy^{-1} = V$. Hence, U is not contained in two distinct conjugate subgroups.

Conversely, suppose that U satisfies the conditions (a) and (b) and put $V = \langle U, x^{-1}Ux \rangle$ for an arbitrary element x of G. Then, we have $U \subseteq V \cap xVx^{-1}$ so that (b) implies $xVx^{-1} = V$. By (a), this shows that $x \in V$ and hence $x \in \langle U, x^{-1}Ux \rangle$. Therefore, U is abnormal in G. This proves the lemma.

Let G be a finite soluble group and let N denote the normalizer of an arbitrary Hall subgroup of G. Then, N is abnormal in G. Indeed, put $K = \langle N, x^{-1}Nx \rangle$ for an arbitrary element x of G. Then, H and $x^{-1}Hx$ are

Hall subgroups of the same order of K. By Theorem 11.1.1, there exists an element y of K such that $x^{-1}Hx = y^{-1}Hy$. So, we have $xy^{-1} \in N$, $x \in K$, which shows that N is abnormal.

We now come to the main theorem of this section:

11.6.1 Carter's Theorem. *Every finite soluble group contains nilpotent self-normalizing subgroups, and all such subgroups are conjugate.*

Proof. Let G be a finite soluble group. We may assume that our theorem is true for all soluble groups of order less than $|G|$.

We first prove the existence of nilpotent self-normalizing subgroups.

Let M be a minimal normal subgroup of G so that $|M| = p^k$ for some prime number p. By the inductive hypothesis, G/M contains a nilpotent self-normalizing subgroup L/M. Then, L has a unique Sylow p-subgroup L_p. Let K_p be a Sylow p-complement of L and C its normalizer in L. It turns out that C is a nilpotent self-normalizing subgroup of G.

Put $C_p = C \cap L_p$. Then, C_p is the unique Sylow p-subgroup of C. For, C_p is normal in C and a p-group; on the other hand,

$$|C: C_p| = |C: C \cap L_p| = |CL_p: L_p| = |L: L_p|,$$

which is prime to p.

We observe that K_p is a normal Sylow p-complement of C, so that $C = C_p \times K_p$. Now, K_p is isomorphic to L/L_p, which is isomorphic to a factor group of the nilpotent group L/M. Therefore, K_p is nilpotent, and hence so is C.

As C is the normalizer of a Hall subgroup of L, it follows from our above remark that C is abnormal in L. In particular, MC is its own normalizer in L, and hence MC/M is its own normalizer in L/M. But, L/M is nilpotent, therefore $MC = L$.

Let x belong to the normalizer of C in G. As M is a normal subgroup of G and $L = MC$, we see that $x^{-1}Lx = L$. But, L is its own normalizer in G, because L/M is self-normalizing in G/M. Therefore, $x \in L$. But, as C is self-normalizing in L, we obtain $x \in C$. Consequently, C is self-normalizing in G.

Next, we prove that any two nilpotent self-normalizing subgroups of G are conjugate. We may assume that this is true for all soluble groups whose order is less than $|G|$.

Let C be a nilpotent self-normalizing subgroup of G. For an arbitrary element x of G, we put $V = \langle C, x^{-1}Cx \rangle$. We shall show that $x \in V$. This is certainly true for $V = G$. Thus, we may assume that V is a proper subgroup of G. Then, the inductive hypothesis shows that the two nilpotent self-normalizing subgroups C and $x^{-1}Cx$ are conjugate in V. Hence, $x^{-1}Cx = v^{-1}Cv$

for a suitable $v \in V$. Therefore, $xv^{-1} \in \mathsf{N}(C) = C$ so that $x \in V$. So, we have $x \in \langle C, x^{-1}Cx \rangle$ for every $x \in G$, and this means that C is abnormal.

Let T be any normal subgroup of G. As C is abnormal, TC is its own normalizer in G, so that TC/T is its own normalizer in G/T. Moreover, TC/T is nilpotent, because it is isomorphic to the factor group $C/T \cap C$ of the nilpotent group C.

Now, let C_1, C_2 be nilpotent self-normalizing subgroups of G and let M be a minimal normal subgroup of G. Then, our previous remark shows that MC_1/M and MC_2/M are nilpotent self-normalizing subgroups of G/M. Thus, by the inductive hypothesis MC_1/M and MC_2/M are conjugate in G/M, which means that MC_1 and MC_2 are conjugate in G. We can, therefore, find an element y of G such that $MC_1 = My^{-1}C_2y$. It is sufficient to prove that C_1 and $y^{-1}C_2y$ are conjugate; in other words, we may assume that $MC_1 = MC_2$. We may also assume that $MC_1 = MC_2 = G$, for otherwise C_1 and C_2 are nilpotent self-normalizing subgroups of the proper subgroup MC_1 of G so that C_1 and C_2 are conjugate in MC_1 by the inductive hypothesis.

Now, $M \cap C_1$ is clearly normal in C_1, but it is also normal in M, because M is abelian. Since $MC_1 = G$, the intersection $M \cap C_1$ is therefore normal in G, so that $M \cap C_1 = M$ or $M \cap C_1 = e$. If $M \cap C_1 = M$, we have $C_1 = MC_1 = G$, and our assertion is trivially true. Thus, we may assume that $M \cap C_1 = M \cap C_2 = e$, so that C_1 and C_2 are complements of M in G.

The order of M is a prime power p^k and $|G:C_1| = |G:C_2| = p^k$. It is easy to see that C_1 and C_2 are maximal subgroups of G. For, if H is a subgroup such that $C_1 \subset H \subset G$, then $M \cap H$ is normal in G and a non-trivial subgroup of M, which is impossible.

Let K_1, K_2 be the Sylow p-complements of C_1, C_2, respectively. Then, K_1 and K_2 are Sylow p-complements of G so that $K_1 = x^{-1}K_2x$ for some $x \in G$.

Suppose that $C_1 \neq x^{-1}C_2x$. Now, K_1 is a Sylow p-complement of the nilpotent groups C_1 and $x^{-1}C_2x$. Hence, K_1 is normal in C_1 and in $x^{-1}C_2x$, and therefore normal in $\langle C_1, x^{-1}C_2x \rangle$. But, $\langle C_1, x^{-1}C_2x \rangle = G$, because C_1 is a maximal subgroup of G. The factor group G/K_1 is a p-group and contains the proper subgroup C_1/K_1, which is self-normalizing. This is a contradiction; hence, we have $C_1 = x^{-1}C_2x$, which proves the theorem.

The nilpotent self-normalizing subgroups of a finite soluble group are called the *Carter subgroups*.

The proof of the last theorem yields the following:

Corollary. Every Carter subgroup is abnormal. If C is a Carter subgroup and T an arbitrary normal subgroup of the finite soluble group G, then CT/T is a Carter subgroup of G/T.

The Carter subgroups have covering and avoidance properties similar to those of the system normalizers as stated in the Theorems 11.4.1 and 11.4.2.

Let C be a Carter subgroup of G. A factor group B/A of G is called a *C-factor* if $C \subseteq \mathsf{N}(A) \cap \mathsf{N}(B)$. A C-factor B/A is called *irreducible* if there is no subgroup B_1 such that $B \supset B_1 \supset A$ and $C \subseteq \mathsf{N}(B_1)$. A C-factor B/A may be regarded as a group with operators, the operators being transformations by the elements of C. An irreducible C-factor is then characterized by the property that it contains no non-trivial admissible subgroup. It is evident that an irreducible C-factor is characteristically simple; hence, it is elementary abelian.

A C-factor B/A is called *central* if $[B, C] \subseteq A$; this means that all the operators act trivially on B/A. Otherwise, B/A is said to be *eccentric.*

11.6.2 *A Carter subgroup C covers all irreducible central C-factors and avoids all irreducible eccentric C-factors.*

Proof. Let B/A be an irreducible C-factor. Then, $A(B \cap C)$ is a subgroup of G between A and B, and C is contained in the normalizer of this subgroup. Therefore, $A(B \cap C) = B$ or $A(B \cap C) = A$, because B/A is irreducible. This shows that C either covers or avoids B/A.

Suppose that C covers B/A, i. e. $A(B \cap C) = B$. Then, we have $|B \cap C : A \cap C| = |B : A|$. Both $B \cap C$ and $A \cap C$ are normal subgroups of C. If there were a normal subgroup T of C such that $B \cap C \supset T \supset A \cap C$, then AT would be a subgroup between B and A that is normalized by C, a contradiction to the irreducibility of B/A. This shows that $B \cap C/A \cap C$ is a chief factor of C.

Now, C is nilpotent so that every chief factor is central. This shows that $[B \cap C, C] \subseteq A \cap C$. An arbitrary element b of B can be written in the form $b = da$, where $d \in B \cap C$ and $a \in A$. For an arbitrary element c of C, we have

$$[b, c] = [da, c] = a^{-1}[d, c]\, c^{-1}ac.$$

Now, $[d, c] \in A$ and $c^{-1}ac \in A$, hence $[b, c] \in A$. So, we see that $[B, C] \subseteq A$, and that B/A is a central C-factor.

Conversely, suppose that C avoids B/A. Then, BC and AC are subgroups of G and $|BC : AC| = |B : A|$. Assume that B/A is a central C-factor so that $[B, C] \subseteq A$. Let $a \in A$, $b \in B$, $c \in C$. Then,

$$b^{-1}acb = b^{-1}abc[c, b].$$

Now, $b^{-1}ab \in A$ and $[c, b] \in A$, therefore $b^{-1}acb \in AC$. Since B belongs to the normalizer of AC, it follows that AC is normal in BC. But, this is a contradiction, because C is abnormal in G. Hence, B/A is an eccentric C-factor.

As we mentioned above, the Carter subgroups belong to a system of classes of conjugate subgroups contained in every finite soluble group. We refer to [24] and [11].

11.7 A Theorem of O. Schmidt

11.7.1 *Let G be a finite group that is not nilpotent but whose proper subgroups are all nilpotent. Then, G is soluble and $|G|$ has precisely two distinct prime divisors.*

For the proof we need the following:

Lemma. Let G be a finite group whose proper subgroups are all nilpotent. Suppose that G contains a maximal subgroup M and a proper subgroup U such that $M \cap U$ is distinct from e and U. Then, G is not simple.

Proof. We assume that the subgroup U, subject to the conditions of the lemma, is chosen in such a way that the order of $D = M \cap U$ is as large as possible. In that case, D turns out to be a normal subgroup of G. By hypothesis, D is a proper subgroup of M as well as of U. Since U and M are nilpotent, the normalizers of D in U and M are larger than D. Denoting the normalizer of D in G by $\mathsf{N}(D)$, we then have

$$D \subset U \cap \mathsf{N}(D), \quad D \subset M \cap \mathsf{N}(D). \tag{11.11}$$

We shall show that $\mathsf{N}(D) = G$. We observe that $\mathsf{N}(D)$ is not contained in M, for otherwise we would have

$$U \cap \mathsf{N}(D) \subseteq U \cap M = D,$$

which contradicts (11.11). Suppose that $\mathsf{N}(D) \neq G$. Then, $\mathsf{N}(D)$ satisfies the conditions on U, but the order of $M \cap \mathsf{N}(D)$ exceeds $|D|$, contradicting the choice of U. So, we have $\mathsf{N}(D) = G$, which proves the lemma.

Proof of Theorem 11.7.1. First, we show that G is soluble. For groups of order 1, there is nothing to prove. We assume that the theorem is true for all groups of order less than $g = |G|$.

If G has a non-trivial normal subgroup T, then both G/T and T are soluble, so that G itself is soluble. Hence, it suffices to show that G is not simple.

Suppose, on the contrary, that G is simple. Then, any maximal subgroup M is self-normalizing. From the lemma, we conclude that every proper subgroup of G that is not contained in M intersects M trivially. It follows that the intersection of any two distinct conjugates of M is the unit element. Moreover, every Sylow subgroup of G is either contained in M, or its intersection with M

is e. Consequently, every Sylow p-subgroup of G for any prime divisor p of $m = |M|$ belongs to some conjugate of M. By Theorem 2.2.3, the conjugates of M cannot contain all the elements of G, so that there is a prime divisor q of g that does not divide m. A Sylow q-subgroup of G is contained in some maximal subgroup M_1 of G, and from what we observed above it follows that the order m_1 of M_1 is relatively prime to m. Note that M_1 too is self-normalizing.

The gm^{-1} conjugates of M contain $(m-1)gm^{-1}$ elements, distinct from each other and from the unit element. Similarly, the gm_1^{-1} conjugates of M_1 contain $(m_1-1)gm_1^{-1}$ distinct elements other than e. Since $(m, m_1) = 1$, any conjugate of M intersects trivially with any conjugate of M_1. This shows that

$$(m-1)\frac{g}{m} + (m_1 - 1)\frac{g}{m_1} \leqq g - 1$$

or

$$1 \leqq g\left(\frac{1}{m} + \frac{1}{m_1} - 1\right),$$

which is obviously false because $m > 1$, $m_1 > 1$. This contradiction shows that G is soluble.

Let $g = p^a q^b \cdots r^b$ be the decomposition of g into powers of distinct prime numbers $p, q, \ldots, r$. We shall prove that g has only two prime divisors p and q. Since G is soluble, it contains a normal subgroup T whose index in G is a prime number. We may choose the notation such that $|G:T| = p$. As T is nilpotent, it is the direct product of its Sylow subgroups:

$$T = P_1 \times Q \times \cdots \times R$$

where $|P_1| = p^{a-1}$, $|Q| = q^b, \ldots, |R| = r^c$. Now, P_1 is contained in a Sylow p-subgroup P of G. Choosing an element x of P not contained in P_1, we have

$$G = T \cup xT \cup \cdots \cup x^{p-1}T.$$

As $Q, \ldots, R$ are characteristic subgroups of T, they are normal in G. In particular, transformation by x maps $Q, \ldots, R$ onto themselves. But, transformation by x does not leave every element of $Q, \ldots, R$ fixed, for otherwise G itself would be nilpotent. Let us assume that Q is not elementwise fixed under transformation by x. Then, $\langle x, Q\rangle$ is not nilpotent, and hence $\langle x, Q\rangle = G$. Consequently, G is of order $p^a q^b$. We observe that the Sylow q-subgroup Q of G is normal in G, and the Sylow p-subgroup $P = \langle x\rangle$ is cyclic; moreover, x^p belongs to the centre of G.

11.8 A Theorem of Frobenius

The Theorem of Frobenius which we shall prove in this section has hardly anything to do with soluble groups. Though the theorem is interesting in itself, remarkably few applications are known so far.

11.8.1 *Let G be any finite group of order g and K a conjugacy class of G containing k elements. Then, for every natural number n, the number of elements x in G such that $x^n \in K$ is divisible by (kn, g).*

Proof. For a given complex C of G, let $F(C, n)$ denote the complex of all those $x \in G$ for which $x^n \in C$; $f(C, n)$ stands for the number of elements in $F(C, n)$.

When $g = 1$, our theorem is trivial. So, we may assume that $g > 1$ and that the theorem is valid for all groups of order less than g. For $n = 1$, we have $f(K, n) = k$, so that our assertion is true. Hence, we may assume that $n > 1$ and that the theorem is proved for all exponents less than n.

If both a and b belong to K, then $b = v^{-1}av$, so that $x^n = a$ implies that $(v^{-1}xv)^n = b$. Thus, we have $f(a, n) = f(b, n)$ and hence $f(K, n) = kf(a, n)$. Now, $x^n = a$ implies that $x^{-1}ax = a$, and this shows that all the solutions x of $x^n = a$ belong to the normalizer $\mathsf{N}(a)$ of a in G. The order of $\mathsf{N}(a)$ is gk^{-1}. Thus, for $k > 1$, we have $|\mathsf{N}(a)| < g$. By the inductive hypothesis, $f(a, n)$ is divisible by (n, gk^{-1}), so that $f(K, n) = kf(a, n)$ is divisible by $k(n, gk^{-1}) = (kn, g)$. This proves our theorem for $k > 1$.

It remains to consider the case $k = 1$. Then, K consists of a single element a belonging to the centre of G.

Suppose that $n = n_1n_2$ is a factorization into natural numbers $n_1 > 1$, $n_2 > 1$ with $(n_1, n_2) = 1$. Writing $D = F(a, n_2)$, we have $F(a, n) = F(D, n_1)$. Obviously, D consists of complete conjugacy classes of G. Thus, it follows from the inductive hypothesis on n that $f(D, n_1) = f(a, n)$ is divisible by (n_1, g). Similarly, we conclude that $f(a, n)$ is divisible by (n_2, g). Now, since $(n_1, n_2) = 1$, (n_1, g) and (n_2, g) are relatively prime, and hence $(n_1, g)(n_2, g) = (n, g)$ is a divisor of $f(a, n)$.

So we finally have to deal with exponents $n = p^m$ that are powers of a prime number p. Let a be of order α. We consider two cases according as p divides α or not.

Suppose that p divides α. All the elements of $F(a, p^m)$ are of order $p^m\alpha$. We choose an element x of $F(a, p^m)$ and determine the number of elements in $\langle x \rangle$ that belong to $F(a, p^m)$. These elements are precisely the products xy, where $y \in \langle x \rangle$ and $y^{p^m} = e$. By Theorem 5.4.5, the number of such elements y is equal to p^m. Moreover, we have $\langle xy \rangle = \langle x \rangle$. This shows that $f(a, p^m)$ is equal to p^m multiplied by the number of distinct cyclic groups that are generated by the elements of $F(a, p^m)$. Thus, $f(a, p^m)$ is divisible by p^m.

Finally, suppose that α and p are relatively prime. We now make use of the fact that a belongs to the centre of G. All the elements of the centre of G whose orders are relatively prime to p form an abelian group Z whose order is not divisible by p. For every pair $z_1, z_2 \in Z$, the equation $z_1 = z_2 y^{p^m}$ has a unique solution $y \in Z$. If $x^{p^m} = z_2$, then $(xy)^{p^m} = z_1$. So, we have $f(z, p^m) = f(a, p^m)$ for all $z \in Z$. Collecting and counting the elements of G with respect to the conjugacy classes of their p^m-th powers, we therefore obtain

$$g = \sum_{C \nsubseteq Z} f(C, p^m) + |Z| f(a, p^m).$$

Here, g and, by what we proved above, all the numbers $f(C, p^m)$ for $C \nsubseteq Z$ are divisible by (p^m, g). Thus, (p^m, g) also divides $|Z|\, f(a, p^m)$. But, since $|Z|$ is relatively prime to p, it follows that (p^m, g) is a divisor of $f(a, p^m)$.

This completes the proof. Taking $K = e$, we obtain:

11.8.2 Frobenius' Theorem. *For every divisor n of the order of the group G, the number of elements x of G such that $x^n = e$ is divisible by n.*

Note that the number in question is at least equal to n, because $x = e$ always satisfies the condition.

Here, the following unsolved question arises. Suppose that, for a divisor n of $|G|$, the equation $x^n = e$ has exactly n solutions $x \in G$. Is it true that all these solutions form a subgroup of G? It is necessary to assume that n is a divisor of $|G|$; for, the equation $x^4 = e$ has four solutions in the symmetric group S_3, but they do not form a subgroup. In the case of a soluble group G, our question can be answered in the affirmative:

11.8.3 *Let G be a finite soluble group. If for a divisor n of $|G|$ the equation $x^n = e$ has precisely n solutions $x \in G$, then these solutions form a characteristic subgroup of G.*

Proof. If the solutions form a subgroup, then it is clearly characteristic. The theorem obviously holds when the order of G is a prime number. We proceed by induction on the group order and assume that our assertion is true for all groups of order less than $|G|$.

As G is soluble, it contains a minimal normal subgroup P, which is elementary abelian. The order of P is a prime power p^k. We consider two cases according as n is divisible by p or not.

I. Suppose that p divides n. Then, all the elements of P are solutions of $x^n = e$. We write

$$n = p^m n_1, \quad g = p^r g_1, \quad \text{where} \quad (n_1, p) = (g_1, p) = 1,$$

and put

$$u = p^{m-k}n_1 \quad \text{if} \quad m \geqq k,$$
$$u = n_1 \quad \text{if} \quad m < k.$$

The order $p^{r-k}g$ of G/P is always divisible by u. It now follows, from Theorem 11.8.2, that the number of elements z in G/P that satisfies the equation $z^u = P/P$ is a multiple of u, say tu. Now, let x be an element of G such that the natural homomorphism of G onto G/P carries x into an element z of G/P for which $z^u = P/P$. This means that $x^u \in P$ and hence that $x^{pu} = e$. As pu divides n, such an element x satisfies the equation $x^n = e$. Moreover, these elements x are representatives of tu cosets of P. Consequently, G contains at least tup^k elements x for which $x^n = e$. Now, $m < k$ implies that $up^k > n$, which contradicts the assumption that $x^n = e$ has precisely n solutions. Hence, we have $m \geqq k$, so that $up^k = n$. As we have already found $tup^k = tn$ solutions of $x^n = e$, it follows that $t = 1$. Thus, the equation $z^u = P/P$ has precisely u solutions $z \in G/P$. By the inductive hypothesis, these u solutions form a subgroup H/P of order u of G/P. Consequently, H is a subgroup of order $up^k = n$ of G whose elements are precisely the solutions of $x^n = e$.

II. Suppose that n and p are relatively prime. In this case, n divides $|G:P|$. By Theorem 11.7.2, the number of elements z in G/P that satisfy the equation $z^n = P/P$ is a multiple tn of n. Let y be an element of G such that the natural homomorphism of G onto G/P carries y into an element z with $z^n = P/P$. Then, we have $y^n \in P$ and hence $y^{pn} = e$. Thus, there are tn cosets of P whose elements satisfy the equation $y^{pn} = e$. The p-th power of such an element y yields a solution of $x^n = e$. We shall show that elements y_1, y_2 of distinct cosets of P yield distinct solutions of x_1, x_2 of $x^n = e$. Indeed, if the cosets Py_1 and Py_2 are distinct, then so are their images z_1 and z_2 under the natural homomorphism of G onto G/P. Now, the assumption $y_1^p = y_2^p$ would imply that $z_1^p = z_2^p$ and this would yield the contradiction $z_1 = z_2$, because we also have $z_1^n = z_2^n$ with n relatively prime to p. From the fact that the equation $z^n = P/P$ has tn solutions $z \in G/P$, we now conclude that $x^n = e$ has at least tn solutions $x \in G$. This shows that $t = 1$. By the inductive hypothesis, G/P contains a subgroup U/P of order n that consists precisely of the solutions of $z^n = P/P$. The subgroup U is of order $p^k n$. As U is soluble, it contains a Sylow p-complement H of order n, and the elements of H are precisely the solutions of $x^n = e$.

11.9 Groups with Cyclic Sylow Subgroups

In this section, we give a survey of all finite groups whose Sylow subgroups are all cyclic. We need the following theorem which is of interest in itself.

11.9.1 *Two successive factor groups $G^{(i)}/G^{(i+1)}$ and $G^{(i+1)}/G^{(i+2)}$ of the commutator chain*

$$G' \supset G'' \supseteq G''' \supseteq \cdots$$

cannot be cyclic unless $G^{(i+1)} = G^{(i+2)}$.

Proof. Without loss of generality, we may assume that $G''' = e$ and that both G'/G'' and G'' are cyclic. Then, we have to prove that $G'' = e$. Let $G'' = \langle a \rangle$. The normalizer of $\langle a \rangle$ is G. If C denotes the centralizer of a in G, then G/C is isomorphic to a group of automorphisms of the cyclic group $\langle a \rangle$. From Theorem 5.4.3, it follows that G/C is abelian. This shows that $G' \subseteq C$, so that G'' belongs to the centre of G'. Since G'/G'' is cyclic, we conclude that G' is abelian, therefore $G'' = e$ as was to be shown.

A group G is called *metacyclic* if both G/G' and G' are cyclic.

11.9.2 *If all the Sylow subgroups of a finite group G are cyclic, then G is cyclic or metacylic. In case G is metacyclic, it is generated by two elements a and b with the defining relations*

$$a^m = e, \quad b^n = e, \ b^{-1}ab = a^s$$

where

$$mn = |G|, \ (m, n) = 1, \ (s - 1, m) = 1, \ s^n \equiv 1 \pmod{m}.$$

Conversely, if a and b satisfy the above relations, then the Sylow subgroups of $G = \langle a, b \rangle$ are cyclic.

Proof. To begin with, we show that G is necessarily soluble if all its Sylow subgroups are cyclic. Let

$$|G| = g = p_1^{a_1} p_2^{a_2} \cdots p_r^{a_r}, \ p_1 < p_2 < \cdots < p_r,$$

be the factorization of g into powers of the prime numbers p_i. We put

$$m = p_i^{c_i} p_{i+1}^{a_{i+1}} \cdots p_r^{a_r}, \qquad 1 \leqq c_i \leqq a_i,$$

and set out to prove that the equation $x^m = e$ has precisely m solutions $x \in G$. This is certainly true if $m = g$. Thus, it is sufficient to prove the following assertion: Let mp be a divisor of g such that p is a prime number and no prime divisor of m is less than p. Suppose that there are precisely mp elements x in G that satisfy the equation $x^{mp} = e$; then, the equation $x^m = e$ has exactly m solutions in G. Let p^{c+1} denote the highest power of p dividing mp. As the Sylow p-subgroups of G are cyclic, there exist elements of order p^{c+1}. This shows that not all the solutions of $x^{mp} = e$ are also solutions of $x^m = e$. By Theorem 11.8.2, the number of solutions of $x^m = e$ is divisible by m, say km. Since these solutions form a proper subset of the solutions of $x^{mp} = e$, we have $1 \leqq k < p$. If an element y of G satisfies the equation $y^{mp} = e$, but not the equation $y^m = e$, then the highest power of p that divides the order t of y is p^{c+1} so that $t = p^{c+1} t_1$ with $(t_1, p) = 1$. An element of $\langle y \rangle$ satisfies the equation $x^{mp} = e$, but not the equation $x^m = e$, if and only if its order is of the form $p^{c+1} t_2$, where t_2 divides t_1. By Theorem 5.4.2, the number $\psi(t)$ of these elements is equal to

$$\psi(t) = \sum_{t_2/t_1} \varphi(p^{c+1} t_2) = p^c(p-1) \sum_{t_2/t_1} \varphi(t_2).$$

This shows that $\psi(t)$ is divisible by $p - 1$. The total number of elements in G satisfying the equation $x^{pm} = e$ but not the equation $x^m = e$ is equal to

$$pm - km = (p - k)\, m,$$

and by our last result this number is divisible by $p - 1$. Since m is not divisible by any prime number less than p, it follows that $p - k$ is divisible by $p - 1$, therefore, $k = 1$. Consequently, $x^m = e$ has precisely m solutions in G.

Taking $m = p_r^{a_r}$, we conclude that G contains precisely $p_r^{a_r}$ elements whose orders are powers of p_r. In other words, G contains a single Sylow p_r-subgroup, and hence the Sylow p_r-subgroup is normal in G. It is now easy to prove that a group G is soluble if all its Sylow subgroups are cyclic. This is certainly true for groups of prime order. We use induction on the order and assume that our assertion is true for all groups of order less than $|G|$. As we just saw, G contains a normal Sylow subgroup S. If the Sylow subgroups of G are cyclic, then so are the Sylow subgroups of G/S. By the inductive hypothesis, G/S is soluble, and as S too is soluble, we conclude that G itself is soluble.

The factor groups of the derived series

$$G \supset G' \supset G'' \supset \cdots$$

are abelian and their Sylow subgroups are cyclic. Hence, the factor groups themselves are cyclic. By Theorem 11.9.1, this implies that $G'' = e$.

If $G' = e$, the group G itself is cyclic.

Now, suppose that $G' \neq e$. Then, $G' = \langle a \rangle$ is a cyclic group of order m, say. Let $G'b$ denote a generating element of the factor group G/G'. Then, a and b generate G. As $\langle a \rangle$ is a normal subgroup, we have $b^{-1}ab = a^s$, where $s \not\equiv 1 \pmod m$; for otherwise G would be abelian. If G/G' is of order n, then $b^n \in G'$ so that

$$a = b^{-n} a b^n = a^{s^n},$$

which shows that $s^n \equiv 1 \pmod m$.

Every element of G can be written in the form $b^j a^i$. By means of (9.3) and (9.4), every commutator $[b^j a^i, b^v a^u]$ can be expressed in terms of the commutators $[a^k, b^l]$, and $[a^k, b^l]$ is a power of $[a, b] = a^{-1}b^{-1}ab = a^{s-1}$. Consequently, a^{s-1} generates G', hence $(s - 1, m) = 1$. Since $b^n \in G$, we have $b^n = a^t$. Thus, b commutes with a^t so that $b^{-1}a^t b = a^t$ and hence $a^{st} = a^t$. Because of $(s - 1, m) = 1$, this shows that $t \equiv 0 (\mathrm{mod}\ m)$ and hence $b^n = e$.

Finally, m and n are relatively prime. For suppose, on the contrary, that p is a common prime divisor of m and n and consider the elements $a_0 = a^{mp^{-1}}$, $b_0 = b^{np^{-1}}$. One easily checks that

$$[a_0, b_0] = a^r, \quad r = mp^{-1}(s^{np^{-1}} - 1).$$

Now, $s^n \equiv 1 (\mathrm{mod}\ p)$ implies that $s^{np^{-1}} \equiv 1 (\mathrm{mod}\ p)$ so that $[a_0, b_0] = e$. It follows that a_0 and b_0 generate a non-cyclic subgroup of order p^2, against the hypothesis.

This proves the first part of the theorem.

Conversely, if a and b satisfy the relations stated in the theorem, then they generate a group G of order mn, which is a splitting extension of a cyclic group of order m by a cyclic group of order n. In this group, every commutator is a power of $a^{-1}b^{-1}ab = a^{s-1}$. From $(s - 1, m) = 1$, it follows that $G' = \langle a \rangle$. Since $(m, n) = 1$, every Sylow subgroup of G is conjugate to a subgroup of $\langle a \rangle$ or of $\langle b \rangle$ and hence cyclic.

This completes the proof. Note that every finite group of square-free order belongs to one of the types described by the theorem.

11.10 A Theorem of Wielandt

Hall's Theorem 11.1.1 is a generalization of Sylow's Theorem 3.2.5. Theorem 11.1.5 shows that the assertion (a) of Sylow's Theorem cannot be generalized to assertion (a) of Hall's Theorem unless the group is soluble. H. Wielandt observed that the assertions (b) and (c) of Sylow's Theorem can also be generalized for Hall subgroups if the solubility condition is replaced by a condition on the Hall subgroup.

11.10.1 Wielandt's Theorem. *Let G be a finite group containing a nilpotent Hall subgroup H. Then, every subgroup of G whose order divides H is contained in a subgroup conjugate to H.*

Proof.[1] Let M be a subgroup of G whose order divides $|H|$. We may assume that $|M| > 1$ and that the theorem holds for all subgroups of G whose orders divide $|H|$ and are less than $|M|$. In particular, we may assume that all proper subgroups of M are contained in conjugates of H and are therefore nilpotent. Thus, it follows from Theorem 11.7.1 that M is soluble. Hence,

[1] The author is indebted to K. H. Müller for this short and elegant proof.

M contains a normal subgroup T such that $|M:T| = p$ is a prime number. Let P be a Sylow p-subgroup of M and K the Sylow p-complement of the nilpotent group T. Clearly, K is a characteristic subgroup of T and hence normal in M. We have $M = KP$ and $P \subseteq \mathsf{N}(K)$, where $\mathsf{N}(K)$ denotes the normalizer of K in G. Let P_1 denote the Sylow p-subgroup and L the Sylow p-complement of H. By the inductive hypothesis, there exists an element x in G such that $x^{-1}Kx \subseteq L$. Then, both $x^{-1}Px$ and P_1 are contained in the normalizer $\mathsf{N}(x^{-1}Kx)$ of $x^{-1}Kx$ in G. By Theorem 3.2.5, there is an element y in $\mathsf{N}(x^{-1}Kx)$ such that $y^{-1}x^{-1}Pxy \subseteq P_1$. This gives

$$y^{-1}x^{-1}Mxy = y^{-1}x^{-1}KPxy = (y^{-1}x^{-1}Kxy)(y^{-1}x^{-1}Pxy)$$
$$= (x^{-1}Kx)(y^{-1}x^{-1}Pxy) \subseteq LP_1 = H,$$

and the theorem is proved.

For generalizations we refer to [2], [33], and [76].

11.11 p-Soluble Groups

For a prime number p, a p'-group is any finite group whose order is not divisible by p.

A finite group G is called *p-soluble* if it has a normal series every factor of which is either a p-group or a p'-group.

If G is finite and soluble, then it is p-soluble for every prime number p.

In a p-soluble group G, the so-called upper p-series

$$e = P_0 \subseteq N_0 \subset P_1 \subset N_1 \subset \cdots \subset P_{l-1} \subset N_{l-1} \subset P_l \subseteq N_l = G$$

is defined as follows:

$$P_0 = e,$$

$$N_i/P_i \text{ is the largest normal } p'\text{-subgroup of } G/P_i,$$

$$P_{i+1}/N_i \text{ is the largest normal } p\text{-subgroup of } G/N_i.$$

It is evident that the P_i and N_i are characteristic subgroups of G. The number $l = l_p = l_p(G)$ is called the *p-length* of G. The p-length has turned out to be an important parameter in the description of the structure of p-soluble groups. In particular, it is closely connected with the structure of the Sylow p-subgroups. We have to confine ourselves to a single theorem on the p-length.

Lemma. The factor group P_i/N_{i-1} contains its centralizer in G/N_{i-1}.

Proof. Let us assume that the centralizer Z/N_{i-1} of P_i/N_{i-1} in G/N_{i-1} is not contained in P_i/N_{i-1}. Then, P_i is a proper subgroup of P_iZ. We choose a normal subgroup M of G of minimal order such that

$$P_i \subset M \subseteq P_i Z.$$

From the definition of P_i, it follows that M/P_i is a p'-group. So, we conclude from Theorem 6.2.2 that there exists a complement Q/N_{i-1} of P_i/N_{i-1} in M/N_{i-1}. Since $Q \subseteq P_i Z$, transformation by an arbitrary element of Q induces an inner automorphism in P_i/N_{i-1}. Since Q/N_{i-1} is a p'-group, it follows that the order of such an automorphism is relatively prime to p. On the other hand, it coincides with an inner automorphism of the p-group P_i/N_{i-1} so that it must be the identity automorphism. Thus, P_i/N_{i-1} and Q/N_{i-1} commute elementwise, therefore

$$M/N_{i-1} = P_i/N_{i-1} \times Q/N_{i-1}.$$

Hence, Q is a normal subgroup of G, which contradicts the definition of N_{i-1}. This proves the lemma.

It follows from the lemma that P_1 contains the centre of an arbitrary Sylow p-Subgroup S of G. Indeed, P_1/N_0 is a p-group so that $P_1 \subseteq SN_0$. Hence, the cosets of N_0 in P_1 can be represented by elements of S. The elements of the centre $\mathsf{Z}(S)$ of S are permutable with these coset representatives so that $\mathsf{Z}(S)N_0/N_0$ belongs to the centralizer of P_1/N_0. Thus, the lemma gives $\mathsf{Z}(S)N_0 \subseteq P_1$ and hence $\mathsf{Z}(S) \subseteq P_1$.

We are now able to establish the following relation between the class of the Sylow p-subgroups and the p-length:

11.11.1 *If the Sylow p-subgroups of G are nilpotent of class c, then* $l_p(G) \leqq c$.

Proof. As our proposition is trivially true for $l_p = 1$, we can proceed by induction on the p-length.

The p-length of G/P_1 is equal to $l_p - 1$, and the Sylow p-subgroups of G/P_1 are isomorphic to $S/S \cap P_1$ where S denotes a Sylow p-subgroup of G. As we observed above, the centre of S is contained in $S \cap P_1$, so that the class of $S/S \cap P_1$ does not exceed $c - 1$. By induction, we see that

$$l_p - 1 \leqq c - 1,$$

which is our assertion.

For further results on the p-length, we refer to [34].

Exercises

The first four exercises provide examples to show that the four assertions of Theorem 11.1.1 are not true for arbitrary finite groups.

1. Show that the alternating group A_5 contains no subgroup of order 15.

2. Let G denote the automorphism group of the elementary abelian group A of order 8. In section 12.1, we shall prove that G is of order 168, and it follows from Theorem 13.7.2

that G is simple. Show that not all the subgroups of order 24 of G are conjugate. (In fact, G contains two classes of conjugate subgroups of order 24. Note that G induces transitive permutation groups on the 7 subgroups of order 2 and on the 7 subgroups of order 4 of A.)

3. The permutations $(1, 2, 3)$ and $(1, 2)(4, 5)$ generate a subgroup U of order 6 of the alternating group A_5. Show that U is not contained in any subgroup of order 12.

4. Prove that A_5 contains 6 Sylow 5-subgroups.

5. Determine the Sylow systems and the system normalizers of the symmetric group S_4. Verify the theorems of section 11.2, 11.3, and 11.4 for this case.

6. Determine the Carter subgroups of S_4 and verify Theorem 11.6.2 for this case.

CHAPTER 12

MISCELLANEOUS TOPICS

12.1 Automorphisms

Let $\mathsf{A}(G)$ denote the automorphism group of the group G. In section 2.3, we observed that the group $\mathsf{J}(G)$ of the inner automorphisms is isomorphic to $G/\mathsf{Z}(G)$.

$\mathsf{J}(G)$ is a normal subgroup of $\mathsf{A}(G)$. For, let φ be an arbitrary automorphism of G and τ_a the inner automorphism induced by $a \in G$, i. e.

$$x\,\tau_a = a^{-1}xa \qquad (x \in G);$$

then, we have

$$x(\varphi^{-1}\,\tau_a\,\varphi) = (a^{-1}(x\,\varphi^{-1})\;a)\;\varphi = (a\,\varphi)^{-1}x(a\,\varphi) = x\,\tau_{a\varphi}.$$

Two non-isomorphic groups may well have isomorphic automorphism groups. For example, the automorphism groups of an infinite cyclic group and of a cyclic group of order 3 are both of order 2.

Not every group occurs as the automorphism group of some group. For example, there is no group G such that $\mathsf{A}(G)$ is cyclic of odd order greater than 1. We prove this under the additional assumption that G is finitely generated. (The proposition is true without this restriction.) First of all, the automorphism group of a non-abelian group G cannot be cyclic; for, otherwise, the subgroup $\mathsf{J}(G)$ of $\mathsf{A}(G)$ would be cyclic; but, since $\mathsf{J}(G) \cong G/\mathsf{Z}(G)$, this is impossible, as we observed in section 2.3. If G is abelian and not of exponent 2, then the mapping

$$x \to x^{-1} \qquad (x \in G)$$

is an automorphism of order 2 so that $|\mathsf{A}(G)|$ is even. Finally, if G is abelian and of exponent 2, then G is the direct product of cyclic groups of order 2; moreover, there must be at least two direct factors unless $|\mathsf{A}(G)| = 1$. But, in this case, we obtain an autormorphism of order 2 by interchanging two direct factors.

12.1.1 *Let T be a normal subgroup of a finite group G. If an automorphism α of G leaves every element of T and every element of the factor group G/T fixed, then the order of α divides $|T|$.*

Proof. We regard α as a permutation of the elements of G and decompose α into the product of disjoint cycles. If the element x of G does not belong to T, then the image of x under α has the form

$$x\alpha = xt, \qquad t \in T.$$

Consequently, α contains the cycle $(x, xt, xt^2, \ldots, xt^{m-1})$, where m denotes the order of t. If these elements do not exhaust the coset xT, then there are more cycles of length m consisting of elements of xT. This shows that m divides $|T|$. The same applies to all cosets of T. The lengths of the cycles may depend on the cosets, but they are always divisors of $|T|$. The order of α is the least common multiple of the lengths of all cycles, i. e. the order of α divides $|T|$.

12.1.2 *If G is a group without centre, then so is $\mathsf{A}(G)$.*

Proof. Let α be an automorphism of G other than the identity, and let y denote an element of G such that $y\alpha \neq y$. Suppose that α belongs to the centre of $\mathsf{A}(G)$. Then, it is permutable with the inner automorphism induced by y, therefore

$$y^{-1}(x\alpha)\, y = (y^{-1}xy)\,\alpha = (y\alpha)^{-1}\,(x\alpha)\,(y\alpha)$$

for every $x \in G$. As $x\alpha$ ranges over the whole group, we conclude that y and $y\alpha$ induce the same inner automorphism so that $y^{-1}(y\alpha)$ belongs to the centre of G. This contradiction proves the theorem.

A group G is said to be *complete* if $\mathsf{Z}(G) = e$ and $\mathsf{A}(G) = \mathsf{J}(G)$. This means that $\mathsf{A}(G) \cong G$. It is easy to check that the symmetric group S_3 is complete. It can be proved that the symmetric groups S_n for $n \neq 6$ are complete (see [7]).

12.1.3 *If G contains a complete normal subgroup T, then $G = T \times C$, where C is the centralizer of T in G.*

Proof. Let

$$G = \bigcup_{y \in R} yT$$

be the decomposition into cosets of T. As T is normal, transformation by an element y of the transversal R induces an automorphism of T. But, since T has only inner automorphisms, the same automorphism is induced by an element t_y of T, i. e.

$$y^{-1}xy = t_y^{-1}xt_y \quad \text{for every} \quad x \in T.$$

Thus, yt_y^{-1} commutes with every element of T. This shows that every coset of T contains an element of the centralizer C of T in G. As the centre of T is trivial, we see that $C \cap T = e$ so that $G = T \times C$, and the theorem is proved.

If a group G is a normal subgroup of some group L, then the transformations of G by the elements of L give rise to automorphisms of G. We shall prove that every finite group G can be embedded in a group L such that G is normal in L and *all* automorphisms of G can be obtained by transforming G by the elements of L. We recall the definition of the right regular permutational representation G^* of G as defined in section 7.2. Owing to the isomorphism $G \cong G^*$, our assertion will be proved if we show that G^* can be embedded as a normal subgroup in a group $H(G)$ of permutations of the elements of G such that all automorphisms of G^* are obtained by transforming G^* by the elements of $H(G)$.

The group G^* consists of the permutations

$$\varrho_a = \begin{pmatrix} x \\ xa \end{pmatrix} \qquad (a, x \in G).$$

Let $S(G)$ denote the group of all permutations of the elements of G. It will turn out that the normalizer $H(G)$ of G^* in $S(G)$ has the required property.

Any automorphism φ of G is a permutation

$$\begin{pmatrix} x \\ x\varphi \end{pmatrix} \qquad (x \in G) \tag{12.1}$$

of the elements of G. If we regard φ as an automorphism of G^*, it carries $\varrho_a \in G^*$ into

$$\varrho_{a\varphi} = \begin{pmatrix} x \\ x(a\varphi) \end{pmatrix} \qquad (x \in G).$$

We have

$$\begin{pmatrix} x \\ x\varphi \end{pmatrix}^{-1} \begin{pmatrix} x \\ xa \end{pmatrix} \begin{pmatrix} x \\ x\varphi \end{pmatrix} = \begin{pmatrix} x\varphi \\ (xa)\varphi \end{pmatrix} = \begin{pmatrix} x\varphi \\ (x\varphi)(a\varphi) \end{pmatrix} = \begin{pmatrix} x \\ x(a\varphi) \end{pmatrix}.$$

Thus, $\varrho_{a\varphi}$ is obtained by transforming ϱ_a by the permutation (12.1). As φ was an arbitrary automorphism, this shows that indeed all automorphisms of G^* are induced by elements of $H(G)$. The permutations of the form (12.1) for all automorphisms of G form a subgroup of $H(G)$, which is isomorphic to $\mathsf{A}(G)$ and will also be denoted by $\mathsf{A}(G)$.

The group $H(G)$ is called the *holomorph* of G.

12.1.4 *The holomorph $H(G)$ is a split extension of G^* by $\mathsf{A}(G)$, i.e.*

$$H(G) = \mathsf{A}(G)\, G^*, \quad G^* \lhd H(G), \quad \mathsf{A}(G) \cap G^* = e^*.$$

Proof. Let

$$\begin{pmatrix} x \\ x' \end{pmatrix} \qquad (x \in G)$$

be an arbitrary permutation of $H(G)$ and let $e' = c$. The permutation

$$\begin{pmatrix} x \\ x' \end{pmatrix}\begin{pmatrix} x \\ xc \end{pmatrix}^{-1} = \begin{pmatrix} x \\ x'' \end{pmatrix}$$

leaves e fixed. If we transform G^* by this permutation, we obtain a certain automorphism of G^*, say φ. Transformation by (12.1) gives rise to the same automorphism, so that

$$\begin{pmatrix} x \\ x'' \end{pmatrix}\begin{pmatrix} x \\ x\varphi \end{pmatrix}^{-1} = \begin{pmatrix} x \\ x' \end{pmatrix}\begin{pmatrix} x \\ xc \end{pmatrix}^{-1}\begin{pmatrix} x \\ x\varphi \end{pmatrix}^{-1} \tag{12.2}$$

is permutable with every element of G^*. It now follows from the results of section 7.2 that (12.2) belongs to the left regular representation $^*G$, and since it leaves e fixed, it is the identity. This shows that

$$\begin{pmatrix} x \\ x' \end{pmatrix} = \begin{pmatrix} x \\ x\varphi \end{pmatrix}\begin{pmatrix} x \\ xc \end{pmatrix}$$

and hence $H(G) = \mathsf{A}(G)G^*$. Clearly, $G^* \lhd H(G)$ by definition. Finally, $\mathsf{A}(G) \cap G^* = e^*$ follows from the fact that G^* is a regular permutation group, whereas $\mathsf{A}(G)$ leaves e fixed. This completes the proof.

The holomorph $H(G)$ had been defined as the normalizer of G^* in $S(G)$. As we saw in section 7.2, the centralizer of G in $S(G)$ is $^*G$. Therefore, $^*G \subset H(G)$. Applying Theorem 12.1.3, we obtain:

12.1.5 *If G is complete, then* $H(G) = {^*G} \times G^*$.

The next two theorems relate the structure of G to properties of $H(G)$ as a permutation group.

12.1.6 *The holomorph $H(G)$ is a primitive permutation group if and only if G is characteristically simple.*

Proof. The stabilizer subgroup of e in $H(G)$ is $\mathsf{A}(G)$. If G^* contains a non-trivial characteristic subgroup C^*, then $\mathsf{A}(G)C^*$ is a subgroup of $H(G)$ such that

$$\mathsf{A}(G) \subset \mathsf{A}(G)\,C^* \subset H(G).$$

Hence, $H(G)$ is imprimitive. Conversely, suppose that there is a subgroup U such that

$$\mathsf{A}(G) \subset U \subset H(G).$$

Then, $C^* = G^* \cap U$ is a non-trivial subgroup of G^* that is normalized by $\mathsf{A}(G)$. This means that C^* is a characteristic subgroup of G^*.

In the proof of the next theorem we make use of the obvious fact that conjugacy is preserved under every automorphism. Indeed, $b = x^{-1}ax$ implies that

$$b\,\varphi = (x\,\varphi)^{-1}(a\,\varphi)(x\,\varphi)$$

for any automorphism φ.

12.1.7 *The holomorph $H(G)$ is doubly transitive if and only if G is an elementary abelian group.*

Proof. It is evident that $H(G)$ is doubly transitive if and only if $\mathsf{A}(G)$ is transitive on the elements $\neq e$ of G. If G is elementary abelian, it follows easily from Theorem 4.2.3 that $\mathsf{A}(G)$ has the property in question. Conversely, suppose that $\mathsf{A}(G)$ is transitive on the elements $\neq e$ of G. Then, all elements $\neq e$ of G are of the same order, and since G certainly contains elements of prime order, that common order must be a prime number. Moreover, because of the transitivity of $\mathsf{A}(G)$, all conjugacy classes $\neq e$ of G contain the same number of elements, h say. This gives $|G| = hr + 1$, where r denotes the number of conjugacy classes $\neq e$ of G. On the other hand, h divides $|G|$ so that $h = 1$. Hence, G is abelian. This completes the proof.

Every inner automorphism carries every element into an element of the same conjugacy class, but there may exist outer automorphisms with the same property. The next theorem deals with such automorphisms.

12.1.8 *The group $\mathsf{K}(G)$ of all automorphisms of G that carry every element into an element of the same conjugacy class of G is a normal subgroup of $\mathsf{A}(G)$. All prime divisors of $|\mathsf{K}(G)|$ divide $|G|$.*

Proof. Every automorphism of G permutes the conjugacy classes. Thus, there is a homomorphism of $\mathsf{A}(G)$ onto a group of permutations of the classes. The kernel of this homomorphism is $\mathsf{K}(G)$. This proves the first part of the theorem.

Let $\varkappa$ be an element of $\mathsf{K}(G)$. Let us assume that the order of $\varkappa$ is a prime number p that does not divide $|G|$. The subgroup $F = \langle G^*, \varkappa\rangle$ of $H(G)$ contains G^* as a normal subgroup of index p. Moreover, $\varkappa$ is not permutable with all the elements of G^* so that the elements of G^* that commute with $\varkappa$ form a proper subgroup U. Those elements of G^* that commute with a fixed element of the class of $\varkappa$ in F form a subgroup conjugate to U. As the conjugates of U cannot contain all the elements of G^*, there exist elements in G^* that are not permutable with any element of the class of $\varkappa$. Let a^* be such an element. Then, $\varkappa$ is not permutable with any element in the conjugacy class of a^* in F. Thus, transformation of a^* by the powers of $\varkappa$ yields p distinct elements. If these p elements do not exhaust the whole class of a^* in F, then we take another element of that class and transform it by the powers of $\varkappa$. This gives a new system of p distinct elements in the class of a^*. This process can be continued until the class of a^* is exhausted. Consequently, the number of elements in the class of a^* is divisible by p. But, this turns out to be impossible, for we shall prove that, on the other hand, the number of elements in the class of a^* in F is a divisor of $|G|$. Indeed, the elements of the class of a^* in F can be obtained as follows. First, we transform a^* by the elements of G^*. This gives rise to the class of a^* in G^*, and the number of elements in this class divides $|G|$. The class of a^* in F is now obtained by transformation by the powers of $\varkappa$. But, this does not lead to new elements, since $\varkappa \in \mathsf{K}(G)$. Thus, the class of a^* in F coincides with the class of a^* in G^*.

Now, let λ be an arbitrary automorphism in $\mathsf{K}(G)$ of order n. If $n = pm$, where the prime number p does not divide $|G|$, then $\varkappa = \lambda^m$ is an automorphism of the kind just considered. This completes the proof.

For the proof of the next theorem we need the following:

Lemma. Let G be a group without centre. If $\mathsf{J}(G)$ is a characteristic subgroup of $\mathsf{A}(G)$, then $\mathsf{A}(G)$ is complete.

Proof. Since the centre of G is trivial, the mapping

$$a \to \begin{pmatrix} x \\ a^{-1}xa \end{pmatrix}$$

is an isomorphism of G onto $\mathsf{J}(G)$. For an arbitrary automorphism φ of G, we have

$$\begin{pmatrix} x \\ x\varphi \end{pmatrix}^{-1} \begin{pmatrix} x \\ a^{-1}xa \end{pmatrix} \begin{pmatrix} x \\ x\varphi \end{pmatrix} = \begin{pmatrix} x\varphi \\ (a\varphi)^{-1}(x\varphi)(a\varphi) \end{pmatrix} = \begin{pmatrix} x \\ (a\varphi)^{-1}x(a\varphi) \end{pmatrix}.$$

This shows that the permutation

$$\begin{pmatrix} x \\ x\varphi \end{pmatrix} \tag{12.3}$$

is not permutable with every element of $\mathsf{J}(G)$, unless φ is the identity automorphism. Thus, the centre of $\mathsf{A}(G)$ is trivial. Moreover, transformation of $\mathsf{J}(G)$ by the permutations (12.3) yields a group of automorphisms of $\mathsf{J}(G)$ isomorphic to $\mathsf{A}(G)$; because of the isomorphism $G \cong \mathsf{J}(G)$, we obtain all the automorphisms of $\mathsf{J}(G)$ in this way. Now, let γ be an element of the holomorph of $\mathsf{A}(G)$ that does not belong to $\mathsf{A}(G)$. Since $\mathsf{J}(G)$ is a characteristic subgroup of $\mathsf{A}(G)$, transformation of $\mathsf{A}(G)$ by γ induces an automorphism of $\mathsf{J}(G)$. But, as we observed, the same automorphism of $\mathsf{J}(G)$ can be obtained by transformation by a certain element α of $\mathsf{A}(G)$. Consequently, $\delta = \gamma\alpha^{-1}$ is permutable with every element of $\mathsf{J}(G)$. Hence, for every $\lambda \in \mathsf{A}(G)$ and every $\xi \in \mathsf{J}(G)$,

$$\delta^{-1}\lambda^{-1}\xi\lambda\delta = \lambda^{-1}\xi\lambda.$$

Moreover, $\delta\xi\delta^{-1} = \xi$, therefore

$$\delta^{-1}\lambda^{-1}\delta\xi\delta^{-1}\lambda\delta = \lambda^{-1}\xi\lambda.$$

Since distinct elements of $\mathsf{A}(G)$ induce distinct automorphisms of $\mathsf{J}(G)$, it follows from the last equation that $\delta^{-1}\lambda\delta = \lambda$. This means that transformation by γ and by α gives rise to the same automorphism of $\mathsf{A}(G)$ so that $\mathsf{A}(G)$ has only inner automorphisms. Thus, $\mathsf{A}(G)$ is complete, which was to be proved.

12.1.9 *Let G be a non-abelian simple group or the direct product of isomorphic non-abelian simple groups. Then, the automorphism group $\mathsf{A}(G)$ of G is complete.*

Proof. We shall verify that G satisfies the conditions of the lemma. Clearly, the centre of G is trivial. Suppose that $\mathsf{J}(G)$ is not a characteristic subgroup of $\mathsf{A}(G)$. Then, there exists an automorphism of $\mathsf{A}(G)$ that carries $\mathsf{J}(G)$ into some subgroup $\mathsf{J}_1(G) \neq \mathsf{J}(G)$. Both $\mathsf{J}(G)$ and $\mathsf{J}_1(G)$ are normal subgroups of $\mathsf{A}(G)$, so that either every element of $\mathsf{J}(G)$ is permutable with every element of $\mathsf{J}_1(G)$ or the intersection $\mathsf{J}(G) \cap \mathsf{J}_1(G)$ is non-trivial. The first possibility cannot occur, for the proof of the lemma showed that $\mathsf{A}(G)$ contains no element other than the identity commuting with every element of $\mathsf{J}(G)$. Thus, $\mathsf{J}(G) \cap \mathsf{J}_1(G)$ is a non-trivial normal subgroup of $\mathsf{A}(G)$ and hence a characteristic subgroup of $\mathsf{J}(G)$. But, this is impossible, because $\mathsf{J}(G)$ is isomorphic to G, which is characteristically simple. Therefore, $\mathsf{J}(G)$ is a characteristic subgroup of $\mathsf{A}(G)$, and hence the theorem follows from the lemma.

In section 10.1, we observed that an elementary abelian group A_r of order p^r has

$$(p^r - 1)(p^r - p) \cdots (p^r - p^{r-1}) \tag{12.4}$$

distinct bases. Every automorphism of A_r carries a basis into a basis. On the other hand, any automorphism of A_r is uniquely determined by its effect on a basis. This shows that (12.4) is equal to the number of automorphisms of A_r.

12.2 Subgroup Lattices

A *lattice* Λ is an algebraic system with two binary operations

$$\alpha \cap \beta \quad \text{and} \quad \alpha \cup \beta \qquad (\alpha, \beta \in \Lambda)$$

satisfying the following laws for arbitrary $\alpha, \beta, \gamma \in \Lambda$:

$$(\alpha \cap \beta) \cap \gamma = \alpha \cap (\beta \cap \gamma) \quad (\alpha \cup \beta) \cup \gamma = \alpha \cup (\beta \cup \gamma) \tag{12.5}$$

$$\alpha \cap \beta = \beta \cap \alpha \quad \alpha \cup \beta = \beta \cup \alpha \tag{12.6}$$

$$\alpha \cap (\alpha \cup \beta) = \alpha \quad \alpha \cup (\alpha \cap \beta) = \alpha. \tag{12.7}$$

From a formula involving the signs $\cap$, $\cup$, and elements of Λ, we obtain the *dual formula* by interchanging $\cap$ and $\cup$. In this sense, the above laws consist of three dual pairs. This yields the following duality principle: If a formula can be deduced from the above laws, then the dual formula is also true and can be obtained in the dual way.

With $\cap$ and $\cup$ in their usual set-theoretical meaning, the subsets of an arbitrary non-empty set form a lattice.

For subgroups U, V of an arbitrary group G, we define $U \cap V$ to mean the set-theoretical intersection and $U \cup V = \langle U, V \rangle$. It is evident that with this definition of $\cap$ and $\cup$ the subgroups of G form a lattice $\Sigma(G)$, the *subgroup lattice* of G. When dealing with subgroups, we always assume that $\cap$ and $\cup$ have the above meanings.

For each element α of an arbitrary lattice, we have

$$\alpha \cap \alpha = \alpha, \quad \alpha \cup \alpha = \alpha. \tag{12.8}$$

For, by the second eq. (12.7), we obtain for an arbitrary element γ

$$\alpha \cap \alpha = \alpha \cap [\alpha \cup (\alpha \cap \gamma)],$$

and now the first eq. (12.7) with $\beta = \alpha \cap \gamma$ gives

$$\alpha \cap [\alpha \cup (\alpha \cap \gamma)] = \alpha.$$

The second relation (12.8) is dual to the first one and can be proved dually.

In an arbitrary lattice, we have

$$\alpha \cap \beta = \alpha \quad \text{if and only if} \quad \alpha \cup \beta = \beta. \tag{12.9}$$

For, $\alpha = \alpha \cap \beta$ implies that

$$\alpha \cup \beta = (\alpha \cap \beta) \cup \beta = \beta \cup (\alpha \cap \beta) = \beta.$$

Similarly, one obtains the converse implication.

Two lattices Λ and Λ' are said to be isomorphic if there is a one-to-one correspondence

$$\alpha \to \alpha' \qquad (\alpha \in \Lambda, \ \alpha' \in \Lambda')$$

between their elements such that

$$(\alpha \cap \beta)' = \alpha' \cap \beta', \quad (\alpha \cup \beta)' = \alpha' \cup \beta'.$$

Clearly, for two isomorphic groups G and H, the subgroup lattices $\Sigma(G)$ and $\Sigma(H)$ are isomorphic. On the other hand, non-isomorphic groups may well have isomorphic subgroup lattices. For example, the subgroup lattices of all groups of prime order are isomorphic, because every group of prime order has only the two trivial subgroups. But, there are also groups whose types are uniquely determined by their subgroup lattices.

A *distributive* lattice is one in which the distributive laws

$$\alpha \cap (\beta \cup \gamma) = (\alpha \cap \beta) \cup (\alpha \cap \gamma) \tag{12.10}$$

$$\alpha \cup (\beta \cap \gamma) = (\alpha \cup \beta) \cap (\alpha \cup \gamma) \tag{12.11}$$

are satisfied for arbitrary elements α, β, γ. Each of the two distributive laws follows from the other. Suppose that (12.10) is satisfied. Then, we have

$$\begin{aligned}(\alpha \cup \beta) \cap (\alpha \cup \gamma) &= [(\alpha \cup \beta) \cap \alpha] \cup [(\alpha \cup \beta) \cap \gamma] \\ &= \alpha \cup [(\alpha \cap \gamma) \cup (\beta \cap \gamma)] = [\alpha \cup (\alpha \cap \gamma)] \cup (\beta \cap \gamma) \\ &= \alpha \cup (\beta \cap \gamma),\end{aligned}$$

which is (12.11). Dually, (12.10) follows from (12.11).

As is well known, the subsets of a non-empty set with the set-theoretical meaning of $\cap$ and $\cup$ form a distributive lattice. The distributive laws are very strong restrictions. One can determine all groups whose subgroup lattices are distributive. For example, if G is finite, $\Sigma(G)$ is distributive if and only if G is cyclic.

Every lattice becomes a partially ordered set by the following definition:

$$\alpha \leqq \beta \quad \text{if and only if} \quad \alpha \cap \beta = \alpha.$$

To prove this we have to show that the relation $\leqq$ is reflexive, antisymmetric, and transitive.

From (12.8), we obtain $\alpha \leqq \alpha$. If $\alpha \leqq \beta$ and $\beta \leqq \alpha$, then $\alpha \cap \beta = \alpha$ and $\beta \cap \alpha = \beta$, so that (12.6) implies $\alpha = \beta$. Finally, $\alpha \leqq \beta$ and $\beta \leqq \gamma$ means that $\alpha \cap \beta = \alpha$ and $\beta \cap \gamma = \beta$; so, we obtain

$$\alpha \cap \gamma = (\alpha \cap \beta) \cap \gamma = \alpha \cap (\beta \cap \gamma) = \alpha \cap \beta = \alpha$$

and hence $\alpha \leqq \gamma$.

A lattice is said to be *modular* if it satisfies the following condition: For any three elements α, β, γ such that $\alpha \leqq \gamma$

$$\alpha \cup (\beta \cap \gamma) = (\alpha \cup \beta) \cap \gamma.$$

Every distributive lattice is modular. For, $\alpha \leqq \gamma$ implies that $\alpha \cap \gamma = \alpha$, and hence $\alpha \cup \gamma = \gamma$; hence by the distributive law,

$$\alpha \cup (\beta \cap \gamma) = (\alpha \cup \beta) \cap (\alpha \cup \gamma) = (\alpha \cup \beta) \cap \gamma.$$

In the subgroup lattice $\Sigma(G)$ of a group G, the operation $\cap$ is the set-theoretical intersection. Hence, two subgroups U and V of G satisfy the relation $U \leqq V$, i. e. $U \cap V = U$, if and only if U is contained in V.

In general, subgroup lattices are not modular; but, we have the following theorem:

12.2.1 *Let U, V, W be three subgroups of an arbitrary group G such that $UV = VU$ and $U \subseteq W$. Then,*

$$U \cup (V \cap W) = (U \cup V) \cap W. \tag{12.12}$$

Proof. Clearly, $U \subseteq W$ implies that

$$U \cup (V \cap W) \subseteq (U \cup V) \cap W. \tag{12.13}$$

On the other hand, let w be an arbitrary element of $(U \cup V) \cap W$. Since $w \in U \cup V$ and $UV = VU$, the element w can be expressed in the form $w = uv$, $u \in U$, $v \in V$. From $U \subseteq W$, we see that $u^{-1}w = v \in W$ so that $v \in V \cap W$. This shows that

$$w = uv \in U \cup (V \cap W).$$

Combining this result with (12.13), we obtain (12.12).

The normal subgroups of G form a sublattice of $\Sigma(G)$. Since any two normal subgroups are permutable, we obtain as a corollary to Theorem 12.2.1:

12.2.2 *The lattice of all normal subgroups of a group is modular.*

Under certain conditions, for example in finite groups, the subnormal subgroups also form a lattice. The main results on this lattice and on subnormal subgroups in general are due to H. Wielandt (cf. [71], [74]).

For further properties of subgroup lattices and their relations to the group structure, we refer to [65].

12.3 The Burnside Problem

Let n be a natural number. A group G is said to be of exponent n if G is periodic and the order of every element of G divides n. The Burnside problem is the question whether a finitely generated group of exponent n is necessarily finite.

The Burnside group $B(n, r)$ is defined as the group generated by r elements in which every element x satisfies the equation $x^n = e$. By Theorem 2.5.3, every group of exponent n with r generators is a homomorphic image of $B(n, r)$. Thus, the solution of the Burnside problem amounts to deciding whether the groups $B(n, r)$ are finite.

Only for a few values of n have the groups $B(n, r)$ been proved finite, and so far the orders of only a small number of the finite groups $B(n, r)$ are known.

The case $n = 2$ is trivial, since every group of exponent 2 is abelian. Thus, $B(2, r)$ is the elementary abelian group of order 2^r.

We shall now prove that the groups $B(3, r)$ are finite. Clearly, $B(3, 1)$ is finite, for it is cyclic of order 3. Let us assume that $B(3, r)$ has the finite order $3^{m(r)}$. We shall show that then also $B(3, r+1)$ is finite. The equation $(xy)^3 = e$ implies that

$$yxy = x^{-1}y^{-1}x^{-1}. \tag{12.14}$$

The group $B(3, r+1)$ can be obtained from $B(3, r)$ by adjoining a further generator z. Thus, every element b of $B(3, r+1)$ can be expressed in the form

$$b = u_1 z^{\varepsilon_1} u_2 z^{\varepsilon_2} u_3 \cdots u_{k-1} z^{\varepsilon_{k-1}} u_k, \tag{12.15}$$

where the u_i belong to $B(3, r)$ and the exponents ε_i are $+1$ or -1. We shall prove that there always exists such an expression of b in which z does not occur more than twice. In case $\varepsilon_{i-1} = \varepsilon_i$, we can apply (12.14) to the corresponding factors on the right-hand side of (12.15) to obtain

$$zu_i z = u_i^{-1} z^{-1} u_i^{-1}$$

or

$$z^{-1} u_i z^{-1} = u_i^{-1} z u_i^{-1}.$$

In both cases, the number of z's on the right-hand side of (12.15) is diminished by 1. By repeated application of this process, we eventually arrive at an expression for b with alternating exponents of z. If z still occurs more than twice, such an expression has the form

$$b = v_1 z v_2 z^{-1} v_3 z \cdots$$

or

$$b = v_1 z^{-1} v_2 z v_3 z^{-1} \cdots$$

where $v_i \in B(3, r)$. In the first case, we use (12.14) to obtain

$$b = v_1 z v_2 z z v_3 z \cdots = v_1 v_2^{-1} z^{-1} v_2^{-1} v_3^{-1} z^{-1} v_3^{-1} \cdots$$

Thus, the number of z's is again diminished by 1. A similar process can be applied in the second case. Consequently, b can be expressed as a product with at most two factors z or z^{-1}. Finally, by (12.14),

$$v_1 z^{-1} v_2 z v_3 = v_1 z^{-1} v_2 z^{-1} z^{-1} v_3 = v_1 v_2^{-1} z v_2^{-1} z^{-1} v_3.$$

This shows that every element of $B(3, r+1)$ can be expressed in one of the forms

$$w_1,\ w_1 z w_2,\ w_1 z^{-1} w_2,\ w_1 z w_2 z^{-1} w_3,$$

where $w_i \in B(3, r)$. Consequently, the number of elements in $B(3, r+1)$ does not exceed

$$3^{m(r)} + 2 \cdot 3^{2m(r)} + 3^{3m(r)}.$$

This number is less than $3^{3m(r)+1}$, hence $m(r+1) \leqq 3m(r)$. So, we have $m(r) \leqq 3^{r-1}$. The true value of $m(r)$, namely

$$m(r) = \binom{r}{3} + \binom{r}{2} + r,$$

is a much deeper result. It is not difficult to show that the groups $B(4, r)$ are finite, but it is not yet known how their orders depend on r.

The so-called restricted Burnside problem is the question whether the following proposition is true:

R_n: For every natural number r, there exists a number $b_{n,r}$ such that the order of every finite group of exponent n with r generators does not exceed $b_{n,r}$.

If R_n is true, there may exist infinite groups of exponent n with r generators; but among the finite groups with this property, there is one of maximal order, $R(n, r)$ say, and every finite group of exponent n with r generators is a homomorphic image of $R(n, r)$. For, every finite group of exponent n with r generators is isomorphic to a factor group F_r/N_i of a free group F_r of rank r with respect to some normal subgroup N_i that contains all the n-th powers. If R_n is true, then there occurs only a finite number of normal subgroups N_i. By Theorem 1.5.2, the intersection N of all these normal subgroups N_i is of finite index in F_r, and $R(n, r) = F_r/N$ has the required property.

Even R_6 is by no means easy to prove. A proof of R_p for every prime number p can be found in [40].

Exercises

1. A group R of automorphisms of a finite group G is called regular if R is a regular permutation group of the elements $\neq e$ of G. Show that G has a regular automorphism group if and only if G is elementary abelian.

2. Let the group G be generated by the elements a, b, c with the defining relations

$$a^8 = b^8 = c^4 = e, \quad b^{-1}ab = a^5, \quad c^{-1}ac = a^5, \quad c^{-1}bc = a^6 b.$$

Show that the mapping

$$a \to a^5, \quad b \to b, \quad c \to c$$

defines an outer automorphism of G that carries every conjugacy class of G into itself.

3. Let G be a non-abelian group of order pq, where p, q are prime numbers and p divides $q - 1$. Show that the subgroup lattice of G is isomorphic to the subgroup lattice of an elementary abelian group of order q^2.

4. Show that the subgroup lattice of an elementary abelian group of order p^2 is not distributive.

5. Show that the subgroup lattice of a finite abelian group is distributive if and only if the group is cyclic. (Hint: Use Exercise 4.)

6. Show that the subgroup lattice of the alternating group A_4 is not modular.

7. Show that the four-group is uniquely determined by its subgroup lattice.

8. Let G and H be two finite cyclic groups. Find a necessary and sufficient condition for $\Sigma(G)$ and $\Sigma(H)$ to be isomorphic.

CHAPTER 13

REPRESENTATIONS

13.1 Representation Modules

A *representation* of a group G by a group L is a homomorphism of G into L. This term is mainly used when L is the group $GL(n, \Omega)$ of all invertible $n \times n$ matrices with elements in a field Ω. Representations of groups by matrices are not only a powerful tool in the theory of groups, but also form the basis for most of the applications of group theory in physics.

A representation Δ of the group G by $GL(n, \Omega)$ assigns to each $a \in G$ an invertible $n \times n$ matrix $\Delta(a)$ with elements in Ω such that

$$\Delta(a)\,\Delta(b) = \Delta(ab)$$

for any two elements a, b of G. It follows that $\Delta(e) = I_n$, where I_n denotes the identity matrix of degree n and

$$\Delta(a^{-1}) = \Delta(a)^{-1}.$$

The number n is called the *degree* of Δ. The representation Δ is said to be *faithful* if the mapping $a \to \Delta(a)$ is an isomorphism of G into $GL(n, \Omega)$.

As is well known, square matrices can be used to describe linear mappings of a vector space into itself. The fact that the description of such a linear mapping by a matrix depends on the basis of the vector space is one of the reasons why the theory of representations becomes more lucid by considering linear mappings of vector spaces rather than matrices. Moreover, this approach will enable us to apply previous results to the present situation.

Let Ω be any field. A *vector space* $\mathfrak{V}$ *of dimension* n over Ω or an Ω*-module* of *rank* n is an algebraic system with the following properties:

I. $\mathfrak{V}$ is an abelian group, written additively.

II. Ω is an operator domain of $\mathfrak{V}$ so that, to any $\alpha \in \Omega$ and any $v \in V$, there corresponds a unique element αv of $\mathfrak{V}$ and

$$\alpha(v + w) = \alpha v + \alpha w \quad (\alpha \in \Omega; \quad v, w \in \mathfrak{V}). \tag{13.1}$$

Moreover, the following conditions are satisfied for arbitrary

$\alpha, \beta \in \Omega$ and $v \in \mathfrak{B}$:

$$(\alpha + \beta)\, v = \alpha v + \beta v \tag{13.2}$$

$$(\alpha\beta)\, v = \alpha(\beta v); \tag{13.3}$$

finally,

$$1v = v \text{ for the unit element } 1 \text{ of } \Omega. \tag{13.4}$$

III. There are n elements $v_1, \ldots, v_n$ of $\mathfrak{B}$ such that any element v of $\mathfrak{B}$ has a unique representation

$$v = \alpha_1 v_1 + \cdots + \alpha_n v_n, \quad \alpha_i \in \Omega.$$

The elements $v_1, \ldots, v_n$ are called a *basis* of $\mathfrak{B}$. We use the notation

$$\mathfrak{B} = \Omega v_1 + \cdots + \Omega v_n.$$

In what follows a *submodule* of $\mathfrak{B}$ always means a subgroup of the additive group $\mathfrak{B}$ that is admissible with respect to Ω.

The reader is assumed to be familiar with the following facts about submodules and bases of $\mathfrak{B}$.[1]

Every submodule $\mathfrak{T}$ of $\mathfrak{B}$ is an Ω-module whose rank m does not exceed n, i. e.

$$\mathfrak{T} = \Omega t_1 + \cdots + \Omega t_m, \quad m \leqq n.$$

$\mathfrak{T}$ is a proper submodule of $\mathfrak{B}$ if and only if $m < n$. Any basis $t_1, \ldots, t_m$ of $\mathfrak{T}$ can be extended to a basis of $\mathfrak{B}$ by adding suitable $n - m$ elements of $\mathfrak{B}$.[2]

From a basis $v_1, \ldots, v_n$ of $\mathfrak{B}$, we obtain any other basis $w_1, \ldots, w_n$ in the form

$$w_i = \sum_{k=1}^{n} \sigma_{ik} v_k \quad (i = 1, \ldots, n), \tag{13.5}$$

where $S = [\sigma_{ik}]$ is an invertible matrix with elements σ_{ik} in Ω. Let

$$\boldsymbol{v} = \begin{bmatrix} v_1 \\ \vdots \\ v_n \end{bmatrix} \quad \boldsymbol{w} = \begin{bmatrix} w_1 \\ \vdots \\ w_n \end{bmatrix};$$

then, the eqs. (13.5) can be written as

$$\boldsymbol{w} = S\boldsymbol{v}.$$

All automorphisms and endomorphisms of $\mathfrak{B}$ that occur in what follows are tacitly assumed to be Ω-automorphisms (Ω-endomorphisms) as defined in section 2.6. An endomorphism γ of $\mathfrak{B}$ is uniquely determined by the

[1] For proofs we refer to any textbook on linear algebra.

[2] cf. Theorem 4.2.3.

images $v_1^\gamma, \ldots, v_n^\gamma$ of a basis of $\mathfrak{V}$. For, since γ is an Ω-endomorphism, we have

$$(\alpha_1 v_1 + \cdots + \alpha_n v_n)^\gamma = (\alpha_1 v_1)^\gamma + \cdots + (\alpha_n v_n)^\gamma = \alpha_1 v_1^\gamma + \cdots + \alpha_n v_n^\gamma .$$

Denoting the column of the elements $v_1^\gamma, \ldots, v_n^\gamma$ by $\boldsymbol{v}^\gamma$, we have

$$\boldsymbol{v}^\gamma = C\boldsymbol{v},$$

where C is an $n \times n$ matrix with elements in Ω. Clearly, γ is an automorphism if and only if the matrix C is invertible. In terms of a given basis $v_1, \ldots, v_n$ of $\mathfrak{V}$, we obtain in this way a one-to-one correspondence $\gamma \leftrightarrow C$ between the endomorphisms of $\mathfrak{V}$ and the $n \times n$ matrices with elements in Ω. For another endomorphism δ and the corresponding matrix D, we find

$$\boldsymbol{v}^{\gamma\delta} = (\boldsymbol{v}^\gamma)^\delta = (C\boldsymbol{v})^\delta = C\boldsymbol{v}^\delta = CD\boldsymbol{v}.$$

Since $\mathfrak{V}$ is abelian, the sum of any two endomorphisms exists in the sense of section 4.2. To the sum $\gamma + \delta$, there corresponds the sum $C + D$. Thus, the endomorphism ring of $\mathfrak{V}$ is isomorphic to the ring of all $n \times n$ matrices with elements in Ω.

We now assume that the elements of a given group G act as operators on $\mathfrak{V}$. We write the elements of G as right operators, so that to every pair $v \in \mathfrak{V}$ and $a \in G$ there corresponds a unique element va of $\mathfrak{V}$ and

$$(v + w)\, a = va + wa \tag{13.6}$$

for arbitrary elements v, w of $\mathfrak{V}$ and a of G.

A module $\mathfrak{V}$ is called a *representation module* of G over Ω if, in addition to (13.6), the following laws are satisfied for arbitrary elements $v \in \mathfrak{V}$, $a, b \in G$:

$$v(ab) = (va)\, b, \tag{13.7}$$

$$ve = v \text{ for the unit element } e \text{ of } G, \tag{13.8}$$

$$(\alpha v)\, a = \alpha(va). \tag{13.9}$$

It follows from (13.7) and (13.8) that a and a^{-1} induce inverse mappings on $\mathfrak{V}$ so that every element of G induces an automorphism of $\mathfrak{V}$.

Every representation module of rank n gives rise to a representation of degree n. For, let $v_1, \ldots, v_n$ be a basis of $\mathfrak{V}$ and denote the column of the elements $v_1 a, \ldots, v_n a$ by $\boldsymbol{v}a$. Then,

$$\boldsymbol{v}a = \Delta(a)\, \boldsymbol{v}, \tag{13.10}$$

where $\Delta(a)$ is an invertible $n \times n$ matrix with elements in Ω. The above laws for a representation module give

$$\boldsymbol{v}(ab) = (\boldsymbol{v}a)\, b = (\Delta(a)\, \boldsymbol{v})\, b = \Delta(a)\, \boldsymbol{v}b = \Delta(a)\, \Delta(b)\, \boldsymbol{v}, \tag{13.11}$$

and, on the other hand,

$$\boldsymbol{v}(ab) = \Delta(ab)\, \boldsymbol{v}; \tag{13.12}$$

hence

$$\Delta(a)\, \Delta(b) = \Delta(ab). \tag{13.13}$$

Conversely, if a representation Δ of G by $GL(n, \Omega)$ is given, then any Ω-module $\mathfrak{V}$ of rank n becomes a representation module of G by defining the effect of the elements of G on $\mathfrak{V}$ as follows: Use (13.10) to define the effect of $a \in G$ on a basis of $\mathfrak{V}$ and require (13.6) and (13.9) to be satisfied. Then, G becomes an operator domain of $\mathfrak{V}$. Now, (13.13) implies (13.11) and (13.12), so that (13.7) is satisfied. Finally, (13.8) follows from $\Delta(e) = I_n$.

In a new basis $\boldsymbol{w} = S\boldsymbol{v}$ of $\mathfrak{V}$, we obtain

$$\boldsymbol{w}\, a = S\, \boldsymbol{v}\, a = S\Delta(a)\, \boldsymbol{v} = S\Delta(a)\, S^{-1} \boldsymbol{w} = \Phi(a)\, \boldsymbol{w}.$$

The matrices $\Phi(a) = S\Delta(a)S^{-1}$ form a representation $\Phi = S\Delta S^{-1}$ of G. The representations Δ and Φ are said to be *equivalent* and will not be regarded as essentially distinct, since they correspond to the same automorphisms of $\mathfrak{V}$ and describe them only in terms of different bases.

Clearly, two representation modules that are operator isomorphic with respect to G as right operator domain yield equivalent representations, and vice versa.

It may happen that a representation module $\mathfrak{V}$ of G contains a submodule $\mathfrak{T}$ that is admissible with respect to G. Let

$$\mathfrak{T} = \Omega t_1 + \cdots + \Omega t_m$$

with $m < n$ be such a submodule. We can find suitable $l = n - m$ elements $u_1, \ldots, u_l$ of $\mathfrak{V}$ such that $t_1, \ldots, t_m, u_1, \ldots, u_l$ is a basis of $\mathfrak{V}$. Using this basis, we obtain a representation $S\Delta S^{-1}$ whose matrices have the following form:

$$S\Delta(a)S^{-1} = \begin{bmatrix} \Delta_1(a) & O \\ \Delta_3(a) & \Delta_2(a) \end{bmatrix}. \tag{13.14}$$

O always denotes a zero matrix of suitable size. In this case O has m rows and l columns. Here, the $m \times m$ matrices $\Delta_1(a)$ form a representation of G with $\mathfrak{T}$ as a representation module. For, if $\boldsymbol{t}$ denotes the column of the elements $t_1, \ldots, t_m$, we have

$$\boldsymbol{t}\, a = \Delta_1(a)\boldsymbol{t},$$

since $\mathfrak{T}$ is admissible. The matrices $\Delta_2(a)$ also form a representation Δ_2 of G. This is easily verified by computing $S\Delta(a)S^{-1}S\Delta(b)S^{-1} = S\Delta(ab)S^{-1}$, or it can be seen as follows: For the column $\boldsymbol{u}$ of the elements $u_1, \ldots, u_l$, we have

$$\boldsymbol{u}a = \Delta_3(a)\,\boldsymbol{t} + \Delta_2(a)\,\boldsymbol{u}; \tag{13.15}$$

hence,

$$\boldsymbol{u}a \equiv \Delta_2(a)\,\boldsymbol{u} \pmod{\mathfrak{T}}.$$

This shows that the factor module $\mathfrak{V}/\mathfrak{T}$ is a representation module of Δ_2.

A representation Δ is called *reducible* if its representation module $\mathfrak{V}$ has a representation submodule $\mathfrak{T}$ other than 0 and $\mathfrak{V}$. Thus, when Δ is reducible, there exists an equivalent representation $S\Delta S^{-1}$ whose matrices are of the form (13.14). Conversely, if there is an equivalent representation $S\Delta S^{-1}$ of the form (13.14), then every representation module of Δ has a non-trivial representation submodule. For, suppose that the matrices $\Delta_1(a)$ form a representation of degree $m < n$. Then, the first m elements of the basis of $\mathfrak{V}$ belonging to $S\Delta S^{-1}$ are the basis of a non-trivial submodule $\mathfrak{T}$ of rank m that is admissible with respect to G. The representation Δ is said to *split* into the *constituents* Δ_1 and Δ_2.

By contrast, a representation is *irreducible* if and only if its representation module contains no non-trivial representation submodule. This means that the representation modules regarded as abelian groups with the operator domains Ω and G are simple.

We now regard $\mathfrak{V}$ as an additive abelian group with the operator domains Ω and G. As the rank of $\mathfrak{V}$ is finite, there exists a composition series

$$\mathfrak{V} = \mathfrak{V}_k \supset \mathfrak{V}_{k-1} \supset \cdots \supset \mathfrak{V}_1 \supset \mathfrak{V}_0 = 0$$

where each $\mathfrak{V}_{i-1}$ is a maximal representation submodule of $\mathfrak{V}_i$ $(i = 1, \ldots, k)$ or, in other words, the factor modules $\mathfrak{V}_i/\mathfrak{V}_{i-1}$ are simple as additive groups with the operator domains Ω and G. We choose a basis of $\mathfrak{V}_1$, then we extend it to a basis of $\mathfrak{V}_2$; after that, we extend this basis of $\mathfrak{V}_2$ to a basis of $\mathfrak{V}_3$, etc. until we finally arrive at a basis of $\mathfrak{V}$. To this particular basis, there belongs a representation of G whose matrices are of the form

$$\begin{bmatrix} \Delta_{11}(a) & 0 & \ldots\ldots & 0 \\ \Delta_{21}(a) & \Delta_{22}(a) & \ldots\ldots & 0 \\ \ldots & \ldots & \ldots & \ldots \\ \Delta_{k1}(a) & \Delta_{k2}(a) & \ldots\ldots & \Delta_{kk}(a) \end{bmatrix}$$

for every $a \in G$. For $i = 1, \ldots, k$, the representation Δ_{ii} arises from the representation module $\mathfrak{V}_i/\mathfrak{V}_{i-1}$, and as these factor modules are simple, it follows that the representations Δ_{ii} are irreducible.

Conversely, each splitting of a representation into irreducible constituents leads to a composition series of the representation module.

By Theorem 2.7.3 for groups with operators, the factor modules of any two composition series of the representation module $\mathfrak{B}$ are operator isomorphic to within order. Since representations are equivalent if and only if their representation modules are operator isomorphic, our previous arguments yield the following theorem:

13.1.1 *The irreducible constituents of any representation are unique up to equivalence and order.*

Let us assume that $\mathfrak{B}$ is a representation module of a reducible representation of G. Then, there exists a non-trivial representation submodule

$$\mathfrak{T} = \Omega t_1 + \cdots + \Omega t_m,$$

and the basis $t_1, \ldots, t_m$ of $\mathfrak{T}$ can be extended to a basis of $\mathfrak{B}$ by suitable elements $u_1, \ldots, u_l$. Writing

$$\mathfrak{U} = \Omega u_1 + \cdots + \Omega u_l,$$

we obtain the direct decomposition $\mathfrak{B} = \mathfrak{T} + \mathfrak{U}$. In general, $\mathfrak{U}$ is not admissible with respect to the operator domain G. Our notation is the same as above, so that (13.15) shows that $\mathfrak{U}$ is admissible if and only if $\Delta_3(a) = O$ for all $a \in G$. If this condition is satisfied, then $\mathfrak{U}$ is also a representation module and belongs to the representation Δ_2. We say that Δ *splits completely* into the constituents Δ_1 and Δ_2.

We shall now show that it is no loss of generality to consider only representations by invertible matrices. Suppose that to every element a of the group G there corresponds a square matrix $\bar{\Delta}(a)$, not necessarily invertible, such that $\bar{\Delta}(a)\bar{\Delta}(b) = \bar{\Delta}(ab)$ for any two elements a and b of G. Clearly, to such a representation in a more general sense, there also corresponds a representation module $\mathfrak{B}$, where now the elements of G as right operators on $\mathfrak{B}$ induce endomorphisms, but not necessarily automorphisms. Let $\mathfrak{B}e$ denote the image of $\mathfrak{B}$ under the unit element e as a right operator. It is evident that $\mathfrak{B}e$ is an Ω-submodule of $\mathfrak{B}$ that is admissible with respect to G and for which e is the identity operator; in other words, $\mathfrak{B}e$ is a representation module in the previous sense. Any element v of $\mathfrak{B}$ can be written as follows:

$$v = ve + (v - ve).$$

So, we obtain a decomposition

$$\mathfrak{B} = \mathfrak{B}e + (\mathfrak{B} - \mathfrak{B}e). \tag{13.16}$$

This decomposition is direct, because e as a right operator carries every element of $\mathfrak{B}e$ into itself and annihilates every element of $\mathfrak{B} - \mathfrak{B}e$. In particular, $\mathfrak{B} - \mathfrak{B}e$ is also admissible with respect to G. By virtue of (13.16), the representation $\bar{\Delta}$ splits completely into the constituents Δ and Δ_0 which belong to the representation modules $\mathfrak{V}e$ and $\mathfrak{V} - \mathfrak{V}e$, respectively. As we observed above, $\mathfrak{V}e$ is a representation module in our previous sense, i. e. all the matrices $\Delta(a)$ are invertible. By contrast, all the elements of $\mathfrak{V} - \mathfrak{V}e$ are annihilated by every element of G, so that $\Delta_0(a) = O$ for each $a \in G$.

A representation is called *completely reducible* if its representation module is the direct sum of simple representation modules. In terms of the representation, this means that there exists an equivalent representation that splits completely into its irreducible constituents.

The next theorem is basic for the representation theory of finite groups.

13.1.2 *If the characteristic of the field Ω is 0 or relatively prime to the order of the finite group G, then every representation of G by matrices with elements in Ω is completely reducible.*

Proof. It is obviously sufficient to prove the following assertion: If a representation module $\mathfrak{V}$ for G contains a non-trivial representation submodule $\mathfrak{T}$, then $\mathfrak{V}$ has a direct decomposition

$$\mathfrak{V} = \mathfrak{T} + \mathfrak{X},$$

where $\mathfrak{X}$ is a representation module for G.

In any case, we have a direct decomposition

$$\mathfrak{V} = \mathfrak{T} + \mathfrak{U},$$

where $\mathfrak{U}$ is not necessarily admissible with respect to G. We shall prove that $\mathfrak{U}$ can be replaced by another direct summand

$$\mathfrak{X} = \Omega x_1 + \cdots + \Omega x_l$$

that is admissible with respect to G. We use the previous notation and write $\boldsymbol{x}$ for the column of the elements $x_1, \ldots, x_l$. We put

$$\boldsymbol{x} = F\boldsymbol{t} + \boldsymbol{u} \tag{13.17}$$

and try to determine the $l \times m$ matrix F with elements in Ω such that $\mathfrak{X}$ is admissible. Note that, for every choice of the matrix F, the elements $x_1, \ldots, x_l$ are linearly independent because $u_1, \ldots, u_l$ are. Moreover, $\mathfrak{T} \cap \mathfrak{X} = 0$ for every choice of F. For, from

$$x = \beta_1 x_1 + \cdots + \beta_l x_l,$$

it follows by (13.17) that

$$x \equiv \beta_1 u_1 + \cdots + \beta_l u_l \pmod{\mathfrak{T}}.$$

Thus, $x \in \mathfrak{T} \cap \mathfrak{X}$ implies that

$$\beta_1 u_1 + \cdots + \beta_l u_l \equiv 0 \pmod{\mathfrak{T}};$$

hence, $\beta_1 = \cdots = \beta_l = 0$, since $\mathfrak{T} \cap \mathfrak{U} = 0$. From $\mathfrak{T} \cap \mathfrak{X} = 0$ and the fact that $\mathfrak{X}$ is an Ω-module of rank l, it follows that for every matrix F in (13.17) we have the direct decomposition

$$\mathfrak{V} = \mathfrak{T} + \mathfrak{X}.$$

From

$$\boldsymbol{t} a = \Delta_1(a)\, \boldsymbol{t}$$

and (13.15) we obtain

$$\begin{aligned}\boldsymbol{x} a &= (F\boldsymbol{t} + \boldsymbol{u})\, a = F\boldsymbol{t} a + \boldsymbol{u} a = F\Delta_1(a)\, \boldsymbol{t} + \Delta_3(a)\, \boldsymbol{t} + \Delta_2(a)\, \boldsymbol{u} \\ &= (F\Delta_1(a) + \Delta_3(a) - \Delta_2(a)\, F)\, \boldsymbol{t} + \Delta_2(a)\, (F\boldsymbol{t} + \boldsymbol{u}) \\ &= (F\Delta_1(a) + \Delta_3(a) - \Delta_2(a)\, F)\, \boldsymbol{t} + \Delta_2(a)\, \boldsymbol{x}.\end{aligned}$$

For $\mathfrak{X}$ to be admissible, the matrix F must satisfy the condition

$$F\Delta_1(a) + \Delta_3(a) - \Delta_2(a)\, F = O \quad \text{for every} \quad a \in G. \tag{13.18}$$

As Δ is a representation, we have

$$\begin{bmatrix} \Delta_1(a) & O \\ \Delta_3(a) & \Delta_2(a) \end{bmatrix} \begin{bmatrix} \Delta_1(b) & O \\ \Delta_3(b) & \Delta_2(b) \end{bmatrix} = \begin{bmatrix} \Delta_1(ab) & O \\ \Delta_3(ab) & \Delta_2(ab) \end{bmatrix}$$

for any two elements a and b of G. This gives

$$\begin{aligned}&\Delta_1(ab) = \Delta_1(a)\, \Delta_1(b), \quad \Delta_2(ab) = \Delta_2(a)\, \Delta_2(b), \\ &\Delta_3(ab) = \Delta_3(a)\, \Delta_1(b) + \Delta_2(a)\, \Delta_3(b).\end{aligned} \tag{13.19}$$

We put $g = |G|$ and

$$F = \frac{1}{g} \sum_{c \in G} \Delta_2(c)^{-1} \Delta_3(c) \tag{13.20}$$

where c ranges over all elements of G. Then,

$$\Delta_2(a)\, F = \frac{1}{g} \sum_{c \in G} \Delta_2(a)\, \Delta_2(c)^{-1}\, \Delta_3(c).$$

If c ranges over all elements of G, then so does $d = ac^{-1}$ for any fixed $a \in G$. By (13.19), we now obtain

$$\begin{aligned}\Delta_2(a)\, F &= \frac{1}{g} \sum_{d \in G} \Delta_2(d)\, \Delta_3(d^{-1}\, a) \\ &= \frac{1}{g} \sum_{d \in G} \Delta_2(d)\, (\Delta_3(d^{-1})\, \Delta_1(a) + \Delta_2(d^{-1})\, \Delta_3(a)) \\ &= \left(\frac{1}{g} \sum_{d \in G} \Delta_2(d)\, \Delta_3(d^{-1})\right) \Delta_1(a) + \Delta_3(a) \\ &= F\Delta_1(a) + \Delta_3(a).\end{aligned}$$

Consequently, the matrix F as defined by (13.20) satisfies the condition (13.18), and this proves the theorem.

If all representations of a given group G are known to be completely reducible, then the problem of finding all representations of G is reduced to the knowledge of all irreducible representations, because all representation modules of G can be obtained as direct sums of simple representation modules.

We shall prove another theorem on the complete reducibility of certain groups of matrices. This theorem deals with matrices whose elements belong to the field of the complex numbers, but it also applies to an important class of infinite groups.

For a matrix M with elements in the field of the complex numbers, let M^* denote the matrix that is obtained by taking the transpose of M and replacing the elements by their complex conjugates. A matrix H is called hermitian if $H^* = H$. By a *hermitian form*, we mean an expression $\xi^* H \xi$, where H is an hermitian $n \times n$ matrix and ξ a column of n variables $\xi_1, \ldots, \xi_n$. We recall some of the basic properties of hermitian forms and matrices; for proofs we refer, for example, to [20]. For arbitrary values of the variables $\xi_1, \ldots, \xi_n$, the expression $\xi^* H \xi$ is always a real number. The hermitian form $\xi^* H \xi$ and the matrix H are said to be positive-definite if always $\xi^* H \xi \geqq 0$ and if $\xi^* H \xi = 0$ implies that $\xi_1 = \cdots = \xi_n = 0$. If H is positive-definite, then the determinant $|H|$ is positive and so are the determinants of all the matrices obtained from H by removing any system of rows and columns with the same suffixes. Let η denote the column of new variables $\eta_1, \ldots, \eta_n$ that are related to $\xi_1, \ldots, \xi_n$ by a linear substitution $\xi = P\eta$ with an invertible $n \times n$ matrix P. We obtain

$$\xi^* H \xi = \eta^* P^* H P \eta .$$

Clearly, if H is positive-definite, then so is $P^* H P$.

A regular $n \times n$ matrix A is called an *automorphic* matrix of the hermitian matrix H if

$$A^* H A = H .$$

If A is automorphic for H, then $P^{-1} A P$ is automorphic for $P^* H P$; for, we have

$$(P^{-1} A P)^* P^* H P (P^{-1} A P) = P^* A^* (P^*)^{-1} P^* H P P^{-1} A P = P^* H P.$$

We now come to the announced theorem on the complete reducibility of certain representations.

13.1.3 *Every representation of a group that consists of automorphic matrices of a positive-definite hermitian form is completely reducible.*

Proof. Let the matrices of the representation Δ of the group G be automorphic matrices of a positive-definite hermitian form. We may assume that there is an equivalent representation $\Phi = S\Delta S^{-1}$ whose matrices are of the form (13.14). The matrices of the representation Φ are automorphic for a positive-definite hermitian matrix H. We partition H into blocks,

$$H = \begin{bmatrix} H_1 & H_2 \\ H_3 & H_4 \end{bmatrix}$$

where H_1 is an $m \times m$ and H_4 an $l \times l$ matrix. Note that the determinant $|H_4|$ does not vanish. As the matrices (13.14) are automorphic for H, we have

$$\begin{bmatrix} \Delta_1(a)^* & \Delta_3(a)^* \\ O & \Delta_2(a)^* \end{bmatrix} \begin{bmatrix} H_1 & H_2 \\ H_3 & H_4 \end{bmatrix} \begin{bmatrix} \Delta_1(a) & O \\ \Delta_3(a) & \Delta_2(a) \end{bmatrix} = \begin{bmatrix} H_1 & H_2 \\ H_3 & H_4 \end{bmatrix}$$

for every $a \in G$. This gives

$$\begin{aligned} \Delta_2(a)^* H_3 \Delta_1(a) + \Delta_2(a)^* H_4 \Delta_3(a) &= H_3, \\ \Delta_2(a)^* H_4 \Delta_2(a) &= H_4. \end{aligned} \tag{13.21}$$

To prove the theorem, we have to find a matrix F satisfying condition (13.18). We shall show that $F = H_4^{-1} H_3$ has the required property. It follows from (13.21) that

$$\Delta_2(a)^* H_3 \Delta_1(a) + \Delta_2(a)^* H_4 \Delta_3(a) - \Delta_2(a)^* H_4 \Delta_2(a) H_4^{-1} H_3 = O.$$

Multiplication on the left by

$$H_4^{-1}(\Delta_2(a)^*)^{-1}$$

gives

$$H_4^{-1} H_3 \Delta_1(a) + \Delta_3(a) - \Delta_2(a) H_4^{-1} H_3 = O$$

or

$$F\Delta_1(a) + \Delta_3(a) - \Delta_2(a) F = O.$$

This completes the proof.

As is well known, for a hermitian matrix H, there always exists a unitary matrix U such that U^*HU is a diagonal matrix. When H is positive-definite, we can therefore find an invertible matrix Q such that $Q^* HQ = I_n$. If A is automorphic for H, then $B = Q^{-1}AQ$ is automorphic for the identity matrix,

$$B^*B = B^* I_n B = I_n,$$

i. e. B is unitary. So, we obtain the following theorem:

13.1.4 *For every group of automorphic matrices of a positive-definite hermitian matrix, there is an equivalent group of unitary matrices.*

Finally, we show that the last two theorems apply to any finite group.

13.1.5 *For every finite group of matrices with elements in the field of the complex numbers, there is a positive-definite hermitian form for which the matrices are automorphic.*

Proof. Let the group consist of the $n \times n$ matrices $M_1, \ldots, M_r$. For an arbitrary positive-definite hermitian square matrix H, we put

$$K = \sum_{i=1}^{r} M_i^* H M_i.$$

Clearly, K is a hermitian matrix. Moreover, K is positive-definite, because the corresponding hermitian form is the sum of the r positive-definite hermitian forms that belong to the matrices $M_i^* H M_i$. For an arbitrary matrix M_k, we obtain

$$M_k^* K M_k = \sum_{i=1}^{r} M_k^* M_i^* H M_i M_k = \sum_{i=1}^{r} (M_i M_k)^* H (M_i M_k) = K,$$

since, for a fixed M_k, the matrices $M_i M_k$, $i = 1, \ldots, r$, range over the whole group. Thus, $M_1, \ldots, M_r$ are automorphic matrices of K.

13.2 Algebras

For the study of representations of finite groups, it is expedient to consider the so-called group algebra, i. e. the ring whose elements are the formal linear combinations of the group elements over some field with multiplication defined in the natural way. Thus, in this section, we shall derive the basic properties of algebras of finite rank over a field.

By an *algebra* A of rank n over a field Ω we mean an associative ring with the following properties:

I. A is a vector space of dimension n over Ω so that in a basis $u_1, \ldots, u_n$

$$A = \Omega u_1 + \cdots + \Omega u_n.$$

II. The products of basis elements are defined by

$$u_i u_j = \sum_{k=1}^{n} \gamma_{ijk} u_k,$$

where the γ_{ijk} are given elements in Ω. The product of two arbitrary elements

$$a = \alpha_1 u_1 + \cdots + \alpha_n u_n, \quad b = \beta_1 u_1 + \cdots + \beta_n u_n$$

is formed according to the distributive laws and under the assumption that the elements of Ω are permutable with the u_i, i. e.

$$ab = \sum_{i=1}^{n} \sum_{j=1}^{n} \alpha_i \beta_j u_i u_j = \sum_{k=1}^{n} \left(\sum_{i=1}^{n} \sum_{j=1}^{n} \alpha_i \beta_j \gamma_{ijk} \right) u_k .$$

The associative law

$$(u_i u_j) u_k = u_i (u_j u_k) \quad (i, j, k = 1, \ldots, n)$$

entails conditions on the elements γ_{ijk}, that can easily be derived.

For our purpose, the most important examples of algebras are the group algebras of finite groups. For a finite group G and a field Ω, the *group algebra* G_Ω *of* G over Ω is the algebra of rank $|G|$ over Ω for which the elements of G form a basis and whose multiplication rule is that of G. Thus, the elements of G_Ω are linear combinations

$$a = \sum_{x \in G} \alpha_x x ,$$

that are multiplied according to the rule

$$ab = \left(\sum_{x \in G} \alpha_x x \right) \left(\sum_{y \in G} \beta_y y \right) = \sum_{x \in G} \sum_{y \in G} \alpha_x \beta_y xy = \sum_{z \in G} \left(\sum_{xy=z} \alpha_x \beta_y \right) z .$$

The associative law is satisfied, since it holds in G.

The reader will be more familiar with another type of algebra that will also play a role in what follows. For a given field Ω, let Ω_n denote the ring of all $n \times n$ matrices with elements in Ω. Clearly, Ω_n is an algebra of rank n^2 over Ω; it is called the *complete matrix algebra of degree n over* Ω. Let E_{ik} denote the $n \times n$ matrix with the unit element 1 of Ω in the i-th row and k-th column and zeros elsewhere. The n^2 matrices E_{ik} $(i, k = 1, \ldots, n)$ obviously form a basis of Ω_n. The familiar rules for multiplication of matrices show that

$$E_{ik} E_{lm} = \begin{cases} E_{im} & \text{for } k = l , \\ O & \text{for } k \neq l . \end{cases}$$

Every system of n^2 elements in an arbitrary algebra that is multiplied according to the same law as the E_{ik} is called a system of *matrix units* of degree n.

By the complete matrix algebra Λ_n of degree n over a skew field Λ, we mean the ring of all $n \times n$ matrices with elements in Λ. If Λ is of finite rank r over the field Ω, then Λ_n is an algebra of rank rn^2 over Ω. For a basis $\lambda_1, \ldots, \lambda_r$ of Λ over Ω, the rn^2 elements

$$\lambda_s E_{ik} \quad (s = 1, \ldots, r; \quad i, k = 1, \ldots, n)$$

form a basis of Λ_n over Ω.

We shall consider only algebras that contain a unit element, i. e. an element e such that $ea = ae = a$ for every element a of the algebra. In a group algebra, the unit element of the group is the unit element.

The *centre* Z of an algebra A consists of all elements $z \in A$ such that

$$az = za \quad \text{for all} \quad a \in A.$$

If A is an algebra over the field Ω, then so is Z. The elements αe, where e is the unit element of A and $\alpha \in \Omega$, certainly belong to Z. It may happen that these elements already exhaust the centre, for example in a complete matrix algebra. For, the centre of Ω_n consists of the matrices αI_n, where I_n denotes the identity matrix of degree n and $\alpha \in \Omega$. The element

$$z = \sum_{i,k=1}^{n} \zeta_{ik} E_{ik} \quad (\zeta_{ik} \in \Omega)$$

belongs to the centre if and only if $zE_{pq} = E_{pq} z$ for $p, q = 1, \ldots, n$. This gives

$$E_{pq} z = \sum_{i,k=1}^{n} \zeta_{ik} E_{pq} E_{ik} = \sum_{k=1}^{n} \zeta_{qk} E_{pk}$$

$$= zE_{pq} = \sum_{i,k=1}^{n} \zeta_{ik} E_{ik} E_{pq} = \sum_{i=1}^{n} \zeta_{ip} E_{iq};$$

hence, $\zeta_{11} = \cdots = \zeta_{nn}$ and $\zeta_{ik} = 0$ for $i \neq k$, i. e.

$$z = \zeta_{11}(E_{11} + E_{22} + \cdots + E_{nn}) = \zeta_{11} I_n.$$

An element c of an algebra is called an *idempotent* if $c^2 = c$ and $c \neq 0$.

An element b of an algebra A is said to be *nilpotent* if $b^k = 0$ for some natural number k. A *properly nilpotent* element of A is an element $v \neq 0$ such that va is nilpotent for every $a \in A$. If va is nilpotent, then so is av, for $(va)^k = 0$ implies that

$$(av)^{k+1} = a(va)^k v = 0.$$

Similarly, from $(av)^k = 0$, it follows that $(va)^{k+1} = 0$.

A subalgebra R of A is called a *right ideal* if $RA = R$. Here, RA stands for the set of all products ra, $r \in R$, $a \in A$. Similarly, a subalgebra L of A is said to be a *left ideal* if $AL = L$. A *two-sided ideal* is one that is at the same time a right and a left ideal. Any algebra A has two trivial two-sided ideals, namely A itself and the so-called zero ideal (0), which consists of the single element 0. Clearly, the intersection of right ideals is a right ideal, and the same applies to left and two-sided ideals.

For any element u of A, let uA denote the totality of all products ua, $a \in A$. It is evident that uA is a right ideal; uA is called the *principal* right ideal generated by u. Similarly, Au is the principal left ideal generated by u.

A right ideal $R \neq (0)$ is said to be *simple* if R contains no right ideal of A other than R and (0). Similarly, we define simple left and two-sided ideals. As every ideal is an Ω-module, we infer that every algebra contains simple one- and two-sided ideals, namely the non-zero ideals that have minimum rank as Ω-modules.

The distributive law

$$(x + y)\, a = xa + ya$$

can be interpreted as follows: If A is regarded as an Ω-module, then for any fixed $a \in A$ the mapping

$$x \to xa \quad (x \in A)$$

is an endomorphism. Thus, A can be regarded as an Ω-module with the right operator domain A. The other distributive law shows that A is also an Ω-module with the left operator domain A. A right ideal is an Ω-submodule admissible with respect to A as right operator domain. Left and two-sided ideals can be characterized in similar a way.

Direct decompositions of A into ideals are to be understood in the sense of section 5.1, where A is regarded as an Ω-module, i.e. an additive group with the operator domain Ω for which A is a domain of right, left, or two-sided operators.

13.2.1 *If*

$$A = R_1 \oplus \cdots \oplus R_m \tag{13.22}$$

is a direct decomposition of the algebra A into a direct sum of right ideals R_i and

$$e = e_1 + \cdots + e_m \quad (e_i \in R_i), \tag{13.23}$$

the corresponding decomposition of the unit element e, then

$$e_i^2 = e_i, \; e_i e_j = 0 \quad \text{for} \quad i \neq j \quad (i, j = 1, \ldots, m). \tag{13.24}$$

Moreover,

$$R_i = e_i A \quad (i = 1, \ldots, m). \tag{13.25}$$

Conversely, if the eqs. (13.23) and (13.24) hold and the R_i are defined by (13.25), then (13.22) is a direct decomposition of A into right ideals.

On account of (13.24), the elements $e_1, \ldots, e_m$ are called a system of *orthogonal idempotents*.

Proof. For an arbitrary element r_i of R_i, we have

$$r_i = e r_i = e_1 r_i + \cdots + e_i r_i + \cdots + e_m r_i,$$

where $e_j \in R_j$ implies that $e_j r_i \in R_j$. On the other hand, the components of r_i in the direct decomposition (13.22) are

$$r_i = 0 + \cdots + 0 + r_i + 0 + \cdots + 0.$$

Comparing the two representations of r_i, we find that

$$e_i r_i = r_i, \quad e_j r_i = 0 \quad \text{for} \quad j \neq i. \tag{13.26}$$

From $e_i r_i = r_i$, we conclude that $R_i \subseteq e_i A$; on the other hand, $e_i \in R_i$ shows that $e_i A \subseteq R_i$ because R_i is a right ideal. This proves (13.25). Taking $r_i = e_i$ in (13.26), we obtain (13.24).

Conversely, suppose that we are given a decomposition (13.23) of the unit element such that (13.24) is satisfied. Define R_i by (13.25). Then, we have

$$a = ea = e_1 a + \cdots + e_m a \tag{13.27}$$

for any $a \in A$. This shows that A is the sum, possibly not direct, of the right ideals $R_1, \ldots, R_m$. To show that the sum is direct, we have to prove that the decomposition (13.27) into R_i-components is unique. Let us assume that

$$a = e_1 b_1 + \cdots + e_m b_m$$

is another decomposition. Then, we obtain

$$0 = e_1 d_1 + \cdots + e_m d_m \tag{13.28}$$

where $d_i = a - b_i$. Multiplication of (13.28) on the left by e_i gives $e_i^2 d_i = e_i d_i = 0$, hence $e_i a = e_i b_i$. Thus, (13.27) is unique and the decomposition (13.22) is direct. This completes the proof.

An algebra is called *semi-simple* if it is the direct sum of simple right ideals.

13.2.2 *An algebra is semi-simple if and only if it contains no properly nilpotent element.*

Proof. If A is semi-simple, then it is completely reducible in the sense of section 4.2 as an additive group with operators. One easily verifies that Theorem 4.2.3 also applies to this case, so that any right ideal $R' \neq (0)$ is a direct summand, i.e.

$$A = R' \oplus R'',$$

where R'' is a right ideal. By Theorem 13.2.1, this direct decomposition leads to a decomposition $e = e' + e''$ into orthogonal idempotents. This shows that every right ideal, other than (0), contains an idempotent.

Assume that A contains a properly nilpotent element v. Then, vA is a right ideal, that is distinct from (0) and consists of nilpotent elements. But, this contradicts the above result that every right ideal $\neq (0)$ contains an idempotent. Thus, there are no properly nilpotent elements in A.

Conversely, suppose that A contains no properly nilpotent elements. We first prove that any simple right ideal R_1 of A contains an idempotent. For an arbitrary element u of R_1, the set uR_1 is a right ideal of A contained in R_1, and since R_1 is simple, we have $uR_1 = (0)$ or $uR_1 = R_1$. It is easy to see that $uR_1 = (0)$ cannot hold for every $u \in R$. For, otherwise, for any element $u_1 \neq 0$ of R_1 and an arbitrary $a \in A$, the product $u_1 a$ would belong to R_1; hence, $u_1 a R_1 = (0)$. This would imply that $(u_1 a)^2 = 0$ for every $a \in A$, which contradicts the assumption that A contains no properly nilpotent elements. This shows that there exists an element c in R_1 such that $cR_1 = R_1$. The elements x in R_1 for which $cx = 0$ form a right ideal contained in R_1, and $cR_1 = R_1 \neq (0)$ shows that this ideal is distinct from R_1. Hence, it is the zero ideal; hence, from $cx = 0$ and $x \in R_1$ it follows that $x = 0$. From $cR_1 = R_1$ and $c \in R_1$, it follows that there is an element e_1 in R_1 such that $ce_1 = c$. This gives $ce_1^2 = ce_1$ or $c(e_1^2 - e_1) = 0$, hence $e_1^2 = e_1$. Thus, R_1 contains the idempotent e_1.

The elements on the right-hand side of

$$e = e_1 + (e - e_1) \tag{13.29}$$

satisfy the equations

$$e_1^2 = e_1, \quad e_1(e - e_1) = (e - e_1)\, e_1 = 0, \; (e - e_1)^2 = e - e_1.$$

Thus, (13.29) is a decomposition of the unit element into orthogonal idempotents. By Theorem 13.2.1, we obtain the direct decomposition

$$A = R_1 \oplus R_1'$$

into the right ideals $R_1 = e_1 A$, $R_1' = (e - e_1)A$. This shows that every simple right ideal of A is a direct summand.

If R_1' is not simple, let R_2 be a simple right ideal of A that is contained in R_1'. As we just saw, A has a direct decomposition $A = R_2 \oplus R_2'$, where R_2' is a right ideal. Taking the intersection with R_1', we obtain $R_1' = R_2 \oplus R''$ whose R'' is a right ideal of A. This shows that

$$A = R_1 \oplus R_2 \oplus R''.$$

If R'' is not simple, we proceed similarly. As A is an Ω-module of finite rank and all right ideals are also Ω-modules, after a finite number of steps we arrive at a direct decomposition of A into simple right ideals. This completes the proof.

13.2.3 *Every right ideal other than* (0) *of a semi-simple algebra* A *is a principal right ideal* $e'A$ *generated by an idempotent* e'. *The ideal* $e'A$ *is two-sided if and only if* e' *belongs to the centre of* A.

Proof. Let $R' \neq (0)$ be a right ideal. The proof of Theorem 13.2.2 shows that there is a right ideal R'' such that

$$A = R' \oplus R''.$$

By Theorem 13.2.1, this gives

$$e = e' + e'',$$

where e' is an idempotent and $R' = e'A$. This proves the first part of the theorem.

Clearly, if e' belongs to the centre of A, then $e'A = Ae'$ is a two-sided ideal.

Now, let J be a two-sided ideal other than (0). As J is also a right ideal, we have $J = e'A$, where e' is an idempotent and $e'x = x$ for every $x \in J$. By Theorem 13.2.2, A contains no properly nilpotent elements. This property is symmetric with respect to right and left. From this, we conclude that A is also a direct sum of simple left ideals and, further, that every left ideal of A is a principal left ideal generated by an idempotent. In particular, we find that $J = Ae^*$, where e^* is an idempotent. Moreover, we have $ye^* = y$ for every $y \in J$. Now, $J = e'A = Ae^*$ shows that $e' = e'e^* = e^*$, so that $J = e'A = Ae'$. For an arbitrary element a of A, we have $e'a \in J$ and hence $e'ae' = e'a$. Similarly, we obtain $e'ae' = ae'$. This gives $e'a = ae'$ for every $a \in A$, hence e' belongs to the centre of A.

Theorem 13.2.3 is now proved.

An algebra A_0 is said to be *simple* if it is semi-simple and contains no two-sided ideal other than A_0 and (0).

13.2.4 *Every semi-simple algebra* A *is a direct sum*

$$A = A_1 \oplus \cdots \oplus A_r$$

of unique simple two-sided ideals A_i. *The* A_i *are simple algebras and* $A_iA_j = 0$ *for* $i \neq j$. *Every right ideal of an* A_i *is a right ideal of* A. *Every simple right ideal of* A *is contained in some* A_i.

Proof. Let A_1 be a simple two-sided ideal of A. By Theorem 13.2.3, we have $A_1 = e_1A$, where e_1 is an idempotent belonging to the centre. Moreover,

$$A = e_1A \oplus (e - e_1)A \tag{13.30}$$

is a direct decomposition into two-sided ideals, since $e - e_1$ as well as e_1 is contained in the centre. As e_1 and $e - e_1$ are orthogonal and belong to the centre, we have

$$e_1a(e - e_1)b = 0$$

for any two elements a and b of A. This shows that not only addition but also multiplication is carried out componentwise with respect to the direct decomposition (13.30). Therefore, a properly nilpotent element of any of the two summands would be a properly nilpotent element of A. Consequently, both e_1A and $(e - e_1)A$ are semi-simple. Moreover, any one- or two-sided ideal of one of the two summands is also an ideal in A. Therefore, $e_1A = A_1$ is a simple algebra. If $(e - e_1)A$ is not simple, we take a simple two-sided ideal A_2 of $(e - e_1)A$ and proceed as above. After a finite number of steps, we arrive at a direct decomposition

$$A = A_1 \oplus \cdots \oplus A_r \tag{13.31}$$

into simple two-sided ideals A_i. For any two elements $a_i \in A_i$, $a_j \in A_j$, and $i \neq j$, we have

$$a_i a_j \in A_i \cap A_j = (0)$$

so that $a_i a_j = 0$. This implies that any properly nilpotent element of some A_i is properly nilpotent in A, and any non-trivial two-sided ideal of some A_i is a two-sided ideal in A. Consequently, the A_i are simple algebras.

For any simple two-sided ideal J of A, not all the intersections $J \cap A_i$ can be equal to (0). Suppose that $J \cap A_j \neq (0)$. Since both J and A_j are simple, this gives $J \cap A_j = A_j = J$. It follows that the simple two-sided ideals A_i are uniquely determined.

For a right ideal R of A, the intersection $R \cap A_j$ is a right ideal of A. If R is simple, there is a unique index i such that $R \cap A_i = R$ and hence $R \subseteq A_i$.

This completes the proof.

Two right ideals R_1 and R_2 are said to be operator isomorphic if they are isomorphic as Ω-modules with the operator domain A, where the effect of an element a of A as an operator is defined as multiplication by a on the right.

13.2.5 *Let*

$$A = e_1A \oplus \cdots \oplus e_rA$$

be a direct decomposition of the semi-simple algebra A into simple two-sided ideals $e_iA = A_i$. Two simple right ideals R' and R'' are operator isomorphic if and only if they are contained in one and the same simple two-sided ideal.

Proof. Suppose that both R' and R'' are contained in A_i. By Theorem 13.2.1, we have $R' = e'A$, $R'' = e''A$ with orthogonal idempotents e', e''. The set of all finite sums of the form

$$a_1e'b_1 + a_2e'b_2 + \cdots + a_me'b_m \quad (a_j, b_j \in A)$$

obviously forms a two-sided ideal of A contained in A_i. This ideal is distinct from (0), because it contains $e_ie'e_i = e'$. Since A_i is simple, it follows that the ideal in question coincides with A_i. Since $e'' \in A_i$, we have

$$e'' = \bar{a}_1 e' \bar{b}_1 + \cdots + \bar{a}_p e' \bar{b}_p$$

with suitable elements $\bar{a}_j$, $\bar{b}_j$ of A. Since e'' is an idempotent, we conclude that there is at least one index k such that

$$e'' \bar{a}_k e' \bar{b}_k \neq 0. \tag{13.32}$$

Therefore, $e'' \bar{a}_k e' \bar{b}_k A$ is a right ideal, which is distinct from (0) and belongs to $e''A = R''$. As R'' is simple, we have

$$e'' \bar{a}_k e' \bar{b}_k A = R''.$$

Moreover, the right ideal $e' \bar{b}_k A$ is distinct from (0) and is contained in the simple right ideal $e'A = R'$, so that

$$e' \bar{b}_k A = R'.$$

Hence, we have

$$cR' = R'',$$

where $c = e'' \bar{a}_k$. The mapping

$$x' \to cx' \quad (x' \in R') \tag{13.33}$$

is therefore a homomorphism of R' onto R'' as Ω-modules. The kernel of this homomorphism consists of all $y' \in R'$ for which $cy' = 0$, so that the kernel is a right ideal of A; moreover, the kernel is distinct from R' because $R'' \neq (0)$. Thus, the kernel is (0), hence (13.33) is an isomorphism. Finally, $c(x'a) = (cx')a$ shows that (13.33) is an operator isomorphism of R' onto R''.

Now, suppose that R' and R'' are contained in distinct two-sided ideals, $R' \subseteq e_i A$, $R'' \subseteq e_j A$, $i \neq j$, say. Then, R' and R'' cannot be operator isomorphic. For, since A_i and A_j annihilate each other, we have

$$r' e_i = r', \quad r' e_j = 0$$

for every $r' \in R'$, and

$$r'' e_i = 0, \quad r'' e_j = r''$$

for every $r'' \in R$. Thus, both parts of the theorem are proved.

The next theorem describes the structure of simple algebras:

13.2.6 *Every simple algebra is a complete matrix algebra over a skew field.*

Proof. A simple algebra A is semi-simple so that A has a direct decomposition

$$A = e_{11} A \oplus e_{22} A \oplus \cdots \oplus e_{nn} A$$

into simple right ideals. The $e_{11}, e_{22}, \ldots, e_{nn}$ are orthogonal idempotents and

$$e = e_{11} + e_{22} + \cdots + e_{nn}.$$

The set of all finite sums

$$a_1 e_{11} b_1 + a_2 e_{11} b_2 + \cdots + a_m e_{11} b_m \quad (a_k, b_k \in A) \tag{13.34}$$

is a two-sided ideal of A and distinct from (0), because $e_{11} = e_{11}e_{11}e_{11}$ is such a sum. Thus, this set coincides with A. It follows that each e_{ii} has a representation of the form (13.34), say

$$e_{ii} = \bar{a}_1 e_{11} \bar{b}_1 + \cdots + \bar{a}_p e_{11} \bar{b}_p.$$

This gives

$$e_{ii} = e_{ii} e_{ii} = \bar{a}_1 e_{11} \bar{b}_1 e_{ii} + \cdots + \bar{a}_p e_{11} \bar{b}_p e_{ii},$$

which shows that for every index i there is an element c_i of A such that $e_{11} c_i e_{ii} \neq 0$.

After these preliminaries, we show that $A_i = e_{ii} A e_{ii}$, i.e. the set of all products $e_{ii} a e_{ii}$, $a \in A$, is a skew field. Clearly, $e_{ii} A e_{ii}$ is a subalgebra of A with the unit element e_{ii}. We have to show that each element $e_{ii} a e_{ii} \neq 0$ has an inverse. Now, $e_{ii} a e_{ii} A$ is a right ideal of A, distinct from (0) and contained in $e_{ii} A$. Since $e_{ii} A$ is simple, we have $e_{ii} a e_{ii} A = e_{ii} A$. Thus, there is an element x of A such that $e_{ii} a e_{ii} x = e_{ii}$. This shows that

$$e_{ii} a e_{ii} e_{ii} x e_{ii} = e_{ii},$$

so that $e_{ii} x e_{ii}$ is the inverse of $e_{ii} a e_{ii}$.

Next, we construct a system of matrix units in A. For each index $i = 2, \ldots, n$, we choose an element c_i such that

$$e_{11} c_i e_{ii} = e_{1i} \neq 0.$$

Then, we have

$$e_{11} e_{1i} = e_{1i} e_{ii} = e_{1i} \qquad (i = 1, \ldots, n).$$

Since $e_{11} A$ is simple and $e_{1i} A \subseteq e_{11} A$, we conclude that $e_{1i} A = e_{11} A$. In particular, we have $e_{1i} d_i = e_{11}$ for suitable $d_i \in A$ and

$$e_{1i} e_{ii} d_i e_{11} = e_{1i} d_i e_{11} = e_{11}.$$

For $i = 2, \ldots, n$, we put

$$e_{i1} = e_{ii} d_i e_{11},$$

so that

$$e_{ii} e_{i1} = e_{i1} e_{11} = e_{i1} \qquad (i = 1, \ldots, n)$$

and

$$e_{1i} e_{i1} = e_{11}.$$

Moreover, we obtain

$$e_{1i} e_{i1} e_{1i} e_{i1} = e_{11}^2 = e_{11}$$

so that $e_{i1} e_{1i} \neq 0$. Since

$$(e_{i1} e_{1i})^2 = e_{i1} e_{1i},$$

$e_{i1}e_{1i}$ is an idempotent of the skew field $\Lambda_i = e_{ii}Ae_{ii}$, and since the unit element is the only idempotent of a skew field, we have

$$e_{i1}e_{1i} = e_{ii} \qquad (i = 1, \ldots, n).$$

Finally, we put

$$e_{i1}e_{1k} = e_{ik}$$

for $i \neq k$. Then, we obtain

$$e_{ik}e_{km} = e_{i1}e_{1k}e_{k1}e_{1m} = e_{i1}e_{11}e_{1m} = e_{im}$$

and for $l \neq k$

$$e_{ik}e_{lm} = e_{ik}e_{kk}e_{ll}e_{lm} = 0.$$

Thus, the n^2 elements e_{ik} $(i, k = 1, \ldots, n)$ form a system of matrix units in A.

For any element λ_1 of the skew field $\Lambda_1 = e_{11}Ae_{11}$, we put

$$\lambda = \lambda_1 + e_{21}\lambda_1 e_{12} + \cdots + e_{n1}\lambda_1 e_{1n}.$$

These elements λ form a skew field Λ, for one readily verifies that the mapping $\lambda_1 \to \lambda$ is an isomorphism of Λ_1 onto Λ. In particular, this isomorphism carries the unit element e_{11} of Λ_1 into

$$e_{11} + e_{21}e_{11}e_{12} + \cdots + e_{n1}e_{11}e_{1n} = e_{11} + e_{22} + \cdots + e_{nn} = e,$$

so that the unit element of Λ coincides with the unit element of A.

Every element λ of Λ is permutable with every matrix unit e_{ik}; for, we have

$$\lambda e_{ik} = e_{i1}\lambda_1 e_{1k} = e_{ik}\lambda.$$

For an arbitrary element a of A, we can put

$$a = eae = \sum_{i,k=1}^{n} e_{ii}ae_{kk}.$$

The summands on the right-hand side can be written as follows:

$$e_{ii}ae_{kk} = e_{i1}e_{1i}ae_{k1}e_{1k}.$$

Here, $e_{1i}ae_{k1} = e_{11}e_{1i}ae_{k1}e_{11}$ is an element of Λ_1. The corresponding element of the skew field Λ is

$$\alpha_{ik} = e_{1i}ae_{k1} + e_{21}e_{1i}ae_{k1}e_{12} + \cdots + e_{n1}e_{1i}ae_{k1}e_{1n}.$$

This gives

$$\alpha_{ik}e_{ik} = e_{ii}ae_{kk}$$

and hence

$$a = \sum_{i,k=1}^{n} \alpha_{ik}e_{ik}.$$

This representation of the elements of A is unique, for

$$\sum_{i,k=1}^{n} \beta_{ik} e_{ik} = 0 \qquad (\beta_{ik} \in \Lambda)$$

implies that

$$\beta_{pq} e_{pq} = e_{pp} \left(\sum_{i,k=1}^{n} \beta_{ik} e_{ik} \right) e_{qq} = 0$$

for $p, q = 1, \ldots, n$. This completes the proof.

If Ω is algebraically closed, we obtain the following description of the semi-simple algebras over Ω:

13.2.7 *Every simple algebra A over an algebraically closed field Ω is a complete matrix algebra over Ω. Every semi-simple algebra over Ω is the direct sum of complete matrix algebras over Ω, and the number of direct summands is equal to the rank of the centre of A over Ω.*

Proof. By Theorem 13.2.6, every simple algebra over Ω is a complete matrix algebra over some skew field Λ. Clearly, the rank of Λ over Ω is finite so that every element of Λ is algebraic over Ω. As Ω is algebraically closed, we see that $\Lambda = \Omega$. This proves the first part of the theorem.

By Theorem 13.2.4, a semi-simple algebra A over Ω is the direct sum

$$A = A_1 \oplus \cdots \oplus A_r \tag{13.35}$$

of simple algebras A_i. By the first part of our theorem, the A_i are complete matrix algebras over Ω. Since $A_i A_j = 0$ for $i \neq j$, multiplication in A is carried out componentwise with respect to the direct decomposition (13.35). Thus, the centre of A is the direct sum of the centres of the A_i. By a previous remark, the centre of a complete matrix algebra over Ω is isomorphic to Ω, i.e. of rank 1 over Ω. Consequently, r is equal to the rank of the centre of A over Ω. This proves our theorem.

Finally, we use Theorem 13.2.6 to obtain a direct decomposition of a simple algebra into simple right ideals.

13.2.8 *Let*

$$\Omega_n = \sum_{i,k=1}^{n} \Omega e_{ik}$$

be the complete matrix algebra over a field Ω. Then, the Ω-modules

$$R_i = \sum_{k=1}^{n} \Omega e_{ik}$$

are simple right ideals of Ω_n and

$$\Omega_n = R_1 \oplus \cdots \oplus R_n$$

is a direct decomposition.

Proof. We have $R_i e_{pq} \subseteq R_i$ for $p, q = 1, \ldots, n$, so that the R_i are right ideals. Let $R \neq (0)$ be a right ideal of Ω_n contained in R_i, and let

$$a = \sum_{k=1}^{n} \alpha_k e_{ik} \qquad (\alpha_k \in \Omega)$$

be an element of R, $a \neq 0$. Then, we have $\alpha_j \neq 0$, say. It follows that $ae_{j1} = \alpha_j e_{i1}$, hence e_{i1} is contained in R. Therefore, R contains the matrix units $e_{i1}e_{1k} = e_{ik}$ for $k = 1, \ldots, n$, hence $R = R_i$. Thus, the R_i are simple right ideals. It is obvious that Ω_n is the direct sum of the R_i.

13.3 The Regular Representation

Let A be an algebra over the field Ω. By a representation of A of degree n over Ω, we mean a ring homomorphism of A into the complete matrix algebra Ω_n of degree n over Ω. This means that to each $a \in A$ there is assigned an $n \times n$ matrix $\Delta(a)$ with elements in Ω such that for arbitrary $a, b \in A$ and $\alpha \in \Omega$

$$\Delta(a + b) = \Delta(a) + \Delta(b),$$

$$\Delta(\alpha a) = \alpha \Delta(a),$$

$$\Delta(ab) = \Delta(a)\Delta(b).$$

A representation Δ of a finite group G gives rise to a representation of the group algebra G_Ω by putting

$$\Delta\left(\sum_{x \in G} \alpha_x x\right) = \sum_{x \in G} \alpha_x \Delta(x).$$

Of course, every representation of G_Ω gives rise to a representation of G.

As in the case of groups, we can introduce representation modules for algebras. The only difference is that condition (13.7) has to be replaced by the three conditions

$$v(ab) = (va)\, b,$$

$$v(a + b) = va + vb, \tag{13.36}$$

$$v(\alpha a) = \alpha(va)$$

for arbitrary elements v of the representation module, a, b of the algebra, and α of Ω. All the results of section 13.1 about the connection between the bases of representation modules and equivalent representations and between reducibility and admissible submodules remain valid for representations of algebras.

Every right ideal of an algebra A over Ω is an Ω-module that is admissible with respect to the operator domain A, where the effect of an element a of A as an operator is defined as multiplication by a on the right. In this case, (13.36), (13.8), and (13.9) are consequences of the laws in A. Thus, every right ideal of A is a representation module for A. The representation induced by a right ideal is irreducible if and only if the right ideal is simple.

The representation, or more precisely, the class of equivalent representations, for which A itself is the representation module is called the *regular representation*. If the elements of the finite group G are taken as a basis of the group algebra G_Ω, then the regular representation of G consists of permutation matrices which correspond to the right regular permutational representation in the sense of section 7.2.

The next theorem is basic for the representations of semi-simple algebras and the so-called ordinary representations of finite groups.

13.3.1 *An algebra A is semi-simple if and only if all representations of A are completely reducible. The representation modules of the irreducible representations of a semi-simple algebra A are operator isomorphic to the simple right ideals of A.*

Proof. First, suppose that A is semi-simple. Let

$$A = e_1 A \oplus \cdots \oplus e_m A$$

be a direct decomposition into simple right ideals and

$$V = \Omega v_1 + \cdots + \Omega v_n$$

a representation module of A. Since the unit element of A is the identity operator on V, we have $V = VA$. Therefore, V is the sum, possibly not direct, of the representation submodules

$$v_i e_k A \qquad (i = 1, \ldots, n;\ k = 1, \ldots, m).$$

If $v_i e_k A \neq 0$, the mapping

$$e_k a \to v_i e_k a \qquad (a \in A)$$

is an operator isomorphism of $e_k A$ onto $v_i e_k A$, for, it is clearly an operator homomorphism. The elements $e_k b$ of $e_k A$ for which $v_i e_k b = 0$ form a right ideal B of A, properly contained in $e_k A$; since $e_k A$ is simple, we have $B = (0)$. Thus, the mapping in question is an operator isomorphism. The representation modules $v_i e_k A$ that are distinct from 0 are therefore operator isomorphic to simple right ideals, hence are simple representation modules. Among the modules $v_i e_k A$, we omit those that are 0. After having arranged the remaining modules in an arbitrary order, we omit those that are contained in the sum of the preceding ones and retain those that are not. Since the modules are simple, we obtain in this way a decomposition of V into a direct sum of suitable modules $v_i e_k A$. Hence, V, and therefore the corresponding representation of A, are completely reducible, and the simple representation submodules of V are operator isomorphic to simple right ideals of A.

Conversely, suppose that every representation of A is completely reducible. Taking the regular representation, we conclude that A is the direct sum of simple Ω-modules that are admissible with respect to multiplication by elements of A as right factors; in other words, A is the direct sum of simple right ideals. Thus, A is semi-simple. This completes the proof.

The simple right ideals of a complete matrix algebra Ω_n yield Ω_n as its own representation. By Theorem 13.2.8, the matrix units

$$e_{r1}, \ldots, e_{rn} \tag{13.37}$$

form a basis of a simple right ideal of Ω_n. For an element

$$a = \sum_{i,k=1}^{n} \alpha_{ik} e_{ik}$$

of Ω_n, we obtain

$$e_{rj} a = \sum_{i,k=1}^{n} \alpha_{ik} e_{rj} e_{ik} = \sum_{k=1}^{n} \alpha_{jk} e_{rk} \qquad (j = 1, \ldots, n).$$

Denoting by $\mathbf{s}$ the column of the elements (13.37), and by $M(a)$ the $n \times n$ matrix $[\alpha_{ik}]$, these equations can be written as follows:

$$\mathbf{s} a = M(a) \mathbf{s}.$$

We observe that the matrix $M(a)$ does not depend on the index r in (13.37), in agreement with the second part of Theorem 13.2.5.

In view of the last theorem, it is important to know whether a group algebra is semi-simple. This question is answered by the following theorem.

13.3.2 *The group algebra G_Ω of a finite group G over a field Ω is semi-simple if and only if the characteristic of Ω does not divide $|G|$.*

Proof. If the characteristic of Ω is not a divisor of $|G|$, then it follows from Theorem 13.1.2 that every representation of G_Ω is completely reducible. By Theorem 13.3.1, G_Ω is therefore semi-simple.

Conversely, suppose that the characteristic of Ω divides $|G|$. We shall show that the element

$$s = \sum_{x \in G} x$$

of G_Ω is properly nilpotent. Then, it follows from Theorem 13.2.2 that G_Ω is not semi-simple.

For any element y of G, we have $sy = s$. For an arbitrary element

$$a = \sum_{y \in G} \alpha_y y$$

of G_Ω, we obtain therefore

$$sa = \Big(\sum_{y \in G} \alpha_y\Big) s,$$

and hence

$$(sa)^2 = \Big(\sum_{y \in G} \alpha_y\Big)^2 s^2 = \Big(\sum_{y \in G} \alpha_y\Big)^2 |G| s = 0,$$

because $|G|$ is a multiple of the characteristic. This proves the theorem.

The representations of a finite group G in $GL(n, \Omega)$ are called *ordinary* if the characteristic of Ω does not divide the order of G. If the characteristic of Ω divides $|G|$, we speak of *modular* representations. In what follows, we deal only with ordinary representations. For an introduction to the more intricate theory of modular representations, we refer to [5] and [13].

An irreducible representation of a group in $GL(n, \Omega)$ may become reducible if Ω is replaced by an extension field. An irreducible representation of a finite group is said to be *absolutely irreducible* if it remains irreducible under an arbitrary algebraic extension of Ω. We shall see later that, in every class of equivalent absolutely irreducible representations of a finite group, there are representations whose coefficients belong to a finite algebraic extension of the corresponding prime field.

By the Theorems 13.2.5, 13.2.7, and 13.3.1, the number of inequivalent absolutely irreducible ordinary representations of a finite group G is equal to the rank of the centre of G_Ω. We shall now prove that the rank of the centre of G_Ω is equal to the number of conjugacy classes in G.

Let $K_1, \ldots, K_t$ denote the conjugacy classes of G. We form the so-called *class sums*

$$k_i = \sum_{x \in K_i} x \qquad (i = 1, \ldots, t)$$

in G_Ω for an arbitrary field Ω.

13.3.3 *The elements $k_1, \ldots, k_t$ of G_Ω are a basis of the centre of G_Ω.*

Proof. An element

$$z = \sum_{x \in G} \zeta_x x \qquad (\zeta_x \in \Omega)$$

belongs to the centre of G_Ω if and only if $y^{-1}zy = z$ for every $y \in G$. This means that

$$z = \sum_{x \in G} \zeta_x x = y^{-1} zy = \sum_{x \in G} \zeta_x y^{-1} xy = \sum_{x \in G} \zeta_{yxy^{-1}} x.$$

Thus, z belongs to the centre if and only if $\zeta_x = \zeta_{yxy^{-1}}$ for any two elements x, y of G, in other words, if conjugate elements of G have equal coefficients in z. Consequently, z can be written in the form

$$z = \sum_{i=1}^{t} \zeta_i k_i.$$

This proves the theorem.

We now return to our proper subject, the representations of finite groups, and summarize the results so far obtained:

13.3.4 *Let G be a finite group.*

(a) *Every ordinary representation of G is completely reducible.*

(b) *The number of classes of equivalent absolutely irreducible representations of G is equal to the number of conjugacy classes in G.*

(c) *Every absolutely irreducible representation class of G occurs among the constituents of the regular representation with a multiplicity equal to its degree.*

(d) *Among the $|G|$ matrices*

$$\Delta(a) \qquad (a \in G)$$

of an absolutely irreducible representation Δ of degree f, there are f^2 linearly independent ones. Thus, only the matrices of the form γI_f, $\gamma \in \Omega$ are permutable with all the matrices $\Delta(a)$, $a \in G$.

(e) *The degrees $f_1, \ldots, f_t$ of the absolutely irreducible representation classes satisfy the equation*

$$f_1^2 + \cdots + f_t^2 = |G|.$$

Proof. (a) is Theorem 13.1.2; (b) follows from the Theorems 13.2.5, 13.2.7, 13.3.1, and 13.3.2. The first part of (c) is a consequence of Theorem 13.3.1. The structure of the group algebra G_Ω over an algebraically closed field Ω is described by Theorem 13.2.7. Let Ω_f be one of the complete matrix algebras that occur as summands in the direct decomposition of G_Ω. By Theorem 13.2.8, Ω_f is the direct sum of f simple right ideals each of which is an Ω-module of rank f. By the second part of Theorem 13.2.5, these f simple right ideals are operator isomorphic so that they yield equivalent absolutely irreducible representations of degree f. Distinct complete matrix algebras that occur as direct summands of G_Ω yield inequivalent representations. This proves the assertion on the multiplicity in (c). As we observed above, the simple right ideals of Ω_f yield Ω_f as its own representation. Thus, the f^2 matrix units of Ω_f are linear combinations of the matrices $\Delta(a)$. This implies (d). By Theorem 13.2.7, the group algebra G_Ω is the direct sum of t complete matrix algebras of the degrees $f_1, \ldots, f_t$. Thus, the rank $|G|$ of G_Ω over Ω is also equal to $f_1^2 + \cdots + f_t^2$, and this gives (e). This completes the proof.

We shall now describe a procedure by which an ordinary representation of a finite group can be decomposed into its irreducible constituents. Though this procedure is hardly practicable except in very simple cases, it provides information on the fields in which the absolutely irreducible representations can be realized.

Let Δ be an ordinary representation of the finite group G in $GL(m,\Lambda)$, where Λ is any field. We consider the subalgebra V of Λ_m consisting of all those matrices that are permutable with every $\Delta(a)$, $a \in G$. It is evident that the rank v of V as a Λ-module depends only on the equivalence class of Δ. By (d) of Theorem 13.3.4, we have $v = 1$ if Δ is absolutely irreducible. Conversely, if Δ is not absolutely irreducible, then it can easily be seen that $v > 1$. For, if Δ is reducible, then the equivalence class of Δ contains a representation of the form

$$\begin{bmatrix} \Delta' & O \\ O & \Delta'' \end{bmatrix}$$

where Δ' and Δ'' are representations of degrees l and $m - l$, respectively. This representation is obviously permutable with all the matrices

$$\begin{bmatrix} \alpha I_l & O \\ O & \beta I_{m-l} \end{bmatrix}.$$

Thus, if Δ is not absolutely irreducible, we can find a matrix M in Λ_m that is not of the form γI_m and that is permutable with all the matrices $\Delta(a)$, $a \in G$. Let μ denote a characteristic root of M. Then, the matrix $W = M - \mu I_m$ is also permutable with every $\Delta(a)$; moreover, the rank w of W is greater than 0 and less than m. The field $\Lambda(\mu)$ is a finite algebraic extension of Λ or coincides with Λ in case $\mu \in \Lambda$. As is well known, there exist two invertible matrices P and Q with elements in $\Lambda(\mu)$ such that

$$PWQ = \begin{bmatrix} I_w & O \\ O & O \end{bmatrix}.$$

From $\Delta(a)W = W\Delta(a)$, we obtain

$$P\Delta(a)P^{-1}PWQ = PWQQ^{-1}\Delta(a)Q.$$

Writing

$$P\Delta(a)P^{-1} = \begin{bmatrix} \Delta_1(a) & \Delta_3(a) \\ \Delta_4(a) & \Delta_2(a) \end{bmatrix} \qquad Q^{-1}\Delta(a)Q = \begin{bmatrix} \bar{\Delta}_1(a) & \bar{\Delta}_3(a) \\ \bar{\Delta}_4(a) & \bar{\Delta}_2(a) \end{bmatrix}$$

with $w \times w$ matrices $\Delta_1(a)$, $\bar{\Delta}_1(a)$, we get

$$\begin{bmatrix} \Delta_1(a) & \Delta_3(a) \\ \Delta_4(a) & \Delta_2(a) \end{bmatrix} \begin{bmatrix} I_w & O \\ O & O \end{bmatrix} = \begin{bmatrix} I_w & O \\ O & O \end{bmatrix} \begin{bmatrix} \bar{\Delta}_1(a) & \bar{\Delta}_3(a) \\ \bar{\Delta}_4(a) & \bar{\Delta}_2(a) \end{bmatrix}.$$

This gives $\bar{\Delta}_3(a) = O$ for all $a \in G$. The proof of Theorem 13.1.2 now shows that the equivalence class of Δ contains a representation whose coefficients belong to $\Lambda(\mu)$ and which splits completely into the constituents $\bar{\Delta}_1$ and $\bar{\Delta}_2$. Thus, if a representation with coefficients in Λ is reducible, then only a finite algebraic extension of Λ is needed to split it completely. The coefficients of the regular representation certainly belong to the prime field, for they are 0 or 1. Our arguments show that the regular representation splits completely into absolutely irreducible constituents after a finite number of finite algebraic extensions. As all representation classes are constituents of the regular representation, we conclude that every class of equivalent representations contains representations whose coefficients belong to a finite algebraic extension of the corresponding prime field.

We cannot discuss the deep problem in what fields all the absolutely irreducible representations of a finite group can be realized, but we refer the reader to [56] and [57].

13.4 Characters

Let G be any group and

$$\mathfrak{V} = \Omega v_1 + \cdots + \Omega v_n, \ \mathfrak{W} = \Omega w_1 + \cdots + \Omega w_m$$

two simple representation modules for G. Let us suppose that there is a G-homomorphism σ of $\mathfrak{V}$ into $\mathfrak{W}$. The image $\mathfrak{V}^\sigma$ of $\mathfrak{V}$ under σ is a G-admissible submodule of $\mathfrak{W}$. As $\mathfrak{W}$ is simple, this means that $\mathfrak{V}^\sigma = \mathfrak{W}$ or $\mathfrak{V}^\sigma = 0$. The kernel of σ is a G-admissible submodule of $\mathfrak{V}$ and hence is equal to 0 or $\mathfrak{V}$. This shows that σ is either a G-isomorphism of $\mathfrak{V}$ onto $\mathfrak{W}$ or carries $\mathfrak{V}$ into the zero of $\mathfrak{W}$. Clearly, the first possibility cannot occur unless $m = n$.

The mapping σ is uniquely determined by the images $v_1^\sigma, \ldots, v_n^\sigma$ of $v_1, \ldots, v_n$. Denoting the columns of the v_i and the w_i by $\boldsymbol{v}$ and $\boldsymbol{w}$, respectively, we have

$$\boldsymbol{v}^\sigma = S\boldsymbol{w}, \tag{13.38}$$

where S is an $n \times m$ matrix with elements in Ω. Our result above means that either S is an invertible matrix or $S = O$.

Let Δ and Φ denote the representations of G obtained from $\mathfrak{V}$ and $\mathfrak{W}$, respectively, so that

$$\boldsymbol{v}a = \Delta(a)\,\boldsymbol{v}, \quad \boldsymbol{w}a = \Phi(a)\,\boldsymbol{w} \tag{13.39}$$

for all $a \in G$. As σ is a G-homomorphism, we have

$$(\boldsymbol{v}^\sigma)\,a = (\boldsymbol{v}a)^\sigma$$

for every $a \in G$. By (13.38) and (13.39), this shows that

$$(\boldsymbol{v}^\sigma)\, a = (S\boldsymbol{w})\, a = S(\boldsymbol{w}a) = S\Phi(a)\,\boldsymbol{w} = (\boldsymbol{v}a)^\sigma = (\Delta(a)\,\boldsymbol{v})^\sigma = \Delta(a)\,\boldsymbol{v}^\sigma = \Delta(a)\, S\boldsymbol{w}\,.$$

So, we have

$$\Delta(a)\, S = S\Phi(a)$$

for every $a \in G$. Conversely, if a matrix S satisfies the last equation for every $a \in G$, then the mapping σ defined by (13.38) is a G-homomorphism of $\mathfrak{V}$ into $\mathfrak{W}$. This leads to the following theorem:

13.4.1 Schur's Lemma. *Let Δ and Φ be representations of a group G that are irreducible over the field Ω. If a matrix S with elements in Ω satisfies the equations*

$$\Delta(a)\, S = S\Phi(a)$$

for every $a \in G$, then S is invertible or $S = O$. In the first case, Δ and Φ are equivalent, namely $S^{-1}\Delta S = \Phi$.

Note that Schur's Lemma holds for an arbitrary field Ω. In particular, it is not necessary to assume that Ω is algebraically closed.

Let

$$\Delta_{ik}(a) \qquad (i, k = 1, \ldots, n)$$

denote the element in the i-th row and k-th column of the matrix $\Delta(a)$, and let $\Phi_{pq}(a)$ have the similar meaning for $\Phi(a)$. We shall now derive a system of basic orthogonality relations for the coefficients of representations of finite groups.

13.4.2 *Let Δ and Φ be inequivalent representations of a finite group G, both irreducible over the field Ω. Then,*

$$\sum_{a \in G} \Delta_{ik}(a)\, \Phi_{pq}(a^{-1}) = 0 \qquad \begin{pmatrix} i, k = 1, \ldots, n; \\ p, q = 1, \ldots, m \end{pmatrix}.$$

Proof. Let

$$U = [u_{ip}] \qquad (i = 1, \ldots, n;\ p = 1, \ldots, m)$$

be an arbitrary $n \times m$ matrix with elements in Ω. The matrix

$$V = \sum_{a \in G} \Delta(a)\, U\Phi(a^{-1})$$

satisfies the equation

$$\Delta(b)\, V\Phi(b^{-1}) = \sum_{a \in G} \Delta(b)\,\Delta(a)\, U\Phi(a^{-1})\,\Phi(b^{-1}) = \sum_{a \in G} \Delta(ba)\, U\Phi(ba)^{-1} = V$$

for every $b \in G$, because when b is fixed, ba ranges over all elements of G as a does. Hence,

$$\Delta(b)\, V = V\Phi(b)$$

for every $b \in G$. Since Δ and Φ are not equivalent, it follows from Theorem 13.4.1 that $V = 0$. This shows that

$$\sum_{a \in G} \sum_{k=1}^{n} \sum_{p=1}^{m} \Delta_{ik}(a)\, u_{kp}\, \Phi_{pq}(a^{-1}) = 0 \qquad \begin{pmatrix} i = 1, \ldots, n; \\ q = 1, \ldots, m \end{pmatrix}.$$

Since this equation holds for every matrix U, we may take a fixed pair k, p of indices and put

$$u_{kp} = 1, \quad u_{lr} = 0 \quad \text{for} \quad l \neq k \quad \text{or} \quad r \neq p.$$

Then,

$$\sum_{a \in G} \Delta_{ik}(a)\, \Phi_{pq}(a^{-1}) = 0,$$

and the theorem is proved.

There is a similar system of relations for the coefficients of one and the same representation. In contrast to the previous theorem, however, it is necessary to assume that the representation is absolutely irreducible.

13.4.3 *The coefficients of an absolutely irreducible representation Δ of degree n of a finite group G satisfy the relations*

$$\sum_{a \in G} \Delta_{ik}(a)\, \Delta_{pq}(a^{-1}) = \frac{g}{n}\, \delta_{iq}\, \delta_{kp} \qquad (i, k, p, q = 1, \ldots, n),$$

where $g = |G|$ and

$$\delta_{ik} = \begin{cases} 1 & \textit{for} \quad i = k, \\ 0 & \textit{for} \quad i \neq k. \end{cases}$$

Proof. Let $U = [\xi_{ik}]$ be an $n \times n$ matrix whose elements ξ_{ik} are indeterminates. The matrix

$$V = \sum_{a \in G} \Delta(a)\, U \Delta(a^{-1}) \tag{13.40}$$

satisfies the equation

$$\Delta(b)\, V \Delta(b^{-1}) = \sum_{a \in G} \Delta(b)\, \Delta(a)\, U \Delta(a^{-1})\, \Delta(b^{-1}) = \sum_{a \in G} \Delta(ba)\, U \Delta(ba)^{-1} = V$$

for every $b \in G$. Since Δ is absolutely irreducible, it follows from Theorem 13.3.4 that

$$V = \varphi(\xi_{11}, \ldots, \xi_{nn})\, I_n. \tag{13.41}$$

Now, (13.40) shows that $\varphi(\xi_{11}, \ldots, \xi_{nn})$ is a linear form in the indeterminates ξ_{ik}, say

$$\varphi(\xi_{11}, \ldots, \xi_{nn}) = \sum_{k=1}^{n} \sum_{p=1}^{n} \gamma_{kp}\, \xi_{kp},$$

where the coefficients γ_{kp} depend on the representation Δ. From (13.40) and (13.41), we obtain

$$\sum_{a\in G}\sum_{k=1}^{n}\sum_{p=1}^{n}\Delta_{ik}(a)\,\xi_{kp}\,\Delta_{pq}(a^{-1}) = \delta_{iq}\sum_{k=1}^{n}\sum_{p=1}^{n}\gamma_{kp}\,\xi_{kp}.$$

As the ξ_{kp} are indeterminates, the coefficients of each ξ_{kp} on both sides of the last equation coincide. Therefore,

$$\sum_{a\in G}\Delta_{ik}(a)\,\Delta_{pq}(a^{-1}) = \delta_{iq}\gamma_{kp}. \tag{13.42}$$

For $q = i$, we obtain

$$\sum_{a\in G}\Delta_{pi}(a^{-1})\,\Delta_{ik}(a) = \gamma_{kp}.$$

The right-hand side is obviously independent of i. Hence, the left-hand side also does not depend on i. Summation over i from 1 to n now yields

$$\sum_{a\in G}\sum_{i=1}^{n}\Delta_{pi}(a^{-1})\,\Delta_{ik}(a) = n\gamma_{kp}. \tag{13.43}$$

But, $\Delta(a^{-1})\Delta(a) = I_n$ or

$$\sum_{i=1}^{n}\Delta_{pi}(a^{-1})\,\Delta_{ik}(a) = \delta_{kp}.$$

Thus, (13.43) gives

$$g\,\delta_{kp} = n\gamma_{kp}.$$

By substituting γ_{kp} in (13.42), we obtain the required relation.

Suppose that G has t conjugacy classes. By Theorem 13.3.4, there are t absolutely irreducible inequivalent representations

$$\Delta^{(1)}, \ldots, \Delta^{(t)}$$

of G whose degrees we denote by $f_1, \ldots, f_t$. Let $\Delta^{(1)}$ be the identity representation, i.e. $\Delta^{(1)}(a) = 1$ for all $a \in G$. Theorems 13.4.2 and 13.4.3 can be summarized in the single relation

$$\sum_{a\in G}\Delta_{ik}^{(r)}(a)\,\Delta_{pq}^{(s)}(a^{-1}) = \frac{g}{f_r}\,\delta_{iq}\,\delta_{kp}\,\delta_{rs} \tag{13.44}$$

$$(i, k = 1, \ldots, f_r;\quad p, q = 1, \ldots, f_s;\quad r, s = 1, \ldots, t).$$

By the *character* χ of a representation Δ of degree n, we mean the traces

$$\chi(a) = \sum_{i=1}^{n}\Delta_{ii}(a)$$

of the matrices $\Delta(a)$. For the unit element e of G, we have $\Delta(e) = I_n$, so that $\chi(e) = n$ is equal to the degree of the representation. Equivalent representations have the same character, since similar matrices $\Delta(a)$ and $S^{-1}\Delta(a)S$ have the same trace. Later, we shall see that the converse is also true, i.e. two representations are equivalent if they have the same character.

For two conjugate elements a and b of G, we have $b = c^{-1}ac$ and hence

$$\Delta(b) = \Delta(c)^{-1}\Delta(a)\,\Delta(c)$$

so that the two similar matrices $\Delta(a)$ and $\Delta(b)$ have the same trace, i.e. $\chi(a) = \chi(b)$. Thus, the character of every representation is a function of the conjugacy classes.

In many applications of group representations, the characters are of greater importance than the representation matrices themselves.

Let

$$\chi^{(1)}, \ldots, \chi^{(t)}$$

denote the characters of the absolutely irreducible representations $\Delta^{(1)}, \ldots, \Delta^{(t)}$. The characters of absolutely irreducible representations are called *simple* characters. The character $\chi^{(1)}$ of the identity representation $\Delta^{(1)}$ is called the *principal character:* $\chi^{(1)}(a) = 1$ for every $a \in G$.

Any ordinary representation Δ of G splits completely into its absolutely irreducible constituents. If precisely m_i of the absolutely irreducible constituents of Δ are equivalent to $\Delta^{(i)}$ $(i = 1, \ldots, t)$, then we write

$$\Delta \sim m_1\Delta^{(1)} + \cdots + m_t\Delta^{(t)}.$$

Since the traces of the matrices $\Delta(a)$ are equal to the sum of the traces of the constituents, we obtain the following expression for the character χ of Δ:

$$\chi = m_1\chi^{(1)} + \cdots + m_t\chi^{(t)}.$$

Henceforth, we confine ourselves to representations whose coefficients belong to the field of all complex numbers.

If a is an element of order m, then $a^m = e$ implies that $\Delta(a)^m = I_n$, so that the characteristic roots $\alpha_1, \ldots, \alpha_n$ of $\Delta(a)$ are m-th roots of unity. Therefore, $\alpha_i\bar{\alpha}_i = 1$ or $\alpha_i^{-1} = \bar{\alpha}_i$, where $\bar{\alpha}$ denotes the complex conjugate of α. The matrix $\Delta(a^{-1}) = \Delta(a)^{-1}$ has the characteristic roots $\alpha_1^{-1}, \ldots, \alpha_n^{-1}$. As the trace is the sum of the characteristic roots, we obtain

$$\chi(a^{-1}) = \alpha_1^{-1} + \cdots + \alpha_n^{-1} = \bar{\alpha}_1 + \cdots + \bar{\alpha}_n = \overline{\chi(a)}.$$

This relation can also be deduced from Theorems 13.1.4 and 13.1.5 which state that every representation of a finite group is equivalent to a representation by unitary matrices.

From the relations (13.44), we derive a system of orthogonality relations for the characters. For $i = k$ and $p = q$, we obtain from (13.44)

$$\sum_{a \in G} \Delta_{ii}^{(r)}(a)\, \Delta_{pp}^{(s)}(a^{-1}) = \frac{g}{f_r}\, \delta_{ip}\, \delta_{rs}.$$

Summation over p from 1 to f_s shows that

$$\sum_{a \in G} \Delta_{ii}^{(r)}(a) \sum_{p=1}^{f_s} \Delta_{pp}^{(s)}(a^{-1}) = \sum_{a \in G} \Delta_{ii}^{(r)}(a)\, \chi^{(s)}(a^{-1}) = \frac{g}{f_r}\, \delta_{rs}$$

and from this we obtain by summation over i from 1 to f_r

$$\sum_{a \in G} \chi^{(r)}(a)\, \chi^{(s)}(a^{-1}) = g\, \delta_{rs}$$

or

$$\sum_{a \in G} \chi^{(r)}(a)\, \overline{\chi^{(s)}(a)} = g\, \delta_{rs}. \tag{13.45}$$

Let $K_1, \ldots, K_t$ denote the conjugacy classes of G and g_i the number of elements in K_i. From each class K_i, we choose an element a_i. Since the value of a character is the same for conjugate elements, (13.45) may be written as follows:

$$\sum_{i=1}^{t} g_i\, \chi^{(r)}(a_i)\, \overline{\chi^{(s)}(a_i)} = g\, \delta_{rs} \tag{13.46}$$

or, what is the same,

$$\sum_{i=1}^{t} \sqrt{\frac{g_i}{g}}\, \chi^{(r)}(a_i) \sqrt{\frac{g_i}{g}}\, \overline{\chi^{(s)}(a_i)} = \delta_{rs}. \tag{13.47}$$

The *character matrix* is defined as the $t \times t$ matrix

$$X = \begin{pmatrix} \sqrt{\frac{g_1}{g}}\, \chi^{(1)}(a_1) & \sqrt{\frac{g_2}{g}}\, \chi^{(1)}(a_2) & \cdots & \sqrt{\frac{g_t}{g}}\, \chi^{(1)}(a_t) \\ \sqrt{\frac{g_1}{g}}\, \chi^{(2)}(a_1) & \sqrt{\frac{g_2}{g}}\, \chi^{(2)}(a_2) & \cdots & \sqrt{\frac{g_t}{g}}\, \chi^{(2)}(a_t) \\ \cdots & \cdots & \cdots & \cdots \\ \sqrt{\frac{g_1}{g}}\, \chi^{(t)}(a_1) & \sqrt{\frac{g_2}{g}}\, \chi^{(t)}(a_2) & \cdots & \sqrt{\frac{g_t}{g}}\, \chi^{(t)}(a_t) \end{pmatrix}.$$

If X^* stands for the complex conjugate transpose of X, the equation

$$XX^* = I_t \tag{13.48}$$

is only another form of writing the relations (13.47). As every matrix is permutable with its inverse, we find that

$$X^*X = I_t,$$

or, in a more explicit notation,

$$\sum_{i=1}^{t} \sqrt{\frac{g_l}{g}}\,\overline{\chi^{(i)}(a_l)}\,\sqrt{\frac{g_m}{g}}\,\chi^{(i)}(a_m) = \delta_{lm}\,. \tag{13.49}$$

13.4.4 *The simple characters in the field of all complex numbers satisfy the orthogonality relations* (13.47) *and* (13.49).

It follows from (13.48) that the matrix X is invertible, so that its rows are linearly independent. Therefore, the expression of an arbitrary character χ in the form

$$\chi = m_1\chi^{(1)} + \cdots + m_t\chi^{(t)}$$

is unique. This shows that the absolutely irreducible constituents of an arbitrary representation are uniquely determined by its character. So, we obtain:

13.4.5 *Two ordinary representations of a finite group are equivalent if and only if their characters coincide.*

Let us consider two representations $\varDelta$ and $\varPhi$ whose absolutely irreducible constituents are given by

$$\varDelta \sim m_1\varDelta^{(1)} + \cdots + m_t\varDelta^{(t)},$$

$$\varPhi \sim n_1\varDelta^{(1)} + \cdots + n_t\varDelta^{(t)}.$$

Denoting the characters of $\varDelta$ and $\varPhi$ by χ and ψ, respectively, we apply (13.45) to obtain

$$\begin{aligned}\sum_{a\in G}\chi(a)\overline{\psi(a)} &= \sum_{a\in G}\sum_{r=1}^{t}\sum_{s=1}^{t} m_r n_s\,\chi^{(r)}(a)\,\overline{\chi^{(s)}(a)}\\ &= \sum_{r=1}^{t}\sum_{s=1}^{t} g m_r n_s\,\delta_{rs} = g\sum_{r=1}^{t} m_r n_r\,.\end{aligned} \tag{13.50}$$

If $\varPhi = \varDelta^{(r)}$ is an absolutely irreducible representation, this gives

$$m_r = \frac{1}{g}\sum_{a\in G}\chi(a)\,\overline{\chi^{(r)}(a)}\,. \tag{13.51}$$

This formula provides an explicit expression for the multiplicity with which the absolutely irreducible representation $\varDelta^{(r)}$ occurs as a constituent in a representation with the character χ. For $\Delta = \Phi$, the relation (13.50) becomes

$$\frac{1}{g} \sum_{a \in G} \chi(a)\overline{\chi(a)} = \frac{1}{g} \sum_{a \in G} |\chi(a)|^2 = \sum_{r=1}^{t} m_r^2 . \tag{13.52}$$

Thus,

$$\frac{1}{g} \sum_{a \in G} |\chi(a)|^2 = 1 \tag{13.53}$$

implies that

$$\sum_{r=1}^{t} m_r^2 = 1 .$$

Hence, precisely one of the multiplicities m_r is equal to 1, whereas the others vanish. This shows that (13.53) is necessary and sufficient for χ to be a simple character.

We now use (13.44) for $r = s$. Multiplication by $\Delta_{ji}^{(r)}(b)$ and summation over i from 1 to f_r yields

$$\sum_{a \in G} \sum_{i=1}^{f_r} \Delta_{ji}^{(r)}(b)\, \Delta_{ik}^{(r)}(a)\, \Delta_{pq}^{(r)}(a^{-1}) = \frac{g}{f_r} \sum_{i=1}^{f_r} \delta_{iq} \delta_{kp} \Delta_{ji}^{(r)}(b),$$

$$\sum_{a \in G} \Delta_{jk}^{(r)}(ba)\, \Delta_{pq}^{(r)}(a^{-1}) = \frac{g}{f_r}\, \delta_{kp} \sum_{i=1}^{f_r} \delta_{iq} \Delta_{ji}^{(r)}(b) ,$$

$$\sum_{a \in G} \Delta_{jk}^{(r)}(ba)\, \Delta_{pq}^{(r)}(a^{-1}) = \frac{g}{f_r}\, \delta_{kp} \Delta_{jq}^{(r)}(b) .$$

For $j = k$ and $p = q$, summation over j and p from 1 to f_r gives

$$\sum_{a \in G} \chi^{(r)}(ba)\, \chi^{(r)}(a^{-1}) = \frac{g}{f_r}\, \chi^{(r)}(b)$$

and, by replacing a^{-1} by a, we obtain

$$\sum_{a \in G} \chi^{(r)}(ba^{-1})\, \chi^{(r)}(a) = \frac{g}{f_r}\, \chi^{(r)}(b) . \tag{13.54}$$

By means of the symbol

$$\delta_{a,b} = \begin{cases} 1 & \text{for} \quad a = b \\ 0 & \text{for} \quad a \neq b \end{cases} \qquad (a, b \in G) ,$$

(13.54) may be written as follows:

$$\sum_{a \in G} \left(\chi^{(r)}(ba^{-1}) - \frac{g}{f_r}\, \delta_{a,b} \right) \chi^{(r)}(a) = 0 .$$

Taking these relations for all elements b of G, we obtain a system of g linear homogeneous equations for the g numbers $\chi^{(r)}(a)$, $a \in G$. Since not all of these numbers vanish, it follows that the determinant of the system vanishes, i.e.

$$\det\left[\chi^{(r)}(ba^{-1}) - \frac{g}{f_r}\,\delta_{a,b}\right] = 0\,,$$

where the index a indicates the rows and b the columns. This equation shows that gf_r^{-1} is a characteristic root of the matrix $[\chi^{(r)}(ba^{-1})]$. The elements of this matrix are sums of roots of unity, hence algebraic integers. Thus, gf_r^{-1} is a root of a monic polynomial whose coefficients are algebraic integers. As is well known, this implies that gf_r^{-1} itself is an algebraic integer. On the other hand, gf_r^{-1} is certainly rational so that it is a natural number. So, we obtain:

13.4.6 *The degree of every absolutely irreducible representation of a finite group G is a divisor of $|G|$.*

From results of section 13.2 and 13.3, we know that the group algebra G_Ω over the field Ω of the complex numbers is the direct sum of t complete matrix algebras whose degrees are $f_1, \ldots, f_t$. Let

$$e_{ik}^{(r)} \quad (r = 1, \ldots, t; \quad i, k = 1, \ldots, f_r)$$

denote a system of matrix units such that

$$a = \sum_{r=1}^{t} \sum_{i,k=1}^{f_r} \Delta_{ik}^{(r)}(a) e_{ik}^{(r)} \tag{13.55}$$

for any element $a \in G$. Using the relations (13.44), we can express the matrix units in terms of the coefficients of the absolutely irreducible representations. Multiplication of (13.55) by $\Delta_{pq}^{(s)}(a^{-1})$ and summation over all $a \in G$ leads to

$$\sum_{a \in G} a\Delta_{pq}^{(s)}(a^{-1}) = \sum_{r=1}^{t} \sum_{i,k=1}^{f_r} \sum_{a \in G} \Delta_{ik}^{(r)}(a)\, \Delta_{pq}^{(s)}(a^{-1})\, e_{ik}^{(r)}$$

$$= \sum_{i,k=1}^{f_s} \delta_{iq}\delta_{kp} \frac{g}{f_s}\, e_{ik}^{(s)} = \frac{g}{f_s}\, e_{pq}^{(s)}\,.$$

Taking $p = q$ and adding over p from 1 to f_s gives us expressions for the unit elements $e^{(s)}$ of the simple direct summands of G_Ω:

$$\frac{g}{f_s}\, e^{(s)} = \frac{g}{f_s} \sum_{p=1}^{f_s} e_{pp}^{(s)} = \sum_{a \in G} \sum_{p=1}^{f_s} a\,\Delta_{pp}^{(s)}(a^{-1}) = \sum_{a \in G} a\chi^{(s)}(a^{-1}).$$

The idempotents $e^{(s)}$, $s = 1, \ldots, t$, form a basis of the centre of G_Ω. The sums

$$k_i = \sum_{x \in K_i} x \quad (i = 1, \ldots, t)$$

of the elements in the conjugacy classes of G are another basis of the centre of G_Ω. Since the centre of G_Ω is an algebra, we have

$$k_l k_m = \sum_{n=1}^{t} c_{lmn} k_n \quad (l, m = 1, \ldots, t), \tag{13.56}$$

where the c_{lmn} are obviously non-negative integers and $c_{lmn} = c_{mln}$. The representation $\Delta^{(s)}$ of G can be extended in the natural way to a representation of G_Ω. By (d) of Theorem 13.3.4, the element k_l is represented by a matrix of the form $\gamma_{sl} I_{f_s}, \gamma_{sl} \in \Omega$. Therefore,

$$\gamma_{sl} I_{f_s} = \sum_{x \in K_l} \Delta^{(s)}(x) .$$

Taking the traces on both sides and using the fact that all the matrices $\Delta^{(s)}(x)$, $x \in K_l$ are similar, we obtain

$$f_s \gamma_{sl} = g_l \chi^{(s)}(a_l) ,$$

where a_l is an arbitrary element of K_l. Hence,

$$\gamma_{sl} = \frac{g_l}{f_s} \chi^{(s)}(a_l) ,$$

so that (13.56) gives

$$\frac{g_l}{f_s} \chi^{(s)}(a_l) \frac{g_m}{f_s} \chi^{(s)}(a_m) = \sum_{n=1}^{t} c_{lmn} \frac{g_n}{f_s} \chi^{(s)}(a_n) . \tag{13.57}$$

13.5 Further Properties of Characters. Examples

The conjugacy classes of abelian groups consist of single elements. Thus, a finite abelian group G has $|G|$ absolutely irreducible representations. By Theorem 13.3.4, all of them are of degree 1. Hence, the absolutely irreducible representations of abelian groups coincide with their characters. In other words,

$$\chi(a)\chi(b) = \chi(ab) \quad (a, b \in G) \tag{13.58}$$

for every simple character χ of a finite abelian group G.

Thus, for an element a of order n, $\chi(a)$ is an n-th root of unity. Conversely, we obtain a representation of the cyclic group $\langle a \rangle$ of order n by assigning to a an arbitrary n-th root of unity.

Suppose that a finite abelian group G is the direct product

$$G = C_1 \times C_2 \times \cdots \times C_s \qquad (13.59)$$

of cyclic groups $C_i = \langle a_i \rangle$ with $|C_i| = n_i$. By (13.58), a simple character χ of G is uniquely determined by the values $\chi(a_i)$; for, in accordance with (13.59), any element a of G has a unique decomposition

$$a = a_1^{k_1} a_2^{k_2} \cdots a_s^{k_s} \quad (k_i \bmod n_i) \qquad (13.60)$$

so that

$$\chi(a) = \chi(a_1)^{k_1}\, \chi(a_2)^{k_2} \cdots \chi(a_s)^{k_s}.$$

For $\chi(a_i)$, we can choose an arbitrary n_i-th root of unity, hence there are n_i possible values of $\chi(a_i)$. Since the $\chi(a_i)$, $i = 1, \ldots, s$, may be chosen independently of each other, we obtain all the $n_1 n_2 \ldots n_s = |G|$ simple characters. For a primitive n_i-th root of unity ε_i, we have $\chi(a_i) = \varepsilon_i^{m_i}$, say. The character χ can then be characterized by the s-tuple

$$(m_1, m_2, \ldots, m_s) \quad (m_i \bmod n_i), \qquad (13.61)$$

where only the residue class mod n_i of the i-th component is relevant.

The product $\chi\chi'$ of two simple characters of G is also a simple character of G, and the same holds for χ^{-1}. This shows that the simple characters of G form a group under multiplication, the *character group* of G. Its order is equal to $|G|$. If χ is given by the s-tuple (13.61) and χ' by $(m_1', m_2', \ldots, m_s')$, then the s-tuple that corresponds to $\chi\chi'$ is

$$(m_1 + m_1', m_2 + m_2', \ldots, m_s + m_s').$$

If the generating elements a_i of the cyclic groups C_i are given, then the expression (13.60) for the elements of G is unique so that a is characterized by the s-tuple

$$(k_1, k_2, \ldots, k_s) \quad (k_i \bmod n_i).$$

The s-tuple of a product is obtained by adding the s-tuples of the factors. This leads to the following theorem:

13.5.1 *The character group of a finite abelian group G is isomorphic to G.*

Now, let G be an arbitrary finite group and N a normal subgroup of G. It is evident that every representation of G/N gives rise to a representation of G. Conversely, a representation of G is a representation of G/N if its character satisfies the condition $\chi(x) = \chi(e)$ for every x of N; for, this equation implies that x is represented by the unit matrix.

Let G' denote the commutator subgroup of G. The factor group G/G' is abelian, so that we have $|G:G'|$ representations of degree 1 of G/G'. Thus,

G has at least $|G:G'|$ representations of degree 1. On the other hand, every character of degree 1 satisfies the equation

$$\chi(a^{-1}b^{-1}ab) = \chi(a)^{-1}\chi(b)^{-1}\chi(a)\,\chi(b) = 1$$

for arbitrary $a, b \in G$, hence it is a representation of G/G'. This gives the following theorem:

13.5.2 *The number of representations of degree 1 of a finite group is equal to the index of its commutator subgroup.*

From two representations Δ and Φ of G, we obtain another representation $\Delta \times \Phi$ by assigning to any $a \in G$ the Kronecker product $\Delta(a) \times \Phi(a)$, for the well known rule for the multiplication of Kronecker products shows that

$$\begin{aligned}(\Delta(a) \times \Phi(a))\,(\Delta(b) \times \Phi(b)) &= \Delta(a)\,\Delta(b) \times \Phi(a)\,\Phi(b)\\ &= \Delta(ab) \times \Phi(ab).\end{aligned}$$

Clearly, the character of $\Delta \times \Phi$ is equal to the product $\chi\psi$ of the characters χ of Δ and ψ of Φ. Let $\Delta^{(1)}, \ldots, \Delta^{(t)}$ denote the absolutely irreducible representations of the finite group G and $\chi^{(1)}, \ldots, \chi^{(t)}$ their characters. The splitting of the products $\Delta^{(i)} \times \Delta^{(k)}$ into their irreducible constituents is indicated by relations of the form

$$\Delta^{(i)} \times \Delta^{(k)} \sim \sum_{l=1}^{t} g_{ikl}\,\Delta^{(l)},$$

where the g_{ikl} are certain non-negative integers with $g_{ikl} = g_{kil}$. The corresponding relations for the characters are

$$\chi^{(i)}\chi^{(k)} = \sum_{l=1}^{t} g_{ikl}\,\chi^{(l)}.$$

Thus, the simple characters of a finite group form a commutative algebra, the so-called character algebra.

Every permutational representation of a finite group G in the sense of section 7.2. gives rise to a representation by permutation matrices, as we already mentioned in section 8.1.

Let H be a subgroup of G. By Theorem 7.2.2, H induces a transitive permutational representation of G and hence a representation Δ_1 of G by permutation matrices. We consider the decomposition

$$G = \bigcup_{i=1}^{n} Hr_i \qquad (Hr_1 = H),$$

of G into right cosets of H. The value $\chi(a)$ of the character χ_1 of Δ_1 for an element a of G is equal to the number of cosets Hr_i with the property Hr_ia

$=Hr_i$; for, $\chi_1(a)$ is equal to the number of entries 1 in the principal diagonal in the permutation matrix $\Delta_1(a)$, i.e. equal to the number of cosets that remain fixed. To compute

$$\sum_{a\in G} \chi_1(a),$$

we can proceed as follows: For each coset Hr_i, we count the number of elements a in G for which $Hr_i a = Hr_i$, and then we add these numbers for all cosets. We have $Hr_i a = Hr_i$ if and only if

$$r_i^{-1}Hr_i a = r_i^{-1}Hr_i;$$

hence, precisely the $|H|$ elements a of $r_i^{-1}Hr_i$ satisfy this condition. This gives

$$\sum_{a\in G} \chi_1(a) = |G:H||H| = |G|.$$

If a permutational representation Δ of G is not transitive but has k orbits, then we can apply the above argument to each transitive constituent of G. For the character χ of Δ, we then obtain

$$\sum_{a\in G} \chi(a) = k|G|. \tag{13.62}$$

This relation can be regarded as (13.51) for the principal character. Thus, k of the absolutely irreducible constituents of Δ are equal to the identity representation.

We now return to the transitive permutational representation Δ_1 of G. Under the permutations that correspond to the elements of the subgroup H, the cosets $Hr_1, \ldots, Hr_n$ are partitioned into, say, s orbits. Clearly, one of these orbits is $Hr_1 = H$ itself, so that $s \geqq 2$. Applying (13.62) to the group H, we obtain

$$\sum_{x\in H} \chi_1(x) = s|H|.$$

Since $\chi_1(r_i^{-1}xr_i) = \chi_1(x)$, this gives

$$\sum_{i=1}^{n} \sum_{x\in r_i^{-1}Hr} \chi_1(x) = ns|H| = s|G|. \tag{13.63}$$

Let us consider a fixed element x of G belonging to some $r_i^{-1}Hr_i$. Then, $\chi_1(x)$ occurs at least once on the left-hand side of (13.63). The multiplicity $m(x)$ with which $\chi_1(x)$ occurs on the left-hand side of (13.63) is obviously equal to the number of indices i for which $r_i^{-1}Hr_i x = r_i^{-1}Hr_i$. From what we observed above, it follows that $m(x) = \chi_1(x)$. For an element a of G that does not belong to any subgroup $r_i^{-1}Hr_i$, we have $\chi_1(a) = 0$. Thus, (13.63) gives

$$\sum_{a \in G} \chi_1(a)^2 = s|G|.$$

As $\chi_1(a)$ is a real number (in fact, a non-negative integer), we may write

$$\sum_{a \in G} \chi_1(a)\,\overline{\chi_1(a)} = s|G|$$

and apply (13.50) to obtain

$$s = \sum_{r=1}^{t} m_r^2,$$

where m_r denotes the multiplicity of the absolutely irreducible representation $\Delta^{(r)}$ of G as a constituent of Δ_1. If $s = 2$, i.e. if the permutational representation is doubly transitive, Δ_1 splits into the identity representation and one absolutely irreducible constituent.

As an example, we now determine all the characters of the symmetric group S_5. The rows of the table below correspond to the characters, the columns to the conjugacy classes of S_5. The columns are indexed by elements of the corresponding classes, and the number of elements in each class is indicated.

	e	$(1, 2)$	$(1,2)(3, 4)$	$(1, 2, 3)$	$(1, 2, 3)(4, 5)$	$(1, 2, 3, 4)$	$(1, 2, 3, 4, 5)$
	1	10	15	20	20	30	24
$\chi^{(1)}$	1	1	1	1	1	1	1
$\chi^{(2)}$	1	-1	1	1	-1	-1	1
$\chi^{(3)}$	4	2	0	1	-1	0	-1
$\chi^{(4)}$	4	-2	0	1	1	0	-1
$\chi^{(5)}$	6	0	-2	0	0	0	1
$\chi^{(6)}$	5	1	1	-1	1	-1	0
$\chi^{(7)}$	5	-1	1	-1	-1	1	0

$\chi^{(1)}$ is the principal character, $\chi^{(2)}$ the so-called alternating character, i.e. the character of the non-identity representation of the factor group S_5/A_5. This is the character we considered in section 7.4. The natural permutational representation of degree 5 of S_5 is certainly doubly transitive. As we observed above, it splits into the identity representation and another absolutely irreducible representation whose character is $\chi^{(3)}$. As $\chi^{(2)}$ is of degree 1, the product $\chi^{(2)}\chi^{(3)} = \chi^{(4)}$ is also a simple character. To obtain the remaining characters we may, for example proceed as follows: By (d) of Theorem 13.3.4, we have

$$f_5^2 + f_6^2 + f_7^2 = 86.$$

Since all the f_i are divisors of 5! they are uniquely determined by this relation, namely with a suitable numbering

$$f_5 = 6, \quad f_6 = 5, \quad f_7 = 5.$$

Equation (13.49) for $l = 1$, $m > 1$ gives the values of the sum

$$6\chi^{(5)} + 5\chi^{(6)} + 5\chi^{(7)}.$$

By means of (13.50), it is easy to verify that the product $\chi^{(3)}\chi^{(3)}$ is the character of a representation whose absolutely irreducible constituents are $\Delta^{(1)}, \Delta^{(3)}$, and two other representations of degrees 5 and 6, respectively. So, we obtain

$$\chi^{(3)}\chi^{(3)} - \chi^{(1)} - \chi^{(3)} = \chi^{(5)} + \chi^{(6)}$$

where $\chi^{(6)}$ denotes the character of degree 5. Similarly, we find

$$\chi^{(3)}\chi^{(4)} - \chi^{(2)} - \chi^{(4)} = \chi^{(5)} + \chi^{(7)}.$$

Since $6\chi^{(5)} + 5\chi^{(6)} + 5\chi^{(7)}$, $\chi^{(5)} + \chi^{(6)}$, and $\chi^{(5)} + \chi^{(7)}$ are known, we can compute $\chi^{(5)}$, $\chi^{(6)}$, and $\chi^{(7)}$, separately.

The computation of characters can be very troublesome. For the characters of some groups, for example the symmetric and the alternating groups, there are explicit formulas. We refer to [3] and [43].

It is no accident that the values of all the characters of S_5 are rational integers. We shall now show that the characters of all symmetric groups have this property. Let a be an element of order m in some symmetric group. For a character χ of degree f, we then obtain

$$\chi(a) = \varepsilon_1 + \varepsilon_2 + \cdots + \varepsilon_f,$$

where the ε_i are m-th roots of unity. For any natural number k, we have

$$\chi(a^k) = \varepsilon_1^k + \varepsilon_2^k + \cdots + \varepsilon_f^k.$$

If k is relatively prime to m, then a and a^k are permutations of the same type so that they belong to the same conjugacy class by Theorem 7.1.1. Hence, $\chi(a) = \chi(a^k)$ for $(k, m) = 1$. On the other hand, the numbers

$$\varepsilon_1^k + \varepsilon_2^k + \cdots + \varepsilon_f^k \quad (k = 1, \ldots, m; \quad (k, m) = 1)$$

are precisely the algebraic conjugates of $\chi(a)$ over the field of the rational numbers. Consequently, $\chi(a)$ is an algebraic integer that coincides with all its conjugates over the field of the rational numbers, i.e. $\chi(a)$ is a rational integer.

We shall now discuss a method of constructing a representation of a finite group G, provided that a representation of a subgroup U of G is given. Let

$$G = \bigcup_{i=1}^{m} Uc_i$$

be the decomposition of G into right cosets of U. Let Γ be a representation of degree n of U. Thus, to every $x \in U$ there corresponds an $n \times n$ matrix $\Gamma(x)$. We define $\Gamma(y)$ to be the $n \times n$ zero matrix fore every element y of G that does not belong to U. For an arbitrary element a of G, we now define

$$\Delta(a) = [\Gamma(c_i a c_k^{-1})] \quad (i, k = 1, \ldots, m).$$

$\Delta(a)$ is an $(mn) \times (mn)$ matrix consisting of m^2 $n \times n$ blocks that are matrices of the representation Γ or zero matrices. For a fixed element a and a fixed index i, there is precisely one k such that $c_i a c_k^{-1} \in U$, and for fixed a and k there is one and only one i such that $c_i a c_k^{-1} \in U$. Thus, $\Delta(a)$ can be obtained from a suitable $m \times m$ permutation matrix by replacing the entries 1 by suitable matrices of the representation Γ and the entries 0 by $n \times n$ zero matrices.

It is easy to verify that the matrices $\Delta(a)$ form a representation of G. Using block multiplication of matrices, we obtain

$$\Delta(a)\,\Delta(b) = \left[\sum_{j=1}^{m} \Gamma(c_i a c_j^{-1})\,\Gamma(c_j b c_k^{-1})\right]. \tag{13.64}$$

A product $\Gamma(c_i a c_j^{-1})\Gamma(c_j b c_k^{-1})$ is distinct from 0 if and only if

$$c_i a c_j^{-1} \in U \quad \text{and} \quad c_j b c_k^{-1} \in U. \tag{13.65}$$

For a fixed index i, there is precisely one index j such that $c_i a c_j^{-1} \in U$, and for every index j, there is precisely one k for which $c_j b c_k^{-1} \in U$. Thus, for any fixed index i, there is precisely one index k for which the sum on the right-hand side of (13.64) is distinct from 0, and this sum consists of a single term, namely the product

$$\Gamma(c_i a c_j^{-1})\,\Gamma(c_j b c_k^{-1}),$$

for which (13.65) is satisfied. Since both $c_i a c_j^{-1}$ and $c_j b c_k^{-1}$ belong to U, we have

$$\Gamma(c_i a c_j^{-1})\,\Gamma(c_j b c_k^{-1}) = \Gamma(c_i a b c_k^{-1}).$$

Therefore,

$$\Delta(a)\,\Delta(b) = \Delta(ab).$$

The representation Δ of G is said to be *induced* by the representation Γ of U.

A representation module V for Δ is the direct sum of m submodules of rank n, and any element of G induces a permutation of these submodules.

If Γ is a permutational representation, we obtain an imprimitive permutation group on the elements of a basis of V. Therefore, for an arbitrary representation Γ, the induced representation is also said to be imprimitive.

When Γ is the identity representation of degree 1 of U, the representation Δ is the permutational representation induced by U in the sense of Theorem 7.2.2 in matrix notation; for, $Uc_i a = Uc_k$ means the same as $c_i a c_k^{-1} \in U$. If Γ is any representation of degree 1, then Δ is a monomial representation of G (see section 8.1). The regular representation of U induces the regular representation of G; this can easily be verified by comparing the characters.

Let ψ denote the character of Γ. In accordance with the definition $\Gamma(y) = 0$ for $y \notin U$, we put

$$\psi(y) = \text{trace of } \Gamma(y) = 0 \quad \text{for} \quad y \notin U. \tag{13.66}$$

For the character χ of Δ, we then obtain

$$\chi(a) = \text{trace of } \Delta(a) = \sum_{i=1}^{m} \psi(c_i a c_i^{-1}).$$

As ψ is a character of U, we have

$$\psi(a) = \psi(xax^{-1})$$

for every $x \in U$, and by (13.66) this also holds if a does not belong to U. Hence,

$$\chi(a) = \sum_{i=1}^{m} \psi(x c_i a c_i^{-1} x^{-1})$$

for every $x \in U$. By summation over all $x \in U$, we obtain

$$|U|\, \chi(a) = \sum_{i=1}^{m} \sum_{x \in U} \psi(x c_i a c_i^{-1} x^{-1}) = \sum_{y \in G} \psi(yay^{-1}), \tag{13.67}$$

where in the last sum y ranges over all elements of G.

Denoting the conjugacy class of a in G by $K(a)$ and the normalizer of a in G by $N(a)$, we get

$$\sum_{y \in G} \psi(yay^{-1}) = |N(a)| \sum_{z \in K(a)} \psi(z).$$

Since $\psi(z) = 0$ for $z \notin U$, it is sufficient to extend the last sum only over those elements z that belong to U. This gives

$$\chi(a) = \frac{|N(a)|}{|U|} \sum_{z \in K(a) \cap U} \psi(z). \tag{13.68}$$

Here, the sum is zero if the intersection $K(a) \cap U$ is empty. The character χ is said to be *induced* by the character ψ of U.

We now come to a remarkable theorem about the absolutely irreducible constituents of those imprimitive representations of G that are induced by the absolutely irreducible representations of U. Let

$$\Gamma^{(1)}, \ldots, \Gamma^{(q)}$$

denote the absolutely irreducible representations of U, and let $\psi^{(k)}$ be the character of $\Gamma^{(k)}$. As before, $\Delta^{(1)}, \ldots, \Delta^{(t)}$ are the absolutely irreducible representations of G and $\chi^{(1)}, \ldots, \chi^{(t)}$ their respective characters. Let the representation Δ of G be induced by the absolutely irreducible representation $\Gamma^{(l)}$ of U. To the decomposition

$$\Delta \sim \sum_{s=1}^{t} m_s \Delta^{(s)}$$

of Δ into absolutely irreducible constituents, there corresponds an expression

$$\chi = \sum_{s=1}^{t} m_s \chi^{(s)} \tag{13.69}$$

for the character χ of Δ. By restricting the absolutely irreducible representation $\Delta^{(r)}$ of G to the subgroup U, we obtain a representation $\Delta_U^{(r)}$ of U. In general, $\Delta_U^{(r)}$ will be reducible, say

$$\Delta_U^{(r)} \sim \sum_{k=1}^{q} n_{rk} \Gamma^{(k)}. \tag{13.70}$$

From (13.68), we obtain

$$\frac{1}{|N(a)|} \chi(a) = \frac{1}{|U|} \sum_{z \in K(a) \cap U} \psi^{(l)}(z).$$

In the notation of section 13.4, the last equation can be written as follows:

$$\frac{g_i}{g} \chi(a_i) = \frac{1}{|U|} \sum_{z \in K_i \cap U} \psi^{(l)}(z).$$

Here, g_i stands for the number of elements in the conjugacy class K_i of G, and a_i is an arbitrary element in K_i. We now multiply the last equation by $\overline{\chi^{(r)}(a_i)}$ and add over a system $a_1, \ldots, a_t$ of representatives of the conjugacy classes of G. By (13.51) and (13.70), we obtain

$$\begin{aligned}
m_r &= \frac{1}{|U|} \sum_{i=1}^{t} \overline{\chi^{(r)}(a_i)} \sum_{z \in K_i \cap U} \psi^{(l)}(z) \\
&= \frac{1}{|U|} \sum_{i=1}^{t} \sum_{k=1}^{q} n_{rk} \overline{\psi^{(k)}(a_i)} \sum_{z \in K_i \cap U} \psi^{(l)}(z) \\
&= \sum_{k=1}^{q} n_{rk} \frac{1}{|U|} \cdot \sum_{z \in U} \overline{\psi^{(k)}(z)}\, \psi^{(l)}(z) \\
&= \sum_{k=1}^{q} n_{rk} \delta_{kl} = n_{rl}.
\end{aligned}$$

This gives the following theorem:

13.5.3 *Suppose that an imprimitive representation Δ of a finite group G is induced by an absolutely irreducible representation $\Gamma^{(l)}$ of a subgroup U of G. Then, the multiplicity of the absolutely irreducible representation $\Delta^{(r)}$ of G as a constituent of Δ is the same as the multiplicity of $\Gamma^{(l)}$ as a constituent of $\Delta^{(r)}$ restricted to U.*

Finally, we determine the characters of the alternating group A_5.

We begin with a general remark on the characters of the symmetric and the alternating groups. Let α denote the alternating character of the symmetric group S_n. If φ is an arbitrary simple character of S_n, then so is $\alpha\varphi$. Since φ is a simple character and real, (13.52) shows that

$$\sum_{a \in S_n} \varphi(a)^2 = n!.$$

If $\alpha\varphi$ is distinct from φ, then it follows from (13.45) that

$$\sum_{a \in S_n} \alpha(a)\ \varphi(a)\ \varphi(a) = 0.$$

Adding the last two equations, we obtain

$$2 \sum_{a \in A_n} \varphi(a)^2 = n!$$

By (13.52), this shows that φ is also a simple character of A_n. Hence, a simple character φ of S_n is also a simple character when restricted to A_n, provided that $\varphi \neq \alpha\varphi$.

As regards the conjugacy classes of A_5, we observe that the classes of S_5 which belong to A_5 are also classes in A_5 except the class of the cycles of length 5, which splits into two classes in A_5.

The table of the characters of A_5 is:

	e 1	(1,2)(3,4) 15	(1,2,3) 20	(1,2,3,4,5) 12	(1,3,5,2,4) 12
$\varphi^{(1)}$	1	1	1	1	1
$\varphi^{(2)}$	4	0	1	-1	-1
$\varphi^{(3)}$	5	1	-1	0	0
$\varphi^{(4)}$	3	-1	0	$\frac{1}{2}(1+\sqrt{5})$	$\frac{1}{2}(1-\sqrt{5})$
$\varphi^{(5)}$	3	-1	0	$\frac{1}{2}(1-\sqrt{5})$	$\frac{1}{2}(1+\sqrt{5})$

The characters $\chi^{(3)}$ and $\chi^{(6)}$ of S_5 have the property $\alpha\chi^{(3)} \neq \chi^{(3)}$, $\alpha\chi^{(6)} \neq \chi^{(6)}$. Therefore, the restrictions $\varphi^{(2)}$ and $\varphi^{(3)}$ of $\chi^{(3)}$ and $\chi^{(6)}$, respectively, to A_5 are simple characters of A_5. The restriction to A_5 of the representation $\Delta^{(5)}$ of S_5, however, is reducible. To obtain another simple character of A_5, we consider an imprimitive representation of A_5 induced by a Sylow 5-subgroup U of A_5. For U, we take the cyclic group generated by $u = (1, 2, 3, 4, 5)$ and for the representation of U, the representation of degree 1 that assigns a primitive fifth root of unity ε to u. Denoting the character of this representation by ψ, we have $\psi(u) = \varepsilon$. By (13.68), we obtain the following values for the character of the induced representation Δ:

$$12 \quad 0 \quad 0 \quad \varrho \quad \varrho',$$

where $\varrho = \varepsilon + \varepsilon^{-1}$, $\varrho' = \varepsilon^2 + \varepsilon^{-2}$ and the conjugacy classes are listed in the same order as in the table above. Note that u and u^{-1} belong to one conjugacy class and u^2 and u^{-2} to the other conjugacy class of order 5. The representation Δ is reducible and splits into the representation of degree 4, the representation of degree 5, and one other absolutely irreducible representation whose character is $\varphi^{(4)}$. For the values of $\varphi^{(4)}$ on the two classes of order 5, we obtain $\varrho + 1$ and $\varrho' + 1$, respectively. Now,

$$\varepsilon^4 + \varepsilon^3 + \varepsilon^2 + \varepsilon + 1 = \varepsilon^{-1} + \varepsilon^{-2} + \varepsilon^2 + \varepsilon + 1 = \varrho + \varrho' + 1 = 0$$

and

$$\varrho^2 = \varepsilon^2 + \varepsilon^{-2} + 2 = \varrho' + 2,$$

so that ϱ is a root of the equation $x^2 + x - 1 = 0$. Taking

$$\varrho = -\frac{1}{2} + \frac{1}{2}\sqrt{5},$$

we obtain the values of $\varphi^{(4)}$ as indicated in the table. The last character is obtained by choosing

$$\varrho = -\frac{1}{2} - \frac{1}{2}\sqrt{5},$$

which means that u is represented by ε^2 instead of ε.

13.6 Applications

In this section, we use the theory of group representations to prove two important theorems on the structure of certain finite groups. One of them, Burnside's Theorem, has only recently been proved independently of representation theory, while no proof of the other theorem without the use of group characters is known. The notation is the same as in the previous sections.

Lemma. Let G be a finite non-abelian simple group. If G has an absolutely irreducible representation $\Delta^{(s)}$ of degree f_s, other than the identity representation, and contains an element $a_l \neq e$ such that the number g_l of elements conjugate to a_l is relatively prime to f_s, then $\chi^{(s)}(a_l) = 0$.

Proof. Since G is simple and $\Delta^{(s)}$ is not the identity representation, it follows that $\Delta^{(s)}$ is faithful. Let us first prove that

$$|\chi^{(s)}(a)| < f_s \tag{13.71}$$

for every $a \neq e$. We know that $\chi^{(s)}(a)$ is the sum of f_s roots of unity:

$$\chi^{(s)}(a) = \varepsilon_1 + \varepsilon_2 + \cdots + \varepsilon_{f_s}.$$

Thus, the vector of $\chi^{(s)}(a)$ in the complex plane is the sum of f_s vectors of length 1. Therefore, $|\chi^{(s)}(a)| \leqq f_s$, where equality occurs only when $\varepsilon_1 = \varepsilon_2 = \cdots = \varepsilon_{f_s}$. In this case, we have $\Delta^{(s)}(a) = \varepsilon_1 I_{f_s}$, so that a belongs to the centre. But, since G is simple and $a \neq e$, this is impossible. Thus, (13.71) is proved.

In (13.57), we keep l fixed and substitute 1, ..., t for m. Then, we obtain t equations, which can be written as follows:

$$\sum_{n=1}^{t} \left(c_{lmn} - \delta_{mn} \frac{g_l}{f_s} \chi^{(s)}(a_l)\right) \frac{g_n}{f_s} \chi^{(s)}(a_n) = 0 \qquad (m = 1, \ldots, t).$$

This is a system of t linear homogeneous equations for the t numbers

$$\frac{g_n}{f_s} \chi^{(s)}(a_n) \qquad (n = 1, \ldots, t).$$

Since not all of these numbers vanish, we conclude that the determinant of the system vanishes. This means that

$$\alpha = \frac{g_l}{f_s} \chi^{(s)}(a_l)$$

is a characteristic root of the matrix

$$[c_{lmn}] \qquad (l \text{ fixed};\ m, n = 1, \ldots, t).$$

As the c_{lmn} are non-negative integers, it follows that α is an algebraic integer.

If g_l and f_s are relatively prime, then there are two integers x and y such that

$$g_l x + f_s y = 1.$$

Therefore,

$$\frac{g_l}{f_s} \chi^{(s)}(a_l) x + \chi^{(s)}(a_l)\, y = \frac{\chi^{(s)}(a_l)}{f_s}.$$

As we just observed, the first term on the left-hand side is an algebraic integer. The second term is also an algebraic integer, because $\chi^{(s)}(a_l)$ is a sum of roots of unity and y is a rational integer. Consequently, $f_s^{-1}\chi^{(s)}(a_l)$ is an algebraic integer.

By (13.71), we have $f_s^{-1}|\chi^{(s)}(a_l)| < 1$. As $\chi^{(s)}(a_l)$ is a sum of roots of unity, it follows that the moduli of the algebraic conjugates of $f_s^{-1}\chi^{(s)}(a_l)$ are also less than n. Thus, $f_s^{-1}\chi^{(s)}(a_l)$ is an algebraic integer whose norm is of modulus less than 1. As is well known, this implies that $f_s^{-1}\chi^{(s)}(a_l) = 0$, and the lemma is proved.

13.6.1 *If the number of elements in a conjugacy class other than e of a finite group G is a power of a prime number, then G is not simple.*

Proof. G is certainly not simple if there is a conjugacy class other than e that contains a single element. We may therefore assume that the number of elements in some class of G is a prime power $p^n > 1$. Let us suppose that, contrary to the assertion of the theorem, G is simple. In (13.49), put $a_m = e$ and let a_l be an element of the class with p^n elements. Then, we obtain

$$\sum_{i=1}^{t} f_i \chi^{(i)}(a_l) = 0. \tag{13.72}$$

For $i = 2, \ldots, t$, i.e. for all characters except the principal character, it follows from the lemma that f_i is divisible by p or $\chi^{(i)}(a_l) = 0$. Moreover, $f_1\chi^{(1)}(a_l) = 1$. Thus, (13.72) shows that

$$1 + p\gamma = 0$$

where γ is an algebraic integer. But, this is a contradiction, because p^{-1} is not an algebraic integer. This completes the proof.

We now come to the first theorem announced above.

13.6.2 Burnside's Theorem. *A group G of order $p^a q^b$ where p and q are prime numbers, is soluble.*

Proof. It is obviously sufficient to show that G is not simple. A Sylow q-subgroup Q of G has a non-trivial centre. Let $a \neq e$ be an element of the centre of Q. Then, the number of elements in the class of a is a power of p. Thus, it follows from Theorem 13.6.1 that G is not simple.

The other main theorem of this section also deals with the existence of a normal subgroup.

13.6.3 Frobenius' Theorem. *Let H be a subgroup of a finite group G such that*

$$H \cap y^{-1}Hy = e$$

for every element y of G that does not belong to H. Then, the elements of G that are not contained in H or any subgroup conjugate to H, together with the unit element, form a normal subgroup of order $|G:H|$.

If the results in section 7.2 are taken into account, it is easy to see that Theorem 13.6.3 is equivalent to the following theorem on permutation groups:

13.6.3′ *Let G be a transitive permutation group of degree n such that all elements of G other than the identity leave at most one symbol fixed. Then, those elements of G that leave no symbol fixed, together with the identity, form a normal subgroup of order n.*

Proof. We prove the theorem in its second form. Let H denote the subgroup of all those elements of G that leave a given symbol fixed. Then, we have $|G{:}H| = n$. We put $|H| = h$. By the hypothesis of the theorem, H and each of the $n - 1$ other conjugates of H intersect trivially. Consequently, there are

$$hn - 1 - n(h - 1) = n - 1$$

elements in G that are distinct from e and belong neither to H nor to any subgroup conjugate to H. Precisely these $n - 1$ elements of G leave no symbol fixed.

In what follows, x denotes an arbitrary element of G that is distinct from e and belongs to H or some subgroup conjugate to H. Further, y denotes an arbitrary element of G that leaves no symbol fixed.

G is a permutation group of degree n, hence we have a representation of G by $n \times n$ permutation matrices. Evidently, the character π of this representation has the following values:

$$\pi(e) = n, \quad \pi(x) = 1, \quad \pi(y) = 0.$$

Since G is transitive, it follows from a result in section 13.5 that the permutational representation contains the identity representation precisely once as a constituent. Thus, by subtracting the principal character from π, we obtain a character π_1 whose values are

$$\pi_1(e) = n - 1, \quad \pi_1(x) = 0, \quad \pi_1(y) = -1. \tag{13.73}$$

Let ϱ_G denote the character of the regular representation of G so that

$$\varrho_G(e) = nh, \quad \varrho_G(x) = 0, \quad \varrho_G(y) = 0.$$

To prove the theorem, it is sufficient to show that $\omega = \varrho_G - h\pi_1$ is a character of G. Indeed, we have

$$\omega(e) = h, \quad \omega(x) = 0, \quad \omega(y) = h.$$

If we know that ω is a character, then the corresponding representation Γ is of degree h. As $\omega(y)$ is a sum of h roots of unity, $\omega(y) = h$ implies that all these roots of unity are equal to 1. Consequently, $\Gamma(y) = I_h$. Hence, the kernel of Γ consists of e and the elements y so that these n elements form a normal subgroup.

It remains to prove that ω is a character of G. Let s be the number of conjugacy classes in H. Then, there are s simple characters $\psi^{(1)}, \ldots, \psi^{(s)}$ of H whose degrees are denoted by $m_1, \ldots, m_s$, respectively. Moreover, we shall denote the character of the regular representation of H by ϱ_H. By Theorem 13.3.4, we have

$$\varrho_H = \sum_{i=1}^{s} m_i \psi^{(i)}, \qquad h = \sum_{i=1}^{s} m_i^2 .$$

For any character ψ of H, let ψ^* denote the induced character of G. As we observed in section 13.5, we have $\varrho_H^* = \varrho_G$. This gives

$$\omega = \varrho_G - h\pi_1 = \sum_{i=1}^{s} m_i(\psi^{(i)*} - m_i \pi_1) .$$

Thus, it suffices to show that for every simple character $\psi^{(i)}$ of degree m_i of H

$$\psi^{(i)*} - m_i \pi_1$$

is a character of G.

To simplify the notation, we write $\psi^{(i)} = \psi$, $m_i = m$, $|G| = g$. Then,

$$\frac{1}{g} \sum_{a \in G} (\psi^*(a) - m\,\pi_1(a)\overline{(\psi^*(a) - m\,\pi_1(a))}$$

$$= \frac{1}{g} \sum_{a \in G} \psi^*(a)\,\overline{\psi^*(a)} - \frac{2m}{g} \sum_{a \in G} \psi^*(a)\,\overline{\pi_1(a)} + \frac{m^2}{g} \sum_{a \in G} \pi_1(a)\,\overline{\pi_1(a)} .$$

The values of ψ^* are

$$\psi^*(e) = mn, \quad \psi^*(x) = \psi(x), \quad \psi^*(y) = 0 .$$

Since $g = nh$, this shows that

$$\frac{1}{g} \sum_{a \in G} \psi^*(a)\,\overline{\psi^*(a)} = \frac{m^2 n}{h} + \frac{1}{nh} \sum_{x} \psi(x)\,\overline{\psi(x)} .$$

In the sum on the right-hand side, x ranges over all elements of the type defined above. By the hypothesis about H and the conjugate subgroups, we can also write

$$\frac{1}{g} \sum_{a \in G} \psi^*(a)\,\overline{\psi^*(a)} = \frac{m^2 n}{h} + \frac{1}{h} \sum_{\substack{x \in H \\ x \neq e}} \psi(x)\,\overline{\psi(x)} ,$$

and from

$$\sum_{z \in H} \psi(z)\,\overline{\psi(z)} = h$$

we derive

$$\frac{1}{g}\sum_{a\in G}\psi^*(a)\,\overline{\psi^*(a)} = \frac{m^2 n}{h} + \frac{1}{h}(h - m^2).$$

From (13.73), we obtain

$$\frac{1}{g}\sum_{a\in G}\psi^*(a)\,\overline{\pi_1(a)} = \frac{1}{g}\,mn(n-1) = \frac{1}{h}\,m(n-1)$$

and

$$\frac{1}{g}\sum_{a\in G}\pi_1(a)\,\overline{\pi_1(a)} = \frac{1}{g}\left((n-1)^2 + n - 1\right) = \frac{n-1}{h}.$$

So, we have

$$\frac{1}{g}\sum_{a\in G}\left(\psi^*(a) - m\,\pi_1(a)\right)\overline{\left(\psi^*(a) - m\,\pi(a)_1\right)} = 1. \tag{13.74}$$

Now, $\psi^* - m\pi_1$ can be expressed in the form

$$\psi^* - m\,\pi_1 = \sum_{j=1}^{t} n_j\,\chi^{(j)}$$

with integral coefficients n_j. It remains to show that the n_j are non-negative. From (13.50) and (13.74), we conclude that

$$\sum_{j=1}^{t} n_j^2 = 1.$$

This shows that precisely one of the n_j is equal to $+1$ or -1, while all the other vanish. Hence, there is a simple character $\chi^{(l)}$ such that

$$\psi^* - m\,\pi_1 = \pm\,\chi^{(l)}.$$

But, for the unit element, we have

$$\psi^*(e) - m\,\pi_1(e) = m > 0$$

so that $\psi^* - m\pi_1 = \chi^{(l)}$. This completes the proof.

13.7 The Projective Special Linear Groups

By $SL(n, \Omega)$, we mean the group of all $n \times n$ matrices with elements in the field Ω whose determinants are equal to 1. This group is called the special linear group of degree n. It is obviously a normal subgroup of $GL(n, \Omega)$.

13.7.1 *The matrices*

$$B_{r,s,\lambda} = I_n + \lambda E_{rs} \qquad \begin{pmatrix} r, s = 1, \ldots, n; \\ r \neq s;\ \lambda \in \Omega \end{pmatrix}$$

generate $SL(n, \Omega)$.

Here, I_n denotes the identity matrix of degree n, and E_{rs} stands for the $n \times n$ matrix with the unit element in the r-th row and s-th column and zeros elsewhere.

Proof. Let A be an arbitrary $n \times n$ matrix with elements in Ω. The product $AB_{r,s,\lambda}$ can be described as follows: The columns of $AB_{r,s,\lambda}$ except the s-th column are the same as those of A, whereas the s-th column of $AB_{r,s,\lambda}$ is obtained from the s-th column of A by adding the r-th column multiplied by λ.

Similarly, $B_{r,s,\lambda}A$ is obtained by adding the s-th row of A, multiplied by λ, to the r-th row, while all the other rows remain unchanged.

Let a_i denote the i-th column of A. It is easy to verify that

$$AB_{r,s,1}B_{s,r,-1}B_{r,s,1}$$

has $-a_s$ as its r-th column and a_r as its s-th column, while all the other columns coincide with those of A. Similarly, the interchange of two rows of A and multiplication of one of them by -1 can be achieved by multiplying A on the left by suitable factors $B_{r,s,\lambda}$.

We observe that the determinants of the $B_{r,s,\lambda}$ are equal to 1, so that the determinant of any matrix remains unchanged under multiplication by an arbitrary $B_{r,s,\lambda}$.

Now, let A be an arbitrary unimodular matrix, i.e. a matrix of $SL(n, \Omega)$. To prove the theorem, we show that A can be transformed into the identity matrix I_n by operations equivalent to multiplications by factors $B_{r,s,\lambda}$ on the right or on the left.

For $n = 1$, there is nothing to prove. We may assume that our assertion is true for all unimodular matrices of degree less than n. We observe that not all elements in the first row of A vanish. In case that the element in the upper left corner of A is 0, we interchange the first and a suitable other column of A and multiply one of them by -1 to obtain a matrix A_1 in which the element α in the upper left corner is distinct from 0. If A itself has a non-zero element in the upper left corner, then we put $A_1 = A$. After that, we add suitable multiples of the first column to the other columns and apply similar operations to the rows such that we arrive at a matrix of the form

$$A_2 = \begin{bmatrix} \alpha & 0 \cdots 0 \\ 0 & \\ \vdots & A_2^* \\ 0 & \end{bmatrix}$$

where A_2^* is an $(n-1) \times (n-1)$ matrix of determinant α^{-1}. If $\alpha \neq 1$, we add the first column to the second, then add the second column multiplied by $(1-\alpha)\alpha^{-1}$ to the first, and finally subtract the first column multiplied by α from the second. So, we obtain a matrix of the form

$$A_3 = \begin{bmatrix} 1 & 0 \cdots 0 \\ 0 & \\ \vdots & A_3^* \\ 0 & \end{bmatrix}$$

where A_3^* is a unimodular matrix of degree $n-1$. By the inductive hypothesis, A_3^* can now be transformed into I_{n-1} without altering the first row and the first column of A_3. This completes the proof.

In section 13.2, we observed that only the matrices γI_n are permutable with all E_{rs}. Thus, the centre of $SL(n, \Omega)$ consists of all matrices ϱI_n with $\varrho^n = 1$.

The factor group of $SL(n, \Omega)$ with respect to its centre is called the *projective special linear group* and is denoted by $PSL(n, \Omega)$. We shall prove that apart from two exceptions the projective special linear groups are simple.

Lemma 1. The group $SL(n, \Omega)$ coincides with its commutator subgroup except for $n = 2$ and $\Omega = GF(2)$ or $\Omega = GF(3)$.[1]

Proof. First, suppose that $n \geqq 3$. For any given pair r and s such that $1 \leqq r \leqq n$, $1 \leqq s \leqq n$, $r \neq s$, we have for $k \neq r$, $k \neq s$

$$B_{r,k,\lambda} B_{k,s,1} B_{r,k,\lambda}^{-1} B_{k,s,1}^{-1} = B_{r,s,\lambda}.$$

Hence, our assertion follows from Theorem 13.7.1.

Next we consider the case $n = 2$. Here,

$$\begin{bmatrix} \lambda & 0 \\ 0 & \lambda^{-1} \end{bmatrix} \begin{bmatrix} 1 & \mu \\ 0 & 1 \end{bmatrix} \begin{bmatrix} \lambda^{-1} & 0 \\ 0 & \lambda \end{bmatrix} \begin{bmatrix} 1 & -\mu \\ 0 & 1 \end{bmatrix} = \begin{bmatrix} 1 & \mu(\lambda^2 - 1) \\ 0 & 1 \end{bmatrix}.$$

If Ω is distinct from $GF(2)$ and $GF(3)$, there exists at least one element λ in Ω such that $\lambda \neq 0$ and $\lambda^2 \neq 1$. Since we can choose μ as an arbitrary element of Ω, it follows that apart from the two exceptions the commutator subgroup of $SL(2, \Omega)$ contains all $B_{1,2,\omega}$, $\omega \in \Omega$. One can prove in a similar way that all $B_{2,1,\omega}$, $\omega \in \Omega$ also belong to the commutator subgroup. Thus, for $n = 2$, the lemma also follows from Theorem 13.7.1.

Let $P_{n-1}(\Omega)$, $n > 1$, denote the $(n-1)$-dimensional projective space over Ω. We choose a fixed coordinate system such that the points of $P_{n-1}(\Omega)$ are given by n-tuples $(x_1, x_2, \ldots, x_n)$ of homogeneous coordinates $x_i \in \Omega$. By

$$x_i' = \sum_{k=1}^{n} \alpha_{ik} x_k \qquad (i = 1, \ldots, n),$$

[1] $GF(p)$ denotes the Galois field with p elements.

the matrices $A = [\alpha_{ik}]$ of $SL(n, \Omega)$ define linear mappings of $P_{n-1}(\Omega)$ onto itself. If A ranges over $SL(n, \Omega)$, we obtain in this way a group T of linear mappings. We consider T as a group of permutations of the points of $P_{n-1}(\Omega)$. As the x_i are homogeneous coordinates, two matrices A and λA give rise to the same mapping. Consequently, T is isomorphic to $PSL(n, \Omega)$.

Lemma 2. The group T is doubly transitive.

Proof. Let

$$\alpha = (\alpha_1, \alpha_2, \ldots, \alpha_n), \quad \beta = (\beta_1, \beta_2, \ldots, \beta_n)$$

be two arbitrary distinct points of $P_{n-1}(\Omega)$. It is sufficient to show that T contains an element τ that carries

$$\varepsilon_1 = (1, 0, 0, \ldots, 0) \quad \text{and} \quad \varepsilon_2 = (0, 1, 0, \ldots, 0)$$

into α and β, respectively. Since $\alpha \neq \beta$, there exists a regular $n \times n$ matrix A_0 such that

$$A_0 = \begin{bmatrix} \alpha_1 & \beta_1 \cdots \\ \alpha_2 & \beta_2 \cdots \\ \vdots & \vdots \\ \alpha_n & \beta_n \cdots \end{bmatrix}.$$

Then, $|A_0|^{-1} A_0$ belongs to $SL(n, \Omega)$ and yields an element τ of T with the required property. This proves the lemma.

Now, let H denote the subgroup of all those elements of T that leave the point ε_1 fixed. The matrices belonging to the elements of H are of the form

$$M = \begin{bmatrix} \mu_1 & \mu_2 \cdots \mu_n \\ 0 & \\ \vdots & M^* \\ 0 & \end{bmatrix} \qquad (\mu_1 \neq 0),$$

where M^* is an $(n-1) \times (n-1)$ matrix with the determinant μ_1^{-1}. Let H_0 denote the set of those elements of H whose matrices have the form

$$M_0 = \begin{bmatrix} 1 & \mu_2 \cdots \mu_n \\ 0 & \\ \vdots & I_{n-1} \\ 0 & \end{bmatrix}.$$

It is easy to see that H_0 is a normal subgroup of H. Note that H_0 is abelian and that all matrices $B_{1,2,\lambda}$ lead to elements of H_0.

13.7.2 *The groups $PSL(n, \Omega)$, $n > 1$, are simple except for $n = 2$, $\Omega = GF(2)$ and $n = 2$, $\Omega = GF(3)$.*

Proof. Since $T \cong PSL(n, \Omega)$, it is sufficient to show that the permutation group T on the points of $P_{n-1}(\Omega)$ is simple. Let N be a normal subgroup of T other than the identity subgroup.

By Lemma 2, T is doubly transitive, hence primitive. It follows from Theorem 7.1.10 that N is transitive. This shows that $T = NH$, where H is defined as above. Evidently, $T_0 = NH_0$ is a normal subgroup of T. It is easy to see that all matrices $B_{r,s,\lambda}$ are conjugate to $B_{1,2,\lambda}$ or $B_{1,2,-\lambda}$ in $SL(n,\Omega)$. Since all matrices $B_{1,2,\lambda}$ lead to elements in H_0, it follows that the matrices $B_{r,s,\lambda}$ belong to elements of T_0. Hence, it follows from Theorem 13.7.1 that $T_0 = T$. Therefore,

$$T/N \cong H_0/N \cap H_0,$$

so that the factor group T/N is abelian. Hence, N contains the commutator subgroup of T, and now Lemma 1 implies that $N = T$, unless we have one of the exceptional cases. This completes the proof.

The exceptional groups are not simple: $PSL(2, GF(2))$ is isomorphic to the symmetric group of degree 3 and $PSL(2, GF(3))$ is isomorphic to the tetrahedral group.

Exercises

1. Find a representation of degree 2 of a cyclic group of prime order p with coefficients in $GF(p)$ that is reducible but not completely reducible.
2. Determine the conditions on the elements γ_{ikl} as defined in section 13.2 for A to be associative.
3. Let T be the algebra of all $n \times n$ triangular matrices with elements in a field. Determine the properly nilpotent elements of T.
4. Determine the idempotents that generate the simple right ideals of the group algebra of S_3 over the field of the rational numbers. Find all absolutely irreducible representations of S_3.
5. Find the idempotent of the group algebra of S_n that belongs to the alternating character.
6. Discuss the representations of the quaternion group.
7. Determine the simple characters of A_4 and S_4.
8. Determine the character algebra of the dihedral group of order 10.
9. Let C_m be the group algebra of a cyclic group of order m over the field of the m-th roots of unity. Find a basis of C_m such that the regular representation splits into m constituents of degree 1.
10. Let G be a finite group with a faithful absolutely irreducible representation. Prove that the centre of G is cyclic (or trivial).
11. Let H be a subgroup of G satisfying the condition of Theorem 13.6.3 so that G contains a normal subgroup of order $n = |G:H|$. Prove that $|H|$ divides $n - 1$.

SUGGESTIONS FOR FURTHER STUDIES

Many of the numerous branches of group theory are not mentioned in this introductory book. We cannot even attempt to give a fairly complete survey of the further parts of group theory. But we should like to suggest some topics for a more detailed and advanced study.

Investigations on group axioms are naturally connected with the theory of algebraic systems more general than groups, such as semi-groups and loops. These topics are treated in [6].

There is an extensive theory of free groups and the presentation of groups as factor groups of free groups. We refer to [46]. In this book, the reader also finds more about the commutator calculus. Reference [12] is an extremely useful survey of generators and defining relations for a large number of special groups.

Chapter 5 contains only the elementary part of the theory of abelian groups. As soon as the restriction to finitely many generators is dropped, the structure of abelian groups cannot be described in a similarly simple way as in Theorem 5.3.1. The main monographs on general abelian groups are [19] and [38]. The first of these books contains a collection of research problems, some of which have been solved by now.

In Chapter 12, we mentioned some connections between lattice theory and group theory. We should like to add that also the theory of direct decompositions can be treated by lattice-theoretical methods. We refer to [41] and [83].

In section 6.3, we introduced the second cohomology group in a purely group theoretical manner. Actually this concept belongs to homological algebra, whose origin is algebraic topology. During the last two decades, homology theory has been developed for its own sake and plays an increasing role in various branches of algebra. The main textbooks on homology theory are [9] and [45]. Reference [16] is an introduction from a group-theoretical point of view. For an important application to the theory of finite groups, we refer to [25].

There are several generalizations of Burnside's Theorem 8.3.1; see for example [66] and [82]. For a generalization of Frobenius' Theorem 13.6.3, we refer to [66] and [75].

As to p-groups, the present book contains only classical results. Further important and deep results are due to P. Hall [29].

In section 11.11, we introduced the concept of p-solubility. More generally, let π be an arbitrary non-empty set of prime numbers. A finite group is said to be π-soluble if every composition index is a prime number in π or not divisible by any prime number in π. One can now ask whether and in what form theorems on soluble groups can be generalized to π-soluble groups. A survey of many interesting results is given in [33]. Other properties referring to the arithmetical structure can also be generalized by assuming that a group has such a property not for all prime numbers but only for those in π.

As is to be expected, much work has been done on permutation groups. For finite permutation groups, we refer to [47], [48], and [80]. Infinite permutation groups have only been studied more recently, and literature on this subject is not extensive. The only monograph on infinite permutation groups seems to be [78].

Much of recent research is connected with the study of certain classes of groups as, for instance, the varieties. A variety is the class of groups in which a given set of identical relations is satisfied. It is evident that a variety is closed under the operations of taking subgroups and homomorphic images and forming cartesian products. Conversely, every class of groups that is closed under these operations turns out to be a variety. Examples of varieties are the class of all abelian groups, the class of all groups of exponent m, and the class of all nilpotent groups whose class does not exceed a given natural number. Reference [51] is an excellent report on this flourishing subject.

One of the methods used in the study of finite soluble groups also depends on certain classes of groups that are defined by closure properties. A formation F (see [24]) is a class of finite soluble groups such that $G \in F$ implies that $G/N \in F$, and $G/N_1 \in F$, $G/N_2 \in F$ implies that $G/N_1 \cap N_2 \in F$ for arbitrary normal subgroups N, N_1, N_2 of G. The theory of formations leads to a system of classes of conjugate subgroups in every finite soluble group. The Carter subgroups are an example of such a class. We refer to [11].

The study of finite groups with fixed-point-free automorphisms of prime order has led to important results. In [67] and [68], it is proved that groups with such automorphisms are nilpotent. It follows that the normal subgroup in Frobenius' Theorem 13.6.3 is nilpotent.

In Chapter 13, we confined ourselves to the study of ordinary representations. As an introduction to the extensive theory of modular representations we recommend [5] and [13].

The main subject of this book is the theory of finite groups. As an excellent introduction to the theory of infinite groups, we strongly recommend [41].

We also mention [69] as a survey of the theory of linear groups over the field of the complex numbers, [14] as a textbook on linear groups over Galois fields, and [70] as a group-theoretical approach to the theory of invariants. Finally, for an introduction to the extensive theory of topological groups, we refer to [55].

BIBLIOGRAPHY

1. Baer, R. Erweiterung von Gruppen und ihren Isomorphismen. *Math. Z.* 38 (1934) 375–416.
2. Baer, R. Verstreute Untergruppen endlicher Gruppen. *Arch. Math.* 9 (1958), 7–17.
3. Boerner, H., *Darstellungen von Gruppen.* Springer-Verlag, Berlin, 1955.
4. Brauer, R. and Fowler, K. A. On groups of even order. *Ann. of Math.* 62 (1955), 565–583.
5. Brauer, R. and Nesbitt, C. On modular characters of groups. *Ann. of Math.* 42 (1941), 556–590.
6. Bruck, R. H. *A survey of binary systems.* Springer-Verlag, Berlin, 1958.
7. Burnside, W. *Theory of groups of finite order.* 2nd ed. Dover Publications, Inc., New York, 1955.
8. Carmichael, R. D. *Introduction to the theory of groups of finite order.* Dover Publications, Inc., New York, 1956.
9. Cartan, H. and Eilenberg, S. *Homological algebra.* Princeton University Press, Princeton, N. J., 1956.
10. Carter, R. W. Nilpotent self-normalizing subgroups of soluble groups. *Math. Z.* 75 (1961), 136–139.
11. Carter, R. W. and Hawkes, T. The F-normalizers of a finite soluble group. *J. of Algebra* 5 (1967), 175–202.
12. Coxeter, H. S. M. and Moser, W. O. J. *Generators and relations for discrete groups.* 2nd ed. Springer-Verlag, Berlin, 1965.
13. Curtis, C. W. and Reiner, I. *Representation theory of finite groups and associative algebras.* Interscience, J. Wiley & Sons, New York, 1962.
14. Dickson, L. E. *Linear groups.* Dover Publications, Inc., New York, 1958.
15. Dixon, J. D. *Problems in group theory.* Blaisdell Publ. Comp., Waltham, Mass., 1967.
16. Eilenberg, S. and MacLane, S. Cohomology theory in abstract groups. I, II. *Ann. of Math.* 48 (1947), 51–78, 326–341.
17. Feit, W. and Thompson, J. G. Solvability of groups of odd order. *Pacific J. Math.* 13 (1963), 775–1029.
18. Fitting, H. Beiträge zur Theorie der Gruppen endlicher Ordnung. *Jber. Deutsche Math.-Verein.* 48 (1938), 77–141.
19. Fuchs, L. *Abelian groups.* Pergamon Press, New York, 1960.
20. Gantmacher, F. R. *The theory of matrices.* I. Chelsea Publishing Co., New York, 1959.
21. Gaschütz, W. Zur Erweiterungstheorie der endlichen Gruppen. *J. reine angew. Math.* 190 (1952), 93–107.

22. Gaschütz, W. Über die Φ-Untergruppen endlicher Gruppen. *Math. Z.* 58 (1953), 160-170.

23. Gaschütz, W. Endliche Gruppen mit treuen absolut-irreduziblen Darstellungen. *Math. Nachr.* 12 (1954), 253-255.

24. Gaschütz, W. Zur Theorie der endlichen auflösbaren Gruppen. *Math. Z.* 80 (1963), 300-305.

25. Gaschütz, W. Kohomologische Trivialitäten und äußere Automorphismen von p-Gruppen. *Math. Z.* 88 (1965), 432-433.

26. Gorenstein, D. *Finite Groups.* Harper & Row, New York, 1968.

27. Hall, M., Jr. *The theory of groups.* Macmillan Comp., New York, 1959.

28. Hall, P. A note on soluble groups. *J. London Math. Soc.* 3 (1928), 98-105.

29. Hall, P., A contribution to the theory of groups of prime-power order. *Proc. London Math. Soc.* 36 (1933), 29-95.

30. Hall, P., A characteristic property of soluble groups. *J. London Math. Soc.* 12 (1937), 198-200.

31. Hall, P., On the Sylow systems of a soluble group. *Proc. London Math. Soc.* 43 (1937), 316-323.

32. Hall, P., On the system normalizers of a soluble group. *Proc. London Math. Soc.* 43 (1937), 507-528.

33. Hall, P., Theorems like Sylows. *Proc. London Math. Soc.* (3) 6 (1956), 286-304.

34. Hall, P., and Higman, G. On the p-length of p-soluble groups and reduction theorems for Burnside's problem. *Proc. London Math. Soc.* (3) 6 (1956), 1-42.

35. Huppert, B. Normalteiler und maximale Untergruppen endlicher Gruppen. *Math. Z.* 60 (1954), 409-434.

36. Huppert, B. *Endliche Gruppen.* I. Springer-Verlag, Berlin, 1967.

37. Kaloujnine, L. La structure des p-groupes de Sylow des groupes symmétriques finis. *Ann. Ecole Norm. Sup.* (3) 65 (1948), 239–276.

38. Kaplansky, I. *Infinite abelian groups.* 2nd ed. University of Michigan Press, 1956.

39. Kegel, O. H. Produkte nilpotenter Gruppen. *Arch. Math.* 12 (1961), 90-93.

40. Kostrikin, A. I. The Burnside problem. *Izv. Akad. Nauk SSSR, Ser. mat.* 23 (1959), 3-34.

41. Kurosh, A. G. *The theory of groups.* I, II. 2nd ed. Chelsea Publ. Comp., New York, 1960.

42. Ledermann, W. *Introduction to the theory of finite groups.* 3rd ed. Oliver & Boyd, London, 1957.

43. Littlewood, D. E. *The theory of group characters and matrix representations of groups.* O. U. P., New York, 1940.

44. MacDuffee, C. C. *An introduction to abstract algebra.* Dover Publications, Inc., New York, 1966.

45. MacLane, S. *Homology.* Springer-Verlag, Berlin, 1963.

46. Magnus, W., Karrass, A. and Solitar, D. *Combinatorial group theory.* Interscience, J. Wiley & Sons, New York, 1966.

47. Manning, W. A., *Primitive groups.* Stanford University Publ. 1921.

48. Miller, G. A., Blichfeldt, H. F. and Dickson, L. E. *Theory and applications of finite groups.* J. Wiley & Sons, New York, 1916.

49. Mostow, G. D., Sampson, J. H. and Meyer, J.-P. *Fundamental structures of algebra.* McGraw-Hill Book Comp., New York, 1963.

50. Neumann, B. H. Permutational products of groups. *J. Austral. Math. Soc.* 1 (1960), 299-310.
51. Neumann, H. *Varieties of groups*. Springer-Verlag, Berlin, 1967.
52. Nomizu, K. *Fundamentals of linear algebra*. McGraw-Hill Book Comp., New York, 1966.
53. Novikov, P. S. Insolubility of the conjugacy problem in the theory of groups. *Izv. Akad. Nauk SSSR, Ser. mat.*, 18 (1954), 485-524.
54. Novikov, P. S. On the algorithmic insolvability of the word problem in group theory. *Amer. Math. Soc. Translat.* (2) 9 (1958), 1-122.
55. Pontryagin, L. S. *Topological groups*. Gordon & Breach Science Publ., Inc., New York, 1966.
56. Roquette, P. Arithmetische Untersuchung des Gruppenringes einer endlichen Gruppe. *J. reine angew. Math.* 190 (1952), 148-168.
57. Roquette, P. Realisierung von Darstellungen endlicher nilpotenter Gruppen. *Arch. Math.* 9 (1958), 241-250.
58. Rotman, J. J. *Theory of groups*. Allyn & Bacon, Boston, 1965.
59. Schenkman, E. *Group theory*. D. van Nostrand Comp., Inc., Princeton, N. J., 1965.
60. Schmidt, O. U. *Abstract theory of groups*. W. H. Freeman & Co., San Francisco, 1966.
61. Scott, W. R. *Group theory*. Prentice-Hall, Inc., Englewood Cliffs, N. J., 1964.
62. Smirnov, V. I. ed. by Silverman, R. A. *Linear algebra and group theory*. McGraw-Hill Book Comp., New York, 1961.
63. Specht, W. *Gruppentheorie*. Springer-Verlag, Berlin, 1956.
64. Speiser, A. *Die Theorie der Gruppen von endlicher Ordnung*. Birkhäuser Verlag, Basel, 1956.
65. Suzuki, M. *Structure of a group and the structure of its lattice of subgroups*. Springer-Verlag, Berlin, 1956.
66. Suzuki, M. On the existence of a Hall normal subgroup. *J. Math. Soc. Japan.* 15 (1963), 387-391.
67. Thompson, J. G. Finite groups with fixed-point-free automorphisms of prime order. *Proc. Nat. Acad. Sci. U.S.A.* 45 (1959), 578-581.
68. Thompson, J. G. Normal complements for finite groups. *Math. Z.* 72 (1960), 332-354.
69. Van der Waerden, B. L. *Gruppen von linearen Transformationen*. Springer-Verlag, Berlin, 1935.
70. Weyl, H. *The classical groups, their invariants and representations*. Princeton University Press, Princeton, N. J., 1946.
71. Wielandt, H. Eine Verallgemeinerung der invarianten Untergruppen. *Math. Z.* 45 (1939), 209-244.
72. Wielandt, H. Zum Satz von Sylow. *Math. Z.* 60 (1954), 407-408.
73. Wielandt, H. Vertauschbare nachinvariante Untergruppen. *Abh. Math. Sem. Univ. Hamburg* 21 (1957), 55-62.
74. Wielandt, H. Sylowgruppen und Kompositionsstruktur. *Abh. Math. Sem. Univ. Hamburg* 22 (1958), 215-228.
75. Wielandt, H. Über die Existenz von Normalteilern in endlichen Gruppen. *Math. Nachr.* 18 (1958), 274-280.
76. Wielandt, H. Zum Satz von Sylow. II. *Math. Z.* 71 (1959), 461-462.
77. Wielandt, H. Ein Beweis für die Existenz von Sylowgruppen. *Arch. Math.* 10 (1959), 401-402.

78. Wielandt, H. *Unendliche Permutationsgruppen.* Mimeographed Lecture Notes, Univ. Tübingen, 1959/1960.
79. Wielandt, H. Gedanken für eine allgemeine Theorie der Permutationsgruppen. *Univ. e Politec. Torino Rend. Sem. Mat.* 21 (1962), 31-39.
80. Wielandt, H. *Finite permutation groups.* Academic Press, New York, 1964.
81. Wielandt, H. and Huppert, B. Arithmetical and normal structure of finite groups. *Proc. Sympos. Pure Math.* 6 (1962), 17-38.
82. Zappa, G. Sur les systèmes distingués de représentants et sur les compléments normaux des sous-groupes de Hall. *Acta Math. Acad. Sci. Hungar.* 13 (1962), 227-230.
83. Zassenhaus, H. *The theory of groups.* 2nd ed. Chelsea Publ. Comp., New York, 1958.

INDEX